BACTERIOLOGY RESEARCH DEVELOPMENTS

NEW TRENDS IN *CORYNEBACTERIUM GLUTAMICUM* BEYOND THE AMINO ACIDS

BACTERIOLOGY RESEARCH DEVELOPMENTS

Additional books in this series can be found on Nova's website
under the Series tab.

Additional e-books in this series can be found on Nova's website
under the e-book tab.

BACTERIOLOGY RESEARCH DEVELOPMENTS

NEW TRENDS IN *CORYNEBACTERIUM GLUTAMICUM*

BEYOND THE AMINO ACIDS

CARLOS BARREIRO
EDITOR

nova publishers

New York

Library of Congress Cataloging-in-Publication Data

ISBN: 978-1-63483-280-9

Library of Congress Control Number: 2015943812

Published by Nova Science Publishers, Inc. † New York

CONTENTS

PREFACE

Five carbons, nine hydrogens, one nitrogen and two couples of oxygen opened an exciting history of manufacturing, scientific and academic discoveries, developments and industrial profits. Glutamic acid ($C_5H_9NO_4$), in addition to its role as a protein constituent, also has a strong taste-enhancing power as Ikeda described it at the very beginning of Twentieth century. The nutrition improvement in Japan was the real cause of this discovery and the dangerous process based on concentrating HCl to obtain glutamate from vegetal raw materials was the reason that led to trail the main character of this book. Thus, the screening of an avian-faeces-contaminated soil sample gathered up from the Ueno Zoo in Tokyo (Japan) was the milestone to finding a natural glutamic acid-producer bacterium initially called *Micrococcus glutamicus* No. 534, which later was re-named *Corynebacterium glutamicum*. A few years later, the amino acid fermentation industry was born. The history of how Kinoshita, Udaka, and Shimono from the Kyowa Hakko Kogyo Company discovered *C. glutamicum* and the subsequent development of amino acid production is an educative reading somehow similar to penicillin discovery by Fleming and its improvement by Florey, Heatley, Chain and Abraham. The strain development (e.g.: random mutagenesis or direct metabolic engineering); large projects around biochemical, molecular biology and genetic characterizations; and a constant aim to update with the latest fashion of technological development (e.g.: genome sequencing, Transcriptomics or Proteomics) have yielded the current knowledge about this bacterium as well as the worldwide amount of amino acids (million tons/year). But, when the production runs successfully and increases year by year, the risk of a decrease in the industrial and scientific efforts could become a possibility. Fortunately, one hundred years later from the glutamic acid discovery, the 21st century has enlarged the industrial hopes for *C. glutamicum*. So, there is life parallel with the "amino acid" production for *C. glutamicum*.

This book presents an update of the current state of *C. glutamicum* in an interesting mixture of traditional and emergent researchers from industries and academia focused on this bacterium. The taxonomical location of *C. glutamicum* opens the book to make room for the update on methodological developments and to conclude with an exhaustive overview of *C. glutamicum* trending products "beyond the amino acids."

Carlos Barreiro

Contributors

Bahillo, Esther
Gadea Biopharma, Parque Tecnológico de León, 24009 León, Spain

Barredo, José-Luis
Gadea Biopharma, Parque Tecnológico de León, 24009 León, Spain

Barreiro, Carlos
INBIOTEC (Instituto de Biotecnología de León)
Proteomics Unit. Avda. Real, 1. 24006 León. Spain
Área de Bioquímica y Biología Molecular. Departamento de Biología Molecular. Universidad de León. Campus de Vegazana s/n, 24071 León, Spain

Cregut, Mickaël
Laboratoire Réactions et Génie des Procédés, CNRS UMR-7274, Université de Lorraine, 2 avenue de la Forêt de Haye, TSA 40602, 54518 Vandoeuvre-lès-Nancy Cedex, France
UMR CNRS 6144 GEPEA CBAC laboratory, University of Nantes 18 Boulevard Gaston Defferre, 85035 La Roche sur Yon, France

Delaunay, Stéphane
Laboratoire Réactions et Génie des Procédés, CNRS UMR-7274, Université de Lorraine, 2 avenue de la Forêt de Haye, TSA 40602, 54518 Vandoeuvre-lès-Nancy Cedex, France

Gil, José A
Area of Microbiology, Department of Molecular Biology, Faculty of Biology
University of León, León-Spain

Guedon, Emmanuel
Laboratoire Réactions et Génie des Procédés, CNRS UMR-7274, Université de Lorraine, 2 avenue de la Forêt de Haye, TSA 40602, 54518 Vandoeuvre-lès-Nancy Cedex, France

Hiraga, Kazumi
Research Institute of Innovative Technology for the Earth, 9-2 Kizugawadai, Kizugawa-shi Kyoto 619-0292, Japan

Inui, Masayuki
Research Institute of Innovative Technology for the Earth, 9-2 Kizugawadai, Kizugawa-shi Kyoto 619-0292, Japan

Junne, Stefan
Technische Universität Berlin, Department of Biotechnology, Chair of Bioprocess Engineering, Ackerstrasse 76 ACK 24, D-13355 Berlin, Germany

Kaboré, Abdoul Karim
Laboratoire Réactions et Génie des Procédés, CNRS UMR-7274, Université de Lorraine, 2 avenue de la Forêt de Haye, TSA 40602, 54518 Vandoeuvre-lès-Nancy Cedex, France

Khuat, Huang Bao Truc
Laboratoire Réactions et Génie des Procédés, CNRS UMR-7274, Université de Lorraine, 2 avenue de la Forêt de Haye, TSA 40602, 54518 Vandoeuvre-lès-Nancy Cedex, France

Lemoine, Anja
Technische Universität Berlin, Department of Biotechnology, Chair of Bioprocess Engineering, Ackerstrasse 76 ACK 24, D-13355 Berlin, Germany

Litsanov, Boris
Swiss Federal Institute of Technology Zurich, Institute of Microbiology, Vladimir-Prelog-Weg 1-5/10, 8093 Zürich, Switzerland

Martín, Juan F.
Área de Microbiología, Departamento de Biología Molecular, Universidad de León, Campus de Vegazana s/n, 24071 León, Spain

Martínez-Cámara, Sonia
Gadea Biopharma, Parque Tecnológico de León, 24009 León, Spain

Martínez-Castro, Miriam
Gadea Biopharma, Parque Tecnológico de León, 24009 León, Spain

Mateos, Luis M
Area of Microbiology, Department of Molecular Biology, Faculty of Biology, University of León, León, Spain

Neubauer, Peter
Technische Universität Berlin, Department of Biotechnology, Chair of Bioprocess Engineering, Ackerstrasse 76 ACK 24, D-13355 Berlin, Germany

Omumasaba, Crispinus A
Research Institute of Innovative Technology for the Earth, 9-2 Kizugawadai, Kizugawa-shi Kyoto 619-0292, Japan

Pérez-Redondo, Rosario
Genomics Unit, Instituto de Biotecnología de León, INBIOTEC,
Avda. Real n°. 1, Parque Científico de León, 24006 León, Spain

Rodríguez-García, Antonio
Genomics Unit, Instituto de Biotecnología de León, INBIOTEC,
Avda. Real n°. 1, Parque Científico de León, 24006 León, Spain
Área de Microbiología, Departamento de Biología Molecular,
Universidad de León, Campus de Vegazana s/n; 24071 León, Spain

Sola-Landa, Alberto
Genomics Unit, Instituto de Biotecnología de León, INBIOTEC,
Avda. Real n°. 1, Parque Científico de León, 24006 León, Spain

Villadangos, Almudena F
Area of Microbiology, Department of Molecular Biology, Faculty of Biology,
University of León, León-Spain

Wieschalka, Stefan
Technical University of Denmark, NNFCfB, Kogle Allé 6, 2970 Hørsholm, Denmark

In: New Trends in *Corynebacterium glutamicum*
Editor: Carlos Barreiro

ISBN: 978-1-63483-280-9

Chapter 1

CORYNEBACTERIUM GENUS, A DUAL GROUP OF CLINICAL AND INDUSTRIAL-RELEVANT BACTERIA

Carlos Barreiro[1,2,*] and Juan F. Martín[3]
[1]INBIOTEC (Instituto de Biotecnología de León), León, Spain
[2]Área de Bioquímica y Biología Molecular
Departamento de Biología Molecular, Universidad de León, Spain
[3]Área de Microbiología. Departamento de Biología Molecular,
Universidad de León, Spain

ASTRACT

The taxonomical situation of living (micro-)organisms is *per se* a dinamyc system, which has enough plasticity as to take advantage of new technologies in order to advance in the positioning of the livings cells. This chapter collects the new taxonomical trends to locate the *Corynebacterium* genus and opens this book with a general overview of one of the top ten prokaryotic (Bacteria and Archaea) genera with regard to its known species number. In addition, a brief review of the dual role (pathogens/commensals and industrial relevant microorganisms) highlighting this genus is also presented. Thus, both the historical outbreaks of *Corynebacterium diphtheriae* and the amino acid production started by *Corynebacterium glutamicum* are reviewed.

INTRODUCTION

The phylum Actinobacteria is divided into six classes, twenty-three orders, fifty-three families and two hundred and twenty-two genus as the volume five of last Bergey's Manual edition presents [1]. This phylum gathers: i) secondary metabolite-producer strains with industrial application as *Streptomyces*, *Nocardia* or *Saccharopolyspora*, which makes it the

[*] Corresponding Author: Carlos Barreiro; c.barreiro@unileon.es

most relevant group in this area overtaking the fungal group [2]; ii) food and feed additive-producer bacterium as *Corynebacterium* [3]; iii) trending bacteria that are used as probiotics, *Bifidobacterium* [4]; and iv) relevant human, animal or plant pathogenic species like *Mycobacterium* [5], *Rhodococcus* [6] or *Renibacterium* [7]. This new Bergey's Manual edition described a new order, Corynebacteriales (ord. nov.; the *Corynebacterium* order), which is included in the Actinobacteria class. Members of this order are found in diverse environments, especially in the soil ecosystem, whereas some strains are described traditionally as pathogens of humans and tamed animals. This new order is divided into six families (formerly seven, before the combination of *Nocardiaceae* and *Gordoniaceae* just in one family [*Nocardiaceae*] with two suborders [8]) gathering thirteen genus.

Table 1. Taxonomical situation of *C. diphtheriae* and *C. glutamicum* as an example of classification inside the phylum Actinobaceria based on the description of Bergey's Manual [1]. Ref: reference

Level	Name	Name description	Reference
Domain	Bacteria	---	[45]
Phylum	Actinobacteria phyl. nov.	Ac.ti.no.bac.te¢ri.a. Gr. n. *actis -inos* a ray, beam, N.L. n. *bacter* a rod; suff. *-ia* ending denoting phylum; N.L. pl. neut. n. *Actinobacteria* actinomycete bacteria with diverse morphologies.	[1]
Class	Actinobacteria	Ac.ti.no.bac.te'ri.a. Gr. n. *actis -inos* a ray, beam; N.L. n. *bacter* a rod; suff. *-ia* ending denoting a class; N.L. pl. neut. n. *Actinobacteria* actinomycete bacteria with diverse morphologies.	[46]
Order	Corynebacteriales ord. nov.	Co.ry.ne.bac.te.ri'a.les. N.L. neut. n. *Corynebacterium* type genus of order; suff. *-ales* ending to denote order; N.L. fem. pl. n. *Corynebacteriales* the *Corynebacterium* order.	[1]
Family	Corynebacteriaceae	Co.ry.ne.bac.te.ri.a.ce'a.e. N.L. neut. n. *Corynebacterium* type genus of the family; suff. -*aceae* ending to denote a family; N.L. fem. pl. n. *Corynebacteriaceae* the *Corynebacterium* family.	[47]
Genus	*Corynebacterium* (Type genus)	Co.ry.ne.bac.te¢ri.um. Gr. n. *coryne* a club; L. neut. n. *bacterium* a rod, and in biology a bacterium (so called because the first ones observed were rod-shaped); N.L. neut. n. *Corynebacterium* a club bacterium.	[48]
Species	*Corynebacterium diphtheriae* (Type species)	diph.the.ri'a.e. Gr. n. *diphthera* leather, skin; N.L. fem. n. *diphtheria* a disease in which a leathery membrane forms in the throat; N.L. gen. n. *diphtheriae* of diphtheria.	[11, 49]
	Corynebacterium glutamicum	glu.ta'mi.c.um. N.L. neut. adj. *glutamicum* pertaining to glutamic acid.	[17, 36]

The genus type of the order Corynebacteriales, which is the core of this chapter, is *Corynebacterium* (Table 1). Under an effective, but informal classification, Michael Bott [9] described the dual features of this genus. On the one side, the "bad guys" are those species relevant to medicine due to their commensalism on the mucous membranes or skin and/or to their pathogenicity in humans and animals, including *Corynebacterium diphtheriae*, *Corynebacterium pseudotuberculosis* or *Corynebacterium jeikeium*. On the other side, the "good guys" are gathering those industrial-interesting species mainly involved in amino acid production such as *Corynebacterium glutamicum*. In a more precise phylogenetic approach, up to 84 different validated species defining the *Corynebacterium* genus are presented in the Bergey's Manual [1]. The collected species on that list are corroborated with the information compiled by Jean P. Euzéby on the website "List of prokaryotic names with standing in nomenclature (LPSN)" (http://www.bacterio.net) [10]. Nowadays, the LPSN presents 117 species and 11 sub-species within the *Corynebacterium* genus.

Initially, the description of the *Corynebacterium* genus by Lehmann and Neumann was based on the accommodation of those pathogenic diphtheroid bacilli or at least on their parasitism in animals [11]. Nowadays, several species appear to be commensals or cause significant infection in animals such as: i) farm animals (cattle, sheep, goats, pigs, horses) causing mastitis, pericarditis, ulcers, pneumonia, pyelonephritis or dermatitis; ii) birds (eagles, storks, penguins); iii) pets (dogs, cats, ferrets, hamsters, mice) by means of otitis or sepsis; and iv) wild animals (seals, tortoises, Scottish wild cats, rats, mice) [1].

In addition, some *Corynebacterium* species detected or recovered from animals have also been documented as a potential infection cause in humans (zoonotic transmission) due to animal handling, animal bites or unknown ways of infection [1]. Besides, some *Corynebacterium* species have been isolated directly from infected humans. Therefore, three close phylogenetically-related species producing the diphtheria exotoxin are the diphtherial agents in humans (*Corynebacterium diphtheriae*, *Corynebacterium ulcerans* and *Corynebacterium pseudotuberculosis*). Other multiple-drug resistant *Corynebacterium* species also related to human diseases causing important and occasionally fatal disease mainly in immunocompromised patients, nosocomial infections at hospitals or nursing homes, have been found. Those bacteria are: i) components of the normal skin flora (*Corynebacterium amycolatum*, *Corynebacterium jeikeium*) and ii) lipophilic species (*Corynebacterium kroppenstedtii*) [12, 13].

The dual character of the *Corynebacterium* genus is due to their industrial relevance on fermentative processes. Therefore, several species recovered from foodstuffs [smear-ripened cheese (*Corynebacterium casei*)], soil (*C. glutamicum*) or water [mucus of sea coral (*Corynebacterium maris*) or sea sediment (*Corynebacterium marinum*)] have been widely used in industrial applications [14–17]. However, not only has the taxonomical evolution of the *Corynebacterium* genus been characterized by the continuous incorporation of new species, rearrangement or redistribution of known species inside previously described ones, but also by the split of species out of the genus. This is the case of the 'plant pathogenic coryneform bacteria'. This group was not considered as true *Corynebacterium* species *sensu stricto* and subsequently reclassified in other genera, mostly in *Curtobacterium* and *Clavibacter* [18–21].

CHARACTERISTIC FEATURES OF THE GENUS *CORYNEBACTERIUM*

As a summary of the *Corynebacterium* genus features, it can be said that the species of this genus are Gram-positive (sometimes unequally stained) [1, 18]. They show straight to slightly curved rods (short or of medium length), which, due to the snapping division, exhibit typical V-shaped or palisade arrangement of cells. They are also non-sporulating, nonmotile, acid-fast stain negative, chemoorganotrophic and catalase positives. Some species present perceptible metachromatic granules (polyphosphate). As for environmental conditions, they can be facultatively anaerobic or aerobic. Many species are acid-producers from glucose or some other sugars in peptone media, whereas several species alkalinize the broth with citrate as sole carbon source. The wall presents A1α type peptidoglycan with directly cross-linked meso-diaminopimelic acid. Arabinose and galactose are the predominant cell wall sugars, but occasionally other sugars are also detected. Short chain mycolic acids 22–36 carbons in length (corynomycolic acids) are present with rare exceptions. Straight-chain saturated or monounsaturated fatty acids are observed. Either eight [MK-8(H2)] and/or nine [MK-9(H2)] isoprene units in the dihydrogenated menaquinones are present. A wide range of G+C content is detected (46 to 71 mol % G+C) in the tDNA, but most species have among 51 and 68 mol% G+C. As a complementary datum, the molecular composition of *C. glutamicum* is presented in Table 2.

IDENTIFICATION OF *CORYNEBACTERIUM* SPECIES

In brief, the present section describes the most relevant technologies used for the identification of *Corynebacterium* species (see detailed description in the fifth volume of Bergey's Manual [1]).

Phylogenetic relatedness among *Corynebacterium* species has been fully inferred by a nearly complete 16S rRNA gene sequence. Since the variance degree of 16S rRNA gene sequencing is of 2% or greater, the majority of the species are discernible from each other without furthergene analysis. Shorter fragments (400–500 bp) of the 16S rRNA gene have been used for rapid identification of *Corynebacterium* species. In contrast, some species are related to such a low degree of variance (≤2%) that it is very difficult to distinguish them. Therefore, additional genes as *rpoB, div*IVA, *dtxR* or the 16S–23S rRNA gene spacer region polymorphisms have had to be used for proper identification [1].

The outbreaks of human disease-related *Corynebacterium* species have been characterized using several molecular typing schemes as: rRNA gene restriction pattern determination (ribotyping), yielding an electromorph type (ET), randomly amplified polymorphic DNA (RAPD), pulsed field gel electrophoresis (PFGE), multilocus enzyme electrophoresis (MEE) or amplified fragment length polymorphism (AFLP) [1].

Over the last years, supported by the development of Bruker Daltonik GmbH through its Biotyper system (http://www.bruker.com), the identification of bacterial and fungal species by MALDI-TOF mass spectrometry has been applied increasingly. Thus, it has been found that 99.1% of the analysed diphtheria causing agents (*C. diphtheriae, C. ulcerans* or *C. pseudotuberculosis*) were properly identified with reliable scores by the MALDI-TOF system. Consequently, this analysis is becoming a routinely-used tool in microbiology laboratories due to the fact that it does not require extensive training on mass spectrometry or chemistry in addition to the affordable consumable cost per sample (excluding the MALDI-TOF instrument itself) [22–24].

Along this chapter, *Corynebacterium* has been presented as a dual genus due to the pathogenic/commensal role of some species in contrast with the industrial-relevant behaviour of other species. Next sections will describe the different microbiological expressions of this genus. On the one hand, the type species of the genus responsible for plagues along human history, *C. diphtheriae*. On the other hand, the well-known L-glutamic acid and L-lysine-producer bacterium *C. glutamicum*, which brought forward the so-called amino acid fermentation industry in the late 1950s.

Table 2. The elemental composition (in percentage) of *C. glutamicum* ATCC13032 cells based on Liebl, 2005 [18]. Those values were obtained from *C. glutamicum* strain ATCC 13032 grown in CGIII (complex medium) containing 4% glucose or in BMCG (synthetic broth). As the author indicated [18], cells were dried after washing with 0.9% NaCl and subsequently analysed by: i) IR spectroscopy (C, H, S after combustion with oxygen; O after combustion to CO_2), ii) heat conductance of N_2 (N), iii) inductively coupled argon plasma mass spectroscopy (ICP-MS) (other elements and trace elements)

Element	CGIII (%)	BMG (%)
C	41.1	40.0
O	36.8	38.4
H	6.09	5.94
N	9.74	7.87
P	1.73	2.13
S	0.31	0.26
K	2.82	3.22
Mg	0.226	0.255
Ca	0.002	0.0101
Al	0.0002	0.0017
Mn	0.00039	0.0108
Fe	0.004	0.031
Co	0.000008	0.000002
Ni	0.000009	0.000046
Cu	0.0003	0.0056
Zn	0.0075	0.0064
Mo	0.00011	0.00009

THE PATHOGENIC SIDE OF *CORYNEBACTERIUM* GENUS: *C. DIPHTHERIAE*

Frequently, the connection between an ancient plague and its etiological agent takes longer than expected. Even though diphtheria is a well-known disease described in the 5th century BC by Hippocrates, it was not until 1883 when Klebs discovered *C. diphtheria*. One year later, Loeffler related this bacterium to the disease. More than fifty years later, Freeman discovered in 1951 that certain bacteriophages could provide non-toxinogenic *C. diphtheriae* with the capacity to produce toxin [21]. Thus, diphtheria is a toxin-mediated disease caused by the bacterium *C. diphtheriae*, whose name derives from the Greek *diphthera*, meaning leather hides (http://www.cdc.gov/vaccines/pubs/pinkbook/dip.html). Diphtheria is an acute infectious disease that affects the upper respiratory tract and, occasionally, the skin. The most typical features when affecting the upper respiratory tract are a membranous pharyngitis (pseudo-membrane), in addition to fever and enlarged anterior cervical lymph nodes and oedema of soft tissue, which produces a 'bull neck' appearance [25]. Thus, the name 'strangling angel of children', which sometimes designates it, refers to the wing shaped pseudomembranes that form in the oropharynx, which is the visual effect of this disease. In the 16th and 17th centuries in Spain, diphtheria was known as 'morbus suffocans' or 'garrotillo', whereas in the 18th century the name was 'throat distemper' in New England (US) [26].

Nowadays, the disease is caused by three toxigenic bacteria that live as commensals in the human pharynx, such as *C. diphtheriae*, which can be detected in four biovars: *gravis, mitis, intermedius* and *belfanti*. Another cause is the presence of zoonotic agents such as *C. pseudotuberculosis* [biovars: i) *equi* (from horses or cattle): a nitrate reducer and ii) *ovis* (from sheep or goats): a non nitrate reducer [27, 28]] and *C. ulcerans*, which is transmitted to humans by contact with contaminated milk, farm animals and pets such as cats and dogs [22, 26, 27].

It has been a major killer of children during epidemics resulting in thousands of deaths. Fortunately, the mortality fell down over the twentieth century due to the improvement of the standards of living and boosted by immunization programmes. In fact, in 1994, as a result of the widespread vaccination programs, the World Health Organization (WHO) proposed diphtheria as a feasible candidate for eradication in the European Region. However, different factors affecting the well-established childhood vaccination programmes such as reductions in vaccination coverage, large-scale population movements by the breakup of the former Soviet Union, the decrease of vaccine and antitoxin supplies and non-properly supported health services provided conditions for its re-emergence. Epidemic proportions were reached in the newly independent states of the former Soviet Union [29, 30]. Thus, in 1995 more than 50000 cases were reported by the WHO European Region, which had decreased by 60% one year later [30].

The incidence of diphtheria epidemics presents clear differences between developed and developing countries. Anyway, an intensive vaccination strategy is the key to bring the disease under control in most countries, even though some endemic transmission still continues [26, 29]. Thus, the WHO reported worldwide 30000 cases and 3000 deaths of diphtheria in 2000 (http://www.who.int/immunization/topics/diphtheria/en/). In 2005, just 500 cases were reported in the WHO European Region, whereas Africa, South America, Southeast Asia and the Indian subcontinent presented outbreaks or large numbers of reported cases. This fact suggests circulation of invasive non-toxigenic strains or partial re-emerging toxigenic genotypic variants [31, 32]. Even though it can be considered as a (very) rare illness, for example, 102 infected patients were reported in the UK from 1986 to 2008 (42 of *C. diphtheria*; 59 of *C. ulcerans*; 1 of *C. pseudotuberculosis*), the Health Authorities continue publishing guides about diphtheria-causing species identification, which shows the current relevance of these minor diseases [e.g.: unvaccinated child from Catalonia (Spain), June,

2015] [25, 27, 33]. The immunotoxins, which are protein toxins linked to a targeting moiety derived from the immune system (e.g.: antibody or antibody fragment), are a positive clinical application of the diphtheria toxin. They are in accordance with the 'magic bullets' originally suggested by Paul Ehrlich. Accordingly, the diphtheria toxin has been the basis for the development of immunotoxins indicated for psoriasis, brain tumour, melanoma, leukaemia ovarian, breast or kidney cancers [34].

THE INDUSTRIAL SIDE OF *CORYNEBACTERIUM* GENUS: *C. GLUTAMICUM*

In 1866, the German chemist Karl Heinrich Ritthausen treating wheat gluten with sulphuric acid discovered a substance that he named as glutamic acid, since it comes from gluten [35]. Almost a hundred years later, a new industry called amino acid fermentation was born. This is the beginning of an amazing history that connects the honest attempt of nutrition improvement at Japan with the development of an impressive technology around a single bacterium, *C. glutamicum*. After the glutamic acid discovery, the next step of the history was mediated by Kikunae Ikeda (University of Tokyo), who found the taste-enhancing power of the monosodium glutamate. He observed the decomposition products of konbu (seaweed), where he found small glutamic acid crystals that presented a sour taste that changed to nice flavour after NaOH addition (monosodium glutamate). The original idea of K. Ikeda was improving nutrition at Japan and achieving a taste enhancement that might then result in a food value increase. This was the basis to start the large scale production of glutamate. Dealing with this challenge was not an easy business, since the glutamate production from konbu, soyabean or gluten required dangerous processes using large amounts of HCl. Notwithstanding, S. Suzuki's created Ajinomoto Co. Inc. and B. Kato set up Kyowa Hakko Co. Inc, both based on the production of glutamate as a food additive. This last one promoted in 1950s a screening program in order to search natural microbial producers that accumulated glutamic acid extracellularly. At this point, a new partner entered, S. Kinoshita, which was responsible for the initial description of *Micrococcus glutamicus* No. 534, which accumulates glutamic acid under biotin limiting conditions [36].

Finally, taxonomical studies led to renaming the bacterium as *C. glutamicum*. This microorganism had also the capability to accumulate others amino acids, such as lysine, arginine, threonine or ornithine. This meant the birth of the so-called amino acid fermentation industry (see the personal description of the processes development by S. Kinoshita on chapter 1 of 'Handbook of *Corynebacterium glutamicum*' [37]).

Nowadays, L-glutamic acid and L-lysine are bulk products (>2.93 and >1.95 million tons/year in 2012, respectively) and the product portfolio of the *Corynebacterium* species has been extended to L-threonine (230 thousand tons/year in 2010), L-tryptophan (4.8 thousand tons/year in 2010), L-valine, L-alanine, L-isoleucine, L-aspartic acid, among other amino acids [38–41]. Due to their variety of nutritional value, taste enhancement or medicinal applications, the amino acids are used as feed supplements, food additives, pharmaceuticals, cosmetics, polymers or agricultural chemicals. Excluding ethanol from the list, the world market for fermentation products was estimated to be of US$14.1 billions in 2004, where amino acids were the second largest product after antibiotics with an average annual growth rate of approximately 9% [42, 43]. Currently, the estimated global market is divided in: 56% feed additives (2004 in 2.5 US$ billions; 2009 in 3.7 US$ billions), 32% food supplements (2004 in 1.4 US$ billions; 2009 in 2.1 US$ billions) and 12% others (2004 in 0.5 US$ billions; 2009 in 0.8 US$ billions) [42, 43]. The world market of amino acids is mainly substantiated by Asian companies, but also by European and North-American corporations: Ajinomoto Co., Inc (Japan), Kyowa Hakko Kogyo Co., Ltd (Japan), Global Bio-chem Technology (China), Cheil Jedang (South Korea); Paik Kwang Industrial co ltd (South Korea); Evonik Industries AG (Germany); Vitalys (Denmark); Archer-Daniels-Midland Company (USA).

This amino acids world market, where *C. glutamicum* is the main actor, has enhanced the R&D devoted to this bacterium. This fact can be traced by the manuscripts published in scientific journals that include *C. glutamicum* (formerly *M. glutamicus*) in their tittles, which go from a couple in the 1950s to more than 450 papers in the 2010s. Based on that knowledge, the beginning of the 21th century has refocused the industrial/scientific point of view about the coryneform bacteria as well as other bacteria and fungi. Accordingly, the development of new products, different from the traditional amino acids, has burst into the coryneform world. Thus, products such as new amino acids, vitamins, alcohols, dicarboxylic acids, decarboxylated amino acids or heterologous proteins have enlarged the portfolio of *Corynebacterium* putative industrial products due to their relevance as chemicals and biodegradable polymers or as ingredients or additives in food, feed, cosmetics and pharmaceuticals [44].

Conclusion

Corynebacterium genus is a large taxonomical group as well as highly relevant due to its dual behaviour as pathogenic and industrial-compound producer. The taxonomical summary presented in the chapter updates the position of the genus within the life tree. In addition, this chapter opens the main door to this book that presents the current trending tools used to enlarge *C. glutamicum* product portfolio.

Acknowledgments

The authors want to thank the European Union programme ERA-IB [SCILS project (EIB.12.057)] through the Spanish Ministry of Economy and Competitiveness (Madrid, Spain) (PCIN-2013-063). Special thanks go to A. Mulero for her excellent critical reading.

References

[1] Goodfellow M, Kämpfer P, Busse H-J, Trujillo ME, Suzuki K, Ludwig W, et al., *Bergey's Manual of Systematic Bacteriology*. The Actinobacteria, part B. Volumen five. Springer; 2012.

[2] Barreiro C, Martín JF, García-Estrada C. *Proteomics Methodology Applied to the Analysis of Filamentous Fungi - New Trends for an Impressive Diverse Group of Organisms.* In: Prasain JK, editor. Tandem Mass Spectrom. - Appl. Princ., Rijeka, Croatia: InTech; 2012.

[3] Vertès AA, Inui M, Yukawa H. Postgenomic approaches to using corynebacteria as biocatalysts. *Annu. Rev. Microbiol.* 2012;66:521–50.

[4] Shiby VK, Mishra HN. Fermented milks and milk products as functional foods-a review. *Crit. Rev. Food Sci. Nutr.* 2013;53:482–96.

[5] Cole ST, Brosch R, Parkhill J, Garnier T, Churcher C, Harris D, et al., Deciphering the biology of *Mycobacterium tuberculosis* from the complete genome sequence. *Nature* 1998;393:537–44.

[6] Letek M, González P, Macarthur I, Rodríguez H, Freeman TC, Valero-Rello A, et al., The genome of a pathogenic rhodococcus: cooptive virulence underpinned by key gene acquisitions. *PLoS. Genet.* 2010;6:e1001145.

[7] Sandell TA, Teel DJ, Fisher J, Beckman B, Jacobson KC. Infections by *Renibacterium salmoninarum* and *Nanophyetus salmincola* Chapin are associated with reduced growth of juvenile Chinook salmon, *Oncorhynchus tshawytscha* (Walbaum), in the Northeast Pacific Ocean. J Fish Dis 2014.

[8] Zhi X-Y, Li W-J, Stackebrandt E. An update of the structure and 16S rRNA gene sequence-based definition of higher ranks of the class Actinobacteria, with the proposal of two new suborders and four new families and emended descriptions of the existing higher taxa. *Int. J. Syst. Evol. Microbiol.* 2009;59:589–608.

[9] Bott M. Corynebacteria: the good guys and the bad guys. *Microbiol. Today* 2007;34:74–7.

[10] Euzéby JP. List of Prokaryotic names with standing in nomenclature. Http://www.bacterio.cict.fr/ 2015.

[11] Lehmann KB, Neumann RO. Atlas und Grundriss der Bakteriologie und Lehrbuch der speciellen bakteriologischen Diagnostik (1st ed.). Munich: Lehmann. Lehmann; 1896.

[12] Funke G, Bernard KA. Coryneform gram-positive rods. In: Murray, editor. *Man. Clin. Microbiol.* 9th ed., ASM Press; 2007, p. 485–514.

[13] Tauch A, Schneider J, Szczepanowski R, Tilker A, Viehoever P, Gartemann K-H, et al., Ultrafast pyrosequencing of *Corynebacterium kroppenstedtii* DSM44385 revealed insights into the physiology of a lipophilic corynebacterium that lacks mycolic acids. *J. Biotechnol.* 2008;136:22–30.

[14] Ben-Dov E, Ben Yosef DZ, Pavlov V, Kushmaro A. *Corynebacterium maris* sp. nov., a marine bacterium isolated from the mucus of the coral *Fungia granulosa*. *Int. J. Syst. Evol. Microbiol.* 2009;59:2458–63.

[15] Brennan NM, Brown R, Goodfellow M, Ward AC, Beresford TP, Simpson PJ, et al., *Corynebacterium mooreparkense* sp. nov. and *Corynebacterium casei* sp. nov., isolated from the surface of a smear-ripened cheese. *Int. J. Syst. Evol. Microbiol.* 2001;51:843–52.

[16] Du Z-J, Jordan EM, Rooney AP, Chen G-J, Austin B. *Corynebacterium marinum* sp. nov. isolated from coastal sediment. *Int. J. Syst. Evol. Microbiol.* 2010;60:1944–7.

[17] Abe S, Takayama K, Kinoshita S. Taxonomic studies on glutamic acid-producing bacteria. *J. Gen. Appl. Microbiol.* 1967;13:279–301.

[18] Liebl W. Corynebacterium Taxonomy. In: Eggeling L, Bott M, editors. Handb. *Corynebacterium glutamicum,* Boca Raton, FL: CRC Press (Taylor & Francis Group); 2005, p. 9–34.

[19] Collins MD, Jones D. Reclassification of Corynebacterium flaccumfaciens, *Corynebacterium betae*, *Corynebacterium oortii* and *Corynebacterium poinsettiae* in the genus *Curtobacterium*, as *Curtobacterium flaccumfaciens* comb. nov. Microbiology 1983;129:3545–8.

[20] Davis MJ, Gillaspie a. G, Vidaver a. K, Harris RW. *Clavibacter*: a new genus containing some phytopathogenic Coryneform bacteria, including *Clavibacter xyli* subsp. *xyli* sp. nov., subsp. nov. and *Clavibacter xyli* subsp. *cynodontis* subsp. nov., pathogens that cause ratoon stunting disease of sugarcane and be. *Int. J. Syst. Bacteriol.* 1984;34:107–17.

[21] Barksdale L. *Corynebacterium diphtheriae* and its relatives. Bacteriol. Rev. 1970;34:378–422.

[22] Bernard K. The genus corynebacterium and other medically relevant coryneform-like bacteria. *J. Clin. Microbiol.* 2012;50:3152–8.

[23] Vila J, Juiz P, Salas C, Almela M, de la Fuente CG, Zboromyrska Y, et al., Identification of clinically relevant *Corynebacterium* spp., *Arcanobacterium haemolyticum*, and *Rhodococcus equi* by matrix-assisted laser desorption ionization-time of flight mass spectrometry. *J. Clin. Microbiol.* 2012;50:1745–7.

[24] Konrad R, Berger a, Huber I, Boschert V, Hörmansdorfer S, Busch U, et al., Matrix-assisted laser desorption/ionisation time-of-flight (MALDI-TOF) mass spectrometry as a tool for rapid diagnosis of potentially toxigenic *Corynebacterium* species in the laboratory management of diphtheria-associated bacteria. *Euro Surveill* 2010;15:1–5.

[25] PHE. Diphtheria: the green book, chapter 15. Immun. against Infect. Dis., Public Health England; 2013, p. 1–17.

[26] Byard RW. Diphtheria - "The strangling angel" of children. *J. Forensic Leg. Med.* 2013;20:65–8.

[27] PHE. UK *Standards for Microbiology Investigations Identification* of *Corynebacterium* species. 2014.

[28] Coyle MB, Lipsky BA. Coryneform bacteria in infectious diseases: clinical and laboratory aspects. *Clin. Microbiol. Rev.* 1990;3:227–46.

[29] Wagner KS, White JM, Lucenko I, Mercer D, Crowcroft NS, Neal S, et al., Diphtheria in the postepidemic period, Europe, 2000-2009. *Emerg. Infect. Dis.* 2012;18:217–25.

[30] Vitek CR, Wharton M. Diphtheria in the former Soviet Union: reemergence of a pandemic disease. Emerg. Infect. Dis. 1998;4:539–50.

[31] Mokrousov I. *Corynebacterium diphtheriae*: genome diversity, population structure and genotyping perspectives. *Infect. Genet. Evol.* 2009;9:1–15.

[32] Zakikhany K, Efstratiou A. Diphtheria in Europe: current problems and new challenges. *Future Microbiol.* 2012;7:595–607.

[33] Wagner KS, White JM, Crowcroft NS, De Martin S, Mann G, Efstratiou A. Diphtheria in the United Kingdom, 1986-2008: the increasing role of *Corynebacterium ulcerans*. *Epidemiol. Infect.* 2010;138:1519–30.

[34] Shapira A, Benhar I. Toxin-based therapeutic approaches. *Toxins (Basel)* 2010;2:2519–83.

[35] Plimmer RHA. The Chemical Constitution of the Proteins: Analysis. In: R.H.A. Plimmer & F.G. Hopkins, editor. *Monogr. Biochem.* 2nd ed., Longmans, Green & Company; 1912, p. 114.

[36] Kinoshita S, Nakayama S, Akita S. Taxonomical study of glutamic acid accumulating bacteria, *Micrococcus glutamicus*, nov. sp. *Bull Agric Chem. Soc. Jpn* 1958;22:176–85.

[37] Kinoshita S. A Short History of the Birth of the Amino Acid Industry in Japan. In: Eggeling L, Bott M, editors. *Handb. Corynebacterium glutamicum*, Boca Raton, FL: CRC Press. T aylor & Francis Group; 2005, p. 1–6.

[38] Hermann T. Industrial production of amino acids by coryneform bacteria. *J. Biotechnol.* 2003;104:155–72.

[39] Ajinomoto. Food products business. 2010.

[40] Ajinomoto. Feed-Use Amino Acids Business. 2011.

[41] Woo HM, Park J-B. Recent progress in development of synthetic biology platforms and metabolic engineering of *Corynebacterium glutamicum. J. Biotechnol.* 2014;180:43–51.

[42] Leuchtenberger W, Huthmacher K, Drauz K. Biotechnological production of amino acids and derivatives: current status and prospects. *Appl. Microbiol. Biotechnol.* 2005;69:1–8.

[43] Ikeda M, Takeno S. Amino Acid Production by *Corynebacterium glutamicum.* In: Yukawa H, Inui M, editors. *Corynebacterium glutamicum. Biol. Biotechnol.*, vol. 23, Berlin, Heidelberg: Springer Berlin Heidelberg; 2013, p. 1–415.

[44] Vasco-Cárdenas MF, Baños S, Ramos A, Martín JF, Barreiro C. Proteome response of *Corynebacterium glutamicum* to high concentration of industrially relevant C4 and C5 dicarboxylic acids. *J. Proteomics* 2013;85:65–88.

[45] Woese CR, Kandler O, Wheelis ML. Towards a natural system of organisms: proposal for the domains Archaea, Bacteria, and Eucarya. *Proc. Natl. Acad. Sci. U S* A 1990;87:4576–9.

[46] Stackebrandt E, Rainey FA, Ward-rainey NL. Proposal for a New Hierarchic Classification System, Actinobacteria classis nov. *Int. J. Syst Bacteriol* 1997;47:479–91.

[47] Zhi X-Y, Li W-J, Stackebrandt E. An update of the structure and 16S rRNA gene sequence-based definition of higher ranks of the class Actinobacteria, with the proposal of two new suborders and four new families and emended descriptions of the existing higher taxa. *Int. J. Syst. Evol. Microbiol.* 2009;59:589–608.

[48] Bernard KA, Wiebe D, Burdz T, Reimer A, Ng B, Singh C, et al., Assignment of *Brevibacterium stationis* (ZoBell and Upham 1944) Breed 1953 to the genus *Corynebacterium*, as *Corynebacterium stationis* comb. nov., and emended description of the genus *Corynebacterium* to include isolates that can alkalinize citrate. *Int. J. Syst. Evol. Microbiol.* 2010;60:874–9.

[49] W. K. *Die Mikroorganismen.* 3rd ed. Leipzig: F.C.W. Vogel; 1886.

In: New Trends in *Corynebacterium glutamicum*
Editor: Carlos Barreiro

ISBN: 978-1-63483-280-9

Chapter 2

DECIPHERING THE DYNAMIC TRANSCRIPTOME EMPLOYING MICROARRAYS

***Rosario Pérez-Redondo*[*1], *Alberto Sola-Landa*[1] and *Antonio Rodríguez-García*[1,2]**

[1] Genomics Unit, Instituto de Biotecnología de León, INBIOTEC, León, Spain
[2] Área de Microbiología, Departamento de Biología Molecular, Universidad de León, León, Spain

ABSTRACT

Over the last decade, microarray technology has boosted comprehension in transcriptional control mechanisms, which allowed fitting different pieces into the puzzle of regulatory networks. This chapter summarizes the current knowledge gained with the use of microarrays on the study of the physiology of *Corynebacterium glutamicum*. Global expression profiles have been obtained from different growing conditions and from mutant and parental strain comparisons. The examples described in this review are related to the metabolism of carbon, nitrogen, and phosphate, as well as of other essential nutrients, the response to stressing agents, and the characterization of regulation in relevant amino acids production pathways. Despite the emerging next-generation sequencing (NGS) applications, microarrays constitute a practical, fast and cheap approach for the study of the transcriptome, and serve to complement this and other techniques.

INTRODUCTION

Corynebacterium glutamicum is one of the most important bacteria with industrial applications, since it is used for the large-scale production of L-amino acids and organic acids. This fact motivated the genome sequencing of the reference strain *C. glutamicum* ATCC 13032 by two different industry-supported groups [1, 2], followed by a number of new

* Corresponding Author: Rosario Pérez-Redondo; r.perez.redondo@unileon.es

genome projects [3-5]. The identification of the different transcriptional regulators, sigma factors and other regulatory mechanisms, and the application of the different 'omics techniques have allowed a deeper understanding of the regulatory networks controlling the *C. glutamicum* metabolism [6-10]. In this chapter, we summarize the main findings obtained by transcriptomic analyses employing microarrays, a powerful tool contributing to this comprehension.

A Complex Regulatory Network Controlling the Central Carbon Metabolism and the Utilization of Alternative Carbon Sources

C. glutamicum has emerged as one of the best studied model microorganisms regarding the carbon metabolism and its regulation [11-13]. As a soil bacterium, it has evolved to be able to grow aerobically on various carbon sources. The repertoire of assimilated carbon sources includes some sugars—like glucose, fructose, ribose, maltose and sucrose [14-17]—, uridine [18], organic acids—like gluconate, acetate, propionate, pyruvate, glutamate, lactate and citrate [14, 19-23]—as well as numerous aromatic compounds [24], and both ethanol and methanol [25-29]. Cultures on mixtures of carbon sources revealed the lack of a stringent carbon catabolite repression, and, generally, glucose is metabolized simultaneously with other carbon sources [12]. Known exceptions are the glucose or fructose repression of the glutamate uptake [19, 30]; the glucose or acetate repression of the alcohol dehydrogenase gene and, hence, of the utilization of ethanol as a carbon source [27, 31]; and the repression of the methanol catabolism in the presence of glucose [29].

The nutritional versatility, in terms of carbon source, of *C. glutamicum*, and the need to adapt the metabolism to the variable conditions of its natural habitat, account for the complex regulatory network that has been revealed through numerous post-genome studies. These studies dealt mainly with transcriptional regulation [12] rather than post-transcriptional regulatory mechanisms [11, 13, 32]. The regulatory network controlling the carbon metabolism shows a hierarchical architecture with three types of transcriptional regulators: local, master, and global [8, 12]. Local regulators control a small number of genes, which are usually clustered with the regulatory gene and are functionally involved in the utilization of alternative carbon sources. This class includes the regulators that control the utilization of fructose, FruR [33]; lactate, LldR [34]; propionate, PrpR [35]; citrate, CitAB [36]; gluconate, GntR1 and GntR2 [28]; uridine, UriR [18]; ribose, RbsR [37]; myo-inositol, IolR [38]; and also the local regulators of catabolic pathways for aromatic compounds [24, 39].

Placed at the top of the regulatory network is GlxR, the only regulator of *C. glutamicum* that can be defined as 'global', accordingly to the criteria of regulating more than 20 genes of different transcriptional units belonging to four or more functional categories [12]. GlxR is a cAMP-binding regulator homologous to cAMP receptor proteins, including the global regulator Crp of *Escherichia coli*. It was firstly characterized as a repressor of the glyoxylate bypass genes *aceA* and *aceB* [40]. Further works established its regulatory role in the control of the alternative carbon source and central carbon metabolisms, mainly as a repressor. This control is exerted through the direct regulation of structural genes, and also through the control of local and master regulators [8, 39-50]. Moreover, the global role of GlxR goes beyond the central carbon metabolism and includes functions related to lipid metabolism [51], anaerobic nitrate respiration [52], phosphate transport [53], and cell resuscitation [42]. Bioinformatic prediction combined with electrophoretic mobility shift assays [41] and recent immunoprecipitation studies [45, 50] identified a plethora of binding sites that underline the global role of GlxR. Indeed, it has been predicted that a total of 14% of *C. glutamicum* genes may be under the direct transcriptional control of GlxR [8].

The transcriptional regulators RamB [54], RamA [55], and SugR [56, 57] function as master regulators of the network module related to carbohydrate metabolism [58]. These master regulators conform a layer of the regulatory network that is placed between the global regulator GlxR and the local regulators. Moreover, they are connected by mutual cross-regulations [12]. SugR is a repressor of genes for sugar uptake and metabolism, including those for phosphotransferase systems, glycolysis, and the fermentative lactate dehydrogenase. In the presence of sugar, the SugR regulon is derepressed by phosphorylated sugar intermediates acting as inducers [56, 57, 59-63].

The utilization of acetate as carbon and energy source requires the expression of genes *aceA* (isocitrate lyase), *aceB* (malate synthase), *pta* (phosphotransacetylase), and *ack* (acetate kinase). The expression of these genes is under the control of two antagonistic regulators, RamA and RamB. On one side, *ramA* is required for acetate utilization and functions as an activator of the aforementioned genes [55]. The *ramA* gene is upregulated when acetate substitutes glucose in the growth medium. It is known that it is self-repressed, repressed by SugR, and activated by GlxR [48, 64]. In turn, RamA activates the transcription of *sugR* [65]. On the other side, RamB was shown to be a repressor of these genes, binding the *pta-ackA* and *aceA/aceB* promoter regions when the cells were cultured in glucose as the sole carbon source. In these culture conditions, RamB is under the positive control of RamA [66]. Beyond the acetate metabolism, several carbon metabolism genes are under the control of RamA, RamB, or both [29, 31, 38, 43, 47, 48, 51, 65, 67-73]. Additionally, RamA, RamB, and GlxR bind the promoter region and control the transcription of *rpf2*, one of the two resuscitation promoting factor genes that are involved in the induction of a growing state in dormant cells [42].

The next sections will cover the role of microarray analyses in the elucidation of the carbon metabolism regulatory network.

Microarray Analysis of the Utilization of Acetate and Propionate

The group of genes that modify its expression in response to an imposed stimulus is named a stimulon. Study of stimulons allows the discovery of the physiological strategies developed to cope with a change in the environment. Among the first microarray studies of *C. glutamicum* were those focused in the acetate stimulon, reviewed by Gerstmeir et al. [74]. Acetate can be used as a carbon source for amino acid production, which guarantees the interest of its metabolism. Important information regarding the regulation of the central metabolic pathways has been gained through comparison of acetate-grown cells with glucose-grown cells. In the earlier genome-wide gene expression studies, three different groups working in parallel detected the same main transcriptional changes due to the presence of acetate [74]. Thus, the most upregulated genes encode for enzymes that activate acetate, for enzymes of the glyoxylate and of the tricarboxylic acid (TCA) cycles, and for the gluconeogenesis enzyme phosphoenolpyruvate (PEP) carboxykinase (the *pck* gene). The set of downregulated genes include those for sugar utilization enzymes.

Activation of both acetate and propionate depends on acetate kinase (AK) and phosphotransacetylase (PTA) enzymes. The respective encoding genes form the *pta-ack* operon and are required for growth on propionate or on acetate as sole carbon sources [75]. Despite this, the mutant strain *C. glutamicum* IN-pta, which lacks both AK and PTA activities, was able to utilize acetate or propionate when cultured on minimal medium containing glucose. A microarray assay of the mutant IN-pta grown on glucose *vs.* on glucose plus acetate, was carried out by Veit et al. [75]. The differential gene expression analysis was aimed at identifying a hypothetical alternative pathway for activation. The results discarded any candidate gene induced in the presence of acetate. This lead to the search of candidates among constitutively and highly expressed genes. For this purpose, RNA extracted from cells cultured on LB and on glucose were Cy3-labelled and hybridized in microarrays together with Cy5-labelled genomic DNA, which served as a reference. This design provides an approximate measure of the absolute transcript abundance [76]. Among the 100 genes with the highest mRNA levels, a gene annotated as a putative acetyl-CoA hydrolase was found. Biochemical characterization of the gene product revealed succinyl-CoA:acetate CoA transferase activity. The role of this gene, named *cat*, in the activation of acetate was confirmed using a mutant deficient in succinyl-CoA transferase and AK-PTA activities. The mutant was not able to grow on acetate as the sole carbon source.

Direct microarray comparisons between the wild type and the RamA- and RamB-deficient mutants were used by Auchter et al. [72] to characterize the regulons of both transcriptional regulators. Culture samples were taken at the mid-exponential growth phase in minimal medium with three formulations of carbon sources: glucose, glucose plus acetate, and acetate. These conditions were applied to all the mutant-parental strain comparisons, except for the RamA-deficient mutant in acetate, since this mutant does not grow in acetate as the sole carbon source. Instead, the *ramA* mutant and the wild type were compared in complex medium and in complex medium with glucose. This analysis confirmed the glucose repression mediated by RamA. Combining the microarray data from the cultures on minimal medium, 253 genes were identified as pertaining to the RamA regulon and 81 genes to the RamB regulon. Some of the observed changes may be explained by an indirect regulation, since 14 and 4 putative regulatory genes were included in the RamA and RamB regulons, respectively. Several functional classes were identified among the differentially-expressed genes. Thus, the regulon genes belong to the carbon metabolism and also to the metabolism of nitrogen, sulphur, iron, and phosphate, to respiration, and to the SOS and stress responses. In relation to the carbon metabolism, the authors concluded that RamB mediates the glucose-dependent repression of several genes—including *pta*, *ack*, *aceA*, *aceB*, and *pck* (acetate utilization and gluconeogenesis)—and that RamB is involved in the transcriptional activation of genes for the uptake of glucose and fructose (*ptsG* and *ptsS*). The regulatory role of RamA is antagonistic to RamB, acting as a repressor of glycolytic genes in the presence of glucose, as an activator of gluconeogenesis in the presence of acetate, and as an activator of TCA cycle genes and genes for acetate and ethanol utilization independent of the carbon source.

Regulation of Carbon Source Uptake

Transport of the carbon source into the cell is the necessary first step for the utilization of polar or charged carbon sources such as carbohydrates and organic acids. The major transporter system for glucose, fructose, and sucrose in *C. glutamicum* is the phosphoenolpyruvate:sugar phosphotransferase system (PTS). This widespread bacterial system is formed by the so-called central part, composed by the enzyme I and the HPr protein (encoded by *ptsI* and *ptsH*, respectively), and the enzyme II, which is the specific component for the transported sugar. Thus, for glucose, fructose, and sucrose the specific components of the system are encoded by *ptsG*, *ptsF,* and *ptsS,* respectively. In this respect, the master regulator SugR was first described as a repressor of sugar uptake [56, 57]. Engels et al. [56] used cells grown on LB to compare the wild type and the mutant Δ*sugR* transcriptomes. The repressor role of SugR was indicated by the higher mRNA levels from *ptsG*, *ptsS,* and also from the clustered genes *fruK* (1-phosphofructokinase) and *ptsF* observed in the mutant. On the other hand, Gaigalat et al. [57] also compared the transcriptomes of SugR-lacking and parental strains, but growing the cells on minimal medium with acetate as a carbon source. Their results confirmed the SugR repression of the previous genes and extended the regulon members to all the clustered genes *ptsI-cg2118-fruK-ptsF-ptsH*. Both research groups confirmed their findings with binding assays and RT-PCR analysis. In a later work, Engels et al. [61] found further target genes for SugR, such as those of the central carbon metabolism, using chromatin immunoprecipitation analyzed by microarray hybridization coupled to comparative microarray expression analysis.

The regulation of transporter systems other than PTS has also been studied using microarrays. These studies include the transport of *myo*-inositol [25] (discussed in the next section), dicarboxylate [77], gluconate [28, 78], uridine [18], ribose [37], lactate [22], and citrate [36]. A common approach is to compare the transcriptomes of the parental strain and of the mutant lacking the transcriptional regulator that controls the transporter genes, such as the UriR repressor of uridine transport and utilization [18, 37], and the RbsR repressor of the ribose transport [37]. Another common approach, combined or not with the previous one, is to compare cells cultivated on different carbon sources. A different strategy was used by Youn et al. [77] to find DccT, the dicarboxylate transporter. *C. glutamicum* does not grow in succinate, fumarate, or malate as the sole carbon source, however, after prolonged incubation spontaneous mutants are able to grow in these compounds. The microarray analysis detected one overexpressed gene, later named *dccT,* which encoded a protein of the divalent anion/Na+ symporter family. The overexpression of this gene proved to be sufficient to allow the growth on medium containing these TCA cycle intermediates as the sole carbon source.

Frunzke et al. [28] used single and double mutants of the *gntR1/gntR2* regulators to characterize the transcriptional regulatory properties of these two paralogous and functionally equivalent regulators. Parental and mutant strains were cultured on minimal medium with either glucose or gluconate, or with both carbon sources. No significant differences were found between the single mutants and the wild type strain in all the medium formulations, which was explained by the mutual complementation between the regulators. However, the double mutant cultivated on glucose showed 26 genes with decreased, and 19 genes with increased mRNA levels. Thus, GntR1 and GntR2 were found to strongly repress the gluconate permease (*gntP*), and the catabolic genes gluconate kinase (*gntK*) and 6-

phosphogluconate dehydrogenase (*gnd*). They also repressed to a lesser extent the cluster of pentose phosphate pathway genes. On the other hand, the *ptsG* gene, encoding the glucose permease component of the PTS system, was activated. *C. glutamicum* R, in contrast to the strain ATCC 13032, only contains the *gntR1* gene. Tanka et al. compared the transcriptomes of this strain and of the *gntR1* deletion derivative to search for new regulon members. The differential expression data were combined with ChIP-on-chip analysis to identify 56 new GntR1 binding sites, including those that control several carbon metabolism genes [78].

The transcriptome comparison of the glutamate-producing *C. glutamicum* strain 2262 and its non-glutamate-producing derivative, 2262NP, revealed the higher expression of the *NCgl2816-lldD* operon for a lactate permease and a quinone-dependent lactate dehydrogenase in the former. The analysis profile of wild type *C. glutamicum* grown on lactate, pyruvate, or glucose revealed that the major difference was the overexpression of the *NCgl2816-lldD* operon on lactate. Twenty genes with higher expression during growth on lactate coded for uncharacterized regulatory, transport, and secretory proteins. As demonstrated by the disruption and complementation experiments, LldD is essential for growth of *C. glutamicum* on lactate [22].

Citrate is the only intermediate of the TCA cycle that supports the growth of *C. glutamicum* as the sole carbon source [23]. A combination of transcriptome and proteomic analyses was used to characterize the citrate utilization in *C. glutamicum.* In the presence of citrate two different citrate-uptake systems encoded by *citH* (originally named *citM*) and *tctCBA* were overexpressed, while genes encoding uptake systems for other carbon sources such as glucose, fructose, sucrose, and gluconate showed decreased expression [23]. Either CitH or TctCBA, each one with a different cation requirement, is sufficient for growth on citrate [36]. The *citH* and *tctCBA* induction exerted by citrate is not subject to glucose-dependent carbon catabolite repression [23]. The transcriptome comparison of a Δ*citAB* mutant and the wild type strain, grown in the presence of citrate and glucose revealed the reduction of *citH* and *tctCBA* transcript levels in the mutant (25- to 100-fold lower). This indicates that the two-component system (TCS) CitAB is responsible for the activation of the genes, and was demonstrated by the in vitro binding of the response regulator CitB to the promoter regions of *citH* and *tctCBA* [36]. A microarray study performed with the wild type strain showed that the response to citrate leads to the overexpression of genes encoding enzymes of the TCA cycle (aconitase, succinyl-CoA synthetase, succinate dehydrogenase, and fumarase), malic enzyme, PEP carboxykinase, gluconeogenic glyceraldehyde-3-phosphate dehydrogenase, and ATP synthase. This probably reflects the higher demand of energy and carbon for gluconeogenesis, which requires the use of citrate as the sole carbon source [23].

Ethanol, Methanol, and *Myo*-Inositol As Carbon Sources

The major enzymes involved in the use of ethanol in *C. glutamicum* are alcohol dehydrogenase—ADH, encoded by *adhA* [31]—and acetaldehyde dehydrogenase—ALDH, encoded by *ald* [70]. Ethanol can be used as a sole carbon source, but, when present on a mixture of glucose and ethanol, the sugar is firstly utilized, and growth exhibits biphasic behavior. ADH and ALDH activities are low during the first growth phase but increase in the second growth phase, indicating that ethanol catabolism in *C. glutamicum* is subject to carbon catabolite control [27]. The master regulators of carbon metabolism, RamA and RamB, control *adhA* [31] and *ald* [70]. As indicated previously in this chapter, RamA and RamB regulons were analyzed by microarrays in RamA- and RamB-deficient strains growing in glucose, glucose plus acetate, or acetate as carbon sources. Sugar uptake, glycolysis, gluconeogenesis, acetate, lactate, or ethanol metabolisms were shown to be controlled by these two regulators [72]. Recently global expression profile has discovered the upregulation of *adhA* and *ald* transcription by methanol. Both enzymes ADH and ALDH are involved in methanol oxidation in *C. glutamicum*. Like ethanol, methanol catabolism is also subject to catabolite repression and depends on RamA [29].

C. glutamicum also uses *myo*-inositol as a carbon and energy source. Microarray analysis comparing growth on glucose and on *myo*-inositol revealed that 31 genes respond to the polyol utilization, with 21 of them being localized in two clusters. Only cluster I was necessary for catabolizing the polyol, being cluster II redundant. A mutant lacking IolT1 and IolT2 *myo*-inositol uptake systems was unable to grow on *myo*-inositol as the sole carbon source [25]. The transcriptional regulator encoded by *cg0196*, adjacent to the *iol* gene cluster I, was attributed to be a transcriptional regulator of *pck* encoding phosphoenolpyruvate carboxykinase [79]. However, a more recent DNA microarray analysis of the Δ*cg0196* mutant and the parental wild type strain revealed the upregulation (>100-fold) of the *iol* genes in the mutant. Moreover, binding sites for the encoded protein were identified in the promoter regions of *iolC*, *iolT1*, and *cg0196* itself, thus the gene has been designated *iolR* [38]. Additionally to the primary function in the control of *myo*-inositol genes, IolR is involved in the complex regulation of *pck* expression for adjusting the metabolic flux at the PEP pyruvate-oxaloacetate node during *myo*-inositol utilization [38].

Other applications of the microarray analysis include the identification of strong promoters inducible by carbon source [80], the ChIP-on-chip analysis of the GlxR global regulator [45], and the analysis of a mutant lacking the F1F0 ATP synthase [81].

The Nitrogen Metabolism

Nitrogen is an essential nutrient and the mechanisms used by bacteria to respond to its limitation are very varied [82]. The effect of nitrogen limitation over the metabolism of *C. glutamicum* has been of a great interest because of the industrial importance of the bacterium to produce amino acids, molecules with a high content in this element. The response of *C. glutamicum* to nitrogen starvation has been studied by transcriptomic and proteomic analyses showing changes in the expression of 248-285 genes [83-86]. Under nitrogen limitation conditions, the cell rearranges the cellular transport, changes the metabolic pathways for nitrogen assimilation, increases the energy generation and the protein stability, and decreases the amino acid and protein synthesis [83, 84]. The highest increase of expression observed corresponds to genes related to nitrogen assimilation, including transport systems and metabolic enzymes such as the ammonium transporters *amtA* and *amtB*; the *gltBD* operon, coding for the glutamate synthase (GOGAT), and the *glnA* gene, for the glutamine synthetase (GS), which allow the assimilation of ammonium via the high-affinity GS/GOGAT pathway; the ABC-type urea uptake system *urtABCDE* and the *ureABCEFGD* operon, encoding the urease enzyme complex, enabling the use of urea as an alternative nitrogen source; and the *crnT* and *codA* genes, for a creatinine permease and a creatinine deaminase, respectively, also allowing the use of creatinine as an alternative nitrogen source. The expression of *glnD* and *glnK* genes, encoding an adenylyl/uridylyl transferase and a P_{II}-type regulatory protein, respectively, which are essential for the *C. glutamicum* nitrogen starvation response, is also increased. All these genes are regulated by AmtR [87-90], the master regulator of the nitrogen control belonging to the TetR family of transcriptional factors [91]. The AmtR operon has been characterized by transcriptomic, metabolomics, and bioinformatic analyses, identifying at least 36 genes directly controlled by this regulator [86, 92, 93]. The mechanism of control over these genes has also been elucidated. AmtR is a negative regulator that in presence of an excess of ammonium stays bound to the AmtR-box, a well-defined palindromic sequence, blocking the expression of the genes [86, 94, 95]. It has been shown that a single highly conserved AmtR binding site is enough for stringent regulation. The non-conserved sites have very limited influence although they are recognized by AmtR [95]. In ammonium limitation conditions, GlnK is adenylated by GlnD [96], and GlnK-AMP interacts with AmtR, releasing AmtR from its DNA targets and allowing the gene expression of repressed genes [86, 97].

THE PHOSPHATE LIMITATION RESPONSE

Phosphate is an indispensable nutrient for all living organisms and the cells have different mechanisms to adapt rapidly to external changes in its concentration. In bacteria, the phosphate response control is exerted by a TCS named PhoR-PhoB in *E. coli* [98] and PhoR-PhoP in *Bacillus subtilis* and *Streptomyces* [99, 100]. *C. glutamicum* utilizes a number of inorganic and organic phosphates as phosphorus source, the inorganic phosphate (Pi) being the preferred form [101]. In a first study of the response of *C. glutamicum* to inorganic phosphate starvation, the genes were hierarchically organized in five subclusters according to their profile of expression [102]. The first subcluster responded rapidly to Pi depletion and kept a high expression. It is formed by genes needed to obtain Pi from organic compounds, including extracellular enzymes and genes encoding Pi transporters, such as the *pstSCAB* operon, coding for the high-affinity phosphate-specific transport system; *glpQ*, a glycerophosphodiester phosphodiesterase required for the utilization of glycerol-phosphodiesters as phosphate and carbon sources; the *ugpAEBC* operon, an ABC transporter for uptake of glycerol 3-phosphate; *nucH*, a putative secreted nuclease probably involved in liberating Pi from extracellular nucleic acids; *ushA*, a secreted 5'-nucleotidase allowing the cell to use nucleotides as phosphorus source [103]; *phoH*, an ATPase of unknown function; an ABC transport system for an unknown compound; and other genes coding for putative proteins [102].

The genes belonging to the second subcluster increased their expression after the Pi shift down, but slowly reached their original level. Two TCS are included in this subcluster, one of them related to copper metabolism and the other one, *cgl2607-cgl2606*, was identified in a subsequent work as *phoS-phoR* (*cgtS3-cgtR3*). PhoSR was found to be the only TCS whose synthesis is rapidly and transiently induced in response to Pi limitation [104]. Phosphorylated PhoR controls the expression of phosphate starvation inducible genes by direct binding to the so-called PHO-boxes present in their promoters, showing the highest affinity for the *pstSCAB* operon [105]. The characterization of the response to a Pi shift down in a Δ*phoSR* mutant showed that none of the genes belonging to the subcluster 1 was induced after Pi limitation [104]. The only exception was the *pstSCAB* operon, although its expression was clearly lower than in the wild type strain, suggesting the control by PhoS-PhoR and one or more additional regulators [104]. More recently, it was shown that the global regulator GlxR binds directly to the *pstSCAB* (and *ugpAEBC*) promoter [8]. The control of GlxR over *pstSCAB* in a phosphate-limited medium depends on carbon source. Activity from a promoter fusion lacking PhoR binding site was twofold when using glucose as the sole carbon source, whereas the activation was minor in acetate medium [53]. Additional factors are probably involved in *pstSCAB* control, because a promoter fusion with both PhoR and GlxR binding sites deleted still showed phosphate starvation induction [53].

The other three subclusters identified in the response to phosphate limitation contained genes apparently not involved in phosphate assimilation. Subcluster 3 (containing genes related to protocatechuate degradation) increased the expression only briefly after the Pi shift down, whereas subclusters 4 (with genes mainly associated to protein synthesis) and 5 (for iron uptake) decreased the expression after Pi limitation [102].

The Metabolism of Other Essential Nutrients: Iron, Sulphur, and Zinc

Iron is an essential element for almost all microorganisms acting as cofactor for many proteins involved in relevant processes such as electron transfer, catalysis, and regulatory processes [106]. However, it possesses problems of toxicity and low solubility [107]. To solve this problem bacteria have evolved different molecular mechanisms to achieve an effective iron homeostasis. In corynebacteria, the global regulator of iron metabolism is DtxR, firstly identified in *Corynebacterium diphtheriae* by its role in controlling the expression of the diphtheria toxin gene [108]. DtxR is a repressor protein that in presence of iron forms a complex that binds to the DtxR-box (a 19 bp palindromic sequence) blocking the expression of genes. In iron limitation conditions, the complex dissociates, derepressing the genes that can then be transcribed [108]. The DtxR regulon has been characterized by transcriptomic and bioinformatic analyses comparing a Δ*dtxR* mutant and the wild type strain. More than 30 transcription units directly controlled by DtxR, both positive and negatively, have been identified. Genes involved in iron acquisition and utilization were overexpressed in the Δ*dtxR* mutant, and therefore repressed by DtxR, whereas genes responsible for iron storage and DNA protection were underexpressed (activated by DtxR) [109, 110]. This dual role is exerted via binding to different regions in the promoters. The DtxR-box in genes repressed by DtxR overlaps with the -35 to -10 promoter region, whereas the DtxR-box in activated genes is located upstream of the -35 box [109].

Among the genes directly regulated by DtxR, are 4 response regulators [109, 110]. RipA is a repressor of the TetR family. Transcriptomic analysis of a Δ*ripA* mutant showed that in iron limitation conditions the expression of different iron-containing proteins is increased in reference to the wild type strain, whereas the *ripA* itself is repressed by DtxR in excess iron conditions [111]. Recently it has been shown that in the presence of high CO_2/HCO_3^- levels, there is a significant activation of the complete DtxR/RipA regulon through the repression of the *ripA* gene, whose expression was decreased six-fold [112]. In these conditions the maximum specific growth rate (μ_{max}) of a Δ*dtxR* mutant was 40% lower compared to the wild type strain, showing the importance of the role of DtxR in presence of increased CO_2/HCO_3^- conditions [112]. A second regulator, directly controlled by DtxR is the TCS *cgtSR11*, named *hrrSA*, was involved in the control of heme homeostasis [113]. Transcriptomic analyses have shown that this TCS activates genes for heme oxygenase and heme-containing proteins, and represses genes involved in heme biosynthesis. When *C. glutamicum* is cultivated with heme as the sole iron source, the response to iron starvation in a Δ*hrrA* mutant is stronger than in the wild type strain: there is a higher expression of several genes belonging to DtxT regulon, and the RipA regulon is underexpressed [113]. A second TCS, named *chrSA* (*cgtSR8*), is involved in heme regulation, being crucial for heme resistance. There is a cross regulation between both TCS, having a balance between heme toxicity and iron acquisition [114]. The other two regulators controlled by DtxR are members of the ArsR family. One of them encodes GlyR, which activates the expression of *glyA*, for a serine hydroxymethyltransferase, with an unclear relationship with iron metabolism [115]. The other ArsR-type regulator is *cg3082*, which controls its own expression and that of *cg3083-cg3085* genes, probably involved in divalent cations transport [12]. A bioinformatic analysis looking for the presence of DtxR-boxes identified new regulators probably directly controlled by DtxR, including *ramB* and *sufR* [110]. RamB is a global regulator of central metabolism (see above) and SufR is the probable regulator of the operon *sufRBDCSUT*, encoding the so-called SUF machinery, involved in the assembly and repair of iron-sulphur clusters [110].

The importance of sulphur metabolism in *C. glutamicum* is related to the production of sulphur-containing amino acids. The master regulator is McbR, a member of the TetR family repressors, detected by its ability to bind to the promoter of *metY*, encoding O-acetyl-L-homoserine sulphydrylase [116]. A proteomic analysis of a Δ*mcbR* mutant allowed identification of 6 genes (*metY*, *metK*, *cysK*, *hom*, *cysI*, and *ssuD*) directly controlled by McbR, all of them key enzymes for the biosynthesis of methionine and cysteine, as well as for the utilization of sulphonate [116]. A subsequent analysis by transcriptomics and bioinformatics identified a total of 86 genes overexpressed in the Δ*mcbR* mutant, 45 of them (clustered in 22 transcription units) directly regulated by McbR, which suggests the presence of the McbR-box in their promoters [117]. These genes are those needed for S-adenosylmethionine (SAM) and cysteine biosynthesis, uptake and utilization of sulphur-containing compounds, and in sulphate reduction. The McbR activity is negatively controlled by the level of intracellular S-adenosyl-L-homocysteine (SAH), which is derived from SAM by demethylation [117]. A different study by proteomic and transcriptomic analysis showed that in a Δ*mcbR* mutant, several oxidative stress proteins were significantly upregulated. In addition, the mutant exhibited a reduced growth rate, reduced glucose uptake rate, and reduced biomass yield, with a reroute of the metabolic flux [118].

Five transcriptional regulators were differentially expressed in the Δ*mcbR* mutant with respect to the wild type strain [117]. McbR controls the expression of *cysR*, a ROK-type regulator controlling the expression of at least seven transcription units belonging to the McbR regulon [119]. CysR can only activate transcription of sulphate assimilation genes when an acceptor for the produced sulphide is available [119]. It has been shown that succinate dehydrogenase subunit A (SdhA) interacts directly with CysR and transcription of *cysR* is diminished in a Δ*sdhA* mutant, although the mechanism of control remains unknown

[120]. The genes directly controlled by CysR include the *fpr2-cysIXHDNYZ* gene cluster, needed for assimilatory sulphate reduction [121], and the regulator *ssuR*, which activates the expression of genes for sulphonate utilization in the absence of sulphate [122]. On the other hand, *ssuR* is also directly controlled by McbR [117]. This double control of McbR over *ssuR,* directly and indirectly through CysR, suggests that genes for sulphonate utilization are only expressed when the source of inorganic sulphur is low [12]. A third regulator directly controlled by McbR is *cg0444*, whose expression was increased in a Δ*mcbR* mutant, compared to the wild type strain [117]. Cg0444 corresponds to RamB, involved in central metabolism (see above). The interaction between carbon and sulphur metabolism is in both directions, since *mcbR* expression was diminished in a Δ*ramA* mutant [72]. The expression of *sufR* is also controlled by McbR [117], in addition to DtxR (see above). The transcription of *suf* operon can occur from a housekeeping promoter, but also from a SigM-dependent promoter under disulphide-stress (see below) [123]. The last regulator controlled by McbR is *cg3373*, an ArsR family transcriptional regulator of unknown function [117].

Like other transition metals, zinc is an essential nutrient, but toxic when present at high concentration. Zur repressor controls the zinc metabolism. In a Δ*zur* mutant, the expression of 23 genes is increased at least two-fold and the expression of 3 genes is decreased in the same ratio [9]. Five of the overexpressed transcription units have a Zur-box in their promoters, including two copies of the ABC-type transporter *znuACB* for zinc uptake. Zur binds to the promoters in a zinc-dependent manner, and when Zn concentration is low, Zur is released allowing the expression of the genes [9]. A different regulation mechanism has been described for *zrf* and *zra* genes, involved in zinc resistance. Zur regulates both genes, but the binding of the repressor to the promoters is inhibited by zinc [124]. In this case, the motif recognized by Zur is a direct repeat sequence [124] instead of the inverted repeat sequence, previously identified [9].

Nitrate Respiration in Anaerobic Conditions

C. glutamicum is a facultative anaerobe able to grow using nitrate as a terminal electron acceptor. In anaerobic conditions, it consumes nitrate and excretes nitrite by expressing the *narKGHJI* gene cluster, coding for a nitrate/nitrite transporter and a respiratory nitrate reductase [125]. Expression of *narKGHIJ* is repressed in aerobic conditions by ArnR, an autoregulated protein that belongs to a novel type of transcriptional regulators [126]. Transcriptomic analyses in aerobic conditions of an Δ*arnR* mutant showed an overexpression of *narKGHIJ* and *hmp*, encoding a flavohemoglobin [126], usually acting as NO detoxifying [127]. In a subsequent study, the expression profile under anaerobic nitrate respiratory conditions was compared to aerobic conditions. The analysis showed the overexpression of *narKGHJI* and *hmp* genes, confirming the previous results, as well as that of several genes involved in the respiratory electron transport chain and others belonging to the SOS regulon [52]. On the other hand, exposure to NO-donating agents under anaerobic conditions induced the expression of *hmp*, but not the expression of the SOS genes [52]. The strong growth defects of a Δ*hmp* mutant under aerobic conditions in the presence of nitrate, nitrite, and the NO-donating reagent sodium nitroprusside, but also in anaerobic nitrate respiration, suggested that Hmp is responsible for nitric oxide conversion to either nitrate or nitrous oxide in *C. glutamicum* [128].

Sigma Factors As Transcriptional Regulators Involved in Stress Conditions

The majority of bacteria species have several different sigma subunits, which can alternatively join the core of the RNA polymerase enzyme. These dissociable subunits (sigma factors) recognize specific promoter DNA sequences and act as transcriptional global regulators, modifying the expression of large gene groups (regulons or sigmulons) in response to environmental stimuli or changes in metabolic conditions. In *C. glutamicum*, SigA and SigB are the primary and the primary-like sigma factors, respectively, whereas SigC, SigD, SigE, SigH, and SigM have been designated extracytoplasmic function (ECF) sigma factors. These ECF are involved in the response to stress conditions, with SigH centered in the cross-regulated network of sigma factors for the control of the concerted stress response [12].

SigA controls the transcription of the majority of the housekeeping genes through its binding to DNA sequences, which are very similar to those recognized by the principal sigma factors of other Gram-positive bacteria, and also by *E. coli* Sig70 [129]. During the transition from the exponential to the stationary growth phase, *sigB* increases its transcription, whereas *sigA* reduces it [130], replacing the sigma subunit in the RNA polymerase. Genes involved in glucose consumption (*pfkA*, *fba*, *tpi*, *gapA*, *pgk*) were found to be controlled by SigB during exponential growth [131]. The SigB dependent genes in transition and stationary phase include those for central metabolism, but also others to deal with the oxidative stress defense, which accompanies the nutrient limitation. Thus, a glyoxilase (*cg1073*), a methionine-R-sulphoxide reductase (*cg2078*), and the nitric oxide-detoxification flavohemoprotein (*hmp*) were found to be upregulated, whereas the universal stress protein (*uspA2*) and the chaperone Hsp70 (*dnaK*) were downregulated, when compared to a *sigB* mutant [130]. Under conditions of oxygen deprivation, the *sigB* gene was upregulated, enhancing glucose consumption. In turn, *sigB* disruption reduced transcription of the glucose metabolism genes [131]. SigB is then accepted to act as a housekeeping sigma factor [10], but also plays a role in the response to various environmental stresses [132]. Other stresses, in which the *sigB* expression increased, are acidic conditions [133] and amino acid starvation [134].

SigC and SigD sigma factors *C. glutamicum* have been poorly studied. The *sigD* gene was found to favor the growth in low oxygen concentration when introduced in a high-oxygen-requiring *C. glutamicum* mutant, suggesting that it is involved in the transcription of genes for adaptation to the microaerobic conditions [135].

The *sigE*-regulated genes have not been described in *C. glutamicum*, but *sigE* is involved in response to cell surface stress caused by antibiotics and detergents. The *sigE* gene is located upstream of *cseE* gene, encoding the anti-sigma factor of SigE. The interaction of both proteins, SigE and CseE, was demonstrated in vitro, although the signal that releases SigE from its antagonist CseE remains unknown [136]. The *sigE* transcript level increases in the nitrogen starvation response [83], in a *rel* mutant unable to induce the stringent response [134], and, as indicated below, during lactic acid adaptation [133].

SigH As a Master Regulator of the Stress Response

Transcription of the *sigM* gene is increased under temperature shocks and disulphide stress caused by diamide treatment. Deletion of *sigM* sensitizes cells to heat or cold and to the presence of the thiol oxidant diamide. A global expression study in a *sigM*-proficient strain revealed the overexpression of *trxB*, *trxB1*, and *trxC* genes for thioredoxins; *suf* operon, along with the transcriptional regulator *sufR,* for the assembly of the Fe-S cluster; and genes for heat shock response [123]. However, this control by *sigM* seems to be confusing since the same genes for heat and disulphide stresses were reported to be *sigH*-dependent in another microarray analysis of the heat shock response [137]. This could be explained as an indirect consequence of the SigH-dependent *sigM* expression. Also, it could be due to the sequence similarity found between the SigM and SigH -10 and -35 consensus motifs allowing the recognition by either RNAP+SigH or RNAP+SigM depending on the conditions [10].

NaOCl produces the chlorination of thiol groups to form sulphenyl chloride. This unstable molecule reacts further with proximal thiols to form disulphides or is overoxidized to Cys-sulphonic acid, thus provoking a disulphide stress response. The expression profile provoked by NaOCl was analyzed by microarrays [138], and fits with that obtained in disulphide stress [137, 139]. The response to NaOCl showed the strong upregulation of the SigH regulon, including genes for the anti-sigma factor (*rshA*) and the regulons of the redox-sensing regulators QorR [137], CyeR [140], and RosR [141]. Moreover, the genes for thioredoxin and mycothiol pathways (*trxB1*, *trxC*, *cg1375*, *trxB*, *mshC*, *mca*, *mtr*) were also overexpressed in the presence of NaOCl [138]. Mycothiol overproduction improved the robustness of *C. glutamicum* to different stresses such as oxidants, antibiotics, ions, or ethanol [142]. The S-mycothiolation process protects redox-sensitive thiols against overoxidation under stress in *Actinomycetes*, as occurs with S-bacillithiolation in *Firmicutes* bacteria [143] or S-glutathionylation in eukaryotic cells, in which also serves to regulate cellular processes by modulating protein function [144].

A heat shock response includes the arrest of cell growth and the increased expression of chaperones, which assist protein folding and repairing, and ATP-dependent proteases, for degrading misfolded proteins. The regulation of the ATP-dependent protease Clp expression was observed to be under a regulatory network involving the transcriptional regulators ClgR, HspR, and the sigma factor SigH [145]. ClgR activates transcription of *clpC* and *clpP1-P2*, and the 3 genes are induced under heat stress independently of *clgR*, through the presence of SigH binding motifs in their promoters. Additionally, HspR binds to *clgR* promoter repressing its expression [145].

A global expression profile of *C. glutamicum* cells subjected to heat shock revealed the upregulation of *sigH* and *hspR*, along with other regulators, and genes for heat shock proteins (HSP) and oxidative stress response [137]. The *sigH* regulon identified 45 genes including the already known *clpP1-P2* and *clpC*, and also *dnaK*, *grpE*, *dnaJ*, *clpB*, *dnaJ2*, *groES*, *groEL1*, and *groEL2*, and allowed the definition of a consensus sequence for SigH binding site. The disruption of *hspR* and *hrcA* also upregulated the expression of some HSP genes, which indicated that SigH, but not SigM nor SigB, regulates heat shock response in concert with HrcA, HspR, and other transcriptional regulators [137].

The activity of SigH is controlled by the anti-sigma factor RshA in *C. glutamicum* as occurs in other actinobacteria. Both *sigH* and *rshA* genes are transcriptionally coupled and expressed from four SigA housekeeping promoters located upstream of *sigH*, whereas the monocistronic *rshA* transcript is under the control of SigH. Transcription of *rshA* from a SigH-dependent promoter turns off the SigH-dependent stress response after the cells have overcome the stress condition [139]. Transcriptome analysis of the *rshA* deletion mutant fits with the already known SigH-dependent genes [137], but also includes the overexpression of SOS and DNA repair genes *uvrA*, *uvrC*, and *uvrD3*. The integration of the DNA damage repair in the SigH network is in accordance with the presence of a SOS box overlapping with the -10 region of the *sigH* P4 promoter [139] for the binding of the LexA repressor [146].

Two promoters SigA- and SigH-dependent were found to drive the transcription of the small RNA ArnA. Heat shock condition was observed to trigger the expression of ArnA, which, as a cis-antisense RNA, could regulate the expression of *gntR2* [147]. Other sRNA can be involved in the response to different stresses. Transcript *cgb_00105* was not detected under heat shock, whereas it was present in cold shock and in chemically induced stresses, although there is no hint of its function [148].

The SOS Response

The sudden increase in DNA damage induces a repair and damage tolerance system (SOS response). The SOS response is controlled by a complex interplay involving LexA and RecA proteins, and it is accepted to function as an important bacterial system in adaptation, but also in mutation and diversification, under a broad range of environmental stresses [149]. To identify the SOS regulon of *C. glutamicum*, a comparison of the transcriptome profiles of the wild type strain growing in control condition or in the presence of the DNA-damaging agent mitomycin C, and that of a *lexA* mutant, was performed. The 48 genes of the SOS regulon detected comprise *recA*, *recX*, *recN*, *lexA*, *uvrA*, and *ruvCAB*, involved in DNA repair; *divS* and *ftsK*, for cell division; and others for general or uncharacterized functions [146]. Seventeen genes out of the 48 belonging to the *C. glutamicum* SOS regulon [146] were upregulated after the shift from aerobic to anaerobic nitrate respiration. The expression pattern of 6 of the upregulated transcripts (*lexA*, *recA*, *uvrA*, *ruvC*, *recN*, and the *divS-nrdR* operon) was found to be controlled by RecA, as tested by qPCR. In that sense, the induction of the SOS regulon plays an important role in tolerance and survival coping with potential nitrosating agents generated from anaerobic nitrate respiration [52].

The Starvation Response

The reaction to nitrogen starvation includes a scavenging of nutrients with a rearrangement of the cellular transport systems, changes in metabolic pathways concerning nitrogen assimilation and amino acid biosynthesis, and a reduced capacity for protein synthesis accompanied by an enhanced proteolysis [83]. To discriminate the genes responding to nitrogen scavenging from those responding to general starvation, both nitrogen and phosphate starvation responses were compared, showing that the decreased expression of genes for ribosomal proteins and translational initiation factors is most likely a general stress effect than a response to nitrogen limitation [85].

The analogue serine hidroxamate acts as amino acid starvation inducer. This stringent response has been analyzed by microarrays in both, Rel-dependent and -independent manner, by means of transcriptome comparison of a control strain and a *rel* deleted mutant. In the Rel-independent control the addition of serine hidroxamate downregulates the expression of 26 genes for ribosomal proteins, *infA* for translation, 3 genes for transcription or DNA replication, and also the *narKGHJI* operon for a nitrate reductase and transport genes. The stringent response induced an oxidative stress answer, with the upregulation of *katA* and *dps* genes encoding a catalase and a DNA protection protein, respectively [134].

The Salt Stress Response

To avoid dehydration in a hyperosmotic environment bacteria accumulate or synthesize compatible solutes. Due to the low energy cost and activation speed, the uptake of compatible solutes—certain amino acids and carbohydrates—is preferred to their biosynthesis. The four *C. glutamicum* osmoregulated uptake systems present overlapping substrate specificities and different transport activities. EctP is considered as a maintenance carrier, whereas the other three carriers, ProB, BetB, and LcoP, allow the optimal regulation depending on the extent of the applied stress [150]. The MtrAB TCS was found to be involved in the osmoprotection as well as in cell wall metabolism. The deletion of the *mtrAB* genes in *C. glutamicum* sensitized cells to penicillin, vancomycin, or lysozyme. Comparison of the global expression profiles of a *mtrAB* mutant and the wild type strain showed that the more underexpressed transcripts corresponded to *betP* and *proP* genes, which encode transporters for the uptake of glycine/betaine and proline, respectively. The increased RNA levels in the Δ*mtrAB* mutant were for *mepA*, *ppmA*, *lpqB*, and *nlpC*, putatively related with the cell wall [151]. MtrA function was studied by chromatine immunoprecipitation-microarray hybridization (ChIP-on-chip) and DNA affinity chromatography assays revealing a dual role, direct activator of *betP* and *proP*, and repressor of *mepA* and *nlpC*, coding for putative cell wall peptidases [152]. The involvement of *proP* and *betP* in osmotic stress defense fits with proteomic and transcriptomic global data obtained in adaptation to salt stress conditions: ProP was the strongest upregulated protein, BetP was also upregulated, and increased levels for both transcripts were detected; in contrast, the other osmolyte carriers, LcoP and EctP, showed no significant regulation in both proteome nor transcriptome data [153].

The pH Adaptation Response

Besides the characterization of the lactate stimulon [22], the adaptation to the low pH induced by lactic acid was studied [133]. The global gene expression profiles of *C. glutamicum* adapted to pH 5.7 with lactic acid and to pH 7.5 without lactic acid were characterized by DNA microarray analysis. Long-term adaptation involved the increased expression of 116 genes and the decreased expression of 90. To characterize the function of 38 upregulated genes, disruption mutants were generated. However, only two of them showed an acid-sensitive growth phenotype. A *sigB* disrupted mutant exhibited decreased growth rate when grown at pH 5.7 and reduced survival after an acid shock exposure. A *sigE* mutant already displayed retarded growth at neutral pH, but grew better at acidic pH than the *sigB* mutant. The lack of acid-sensitive phenotypes in 34 out of 36 disrupted genes was considered either a redundancy in acid adaptation response or coincidental effects. Among the upregulated genes in acidic condition, the enhanced expression of the *nrd* (ribonucleotide reductase) operon and that of a gene encoding a DNA ATPase repair protein accounts for the induction of a repair system for acid-induced DNA damage [133].

In *C. glutamicum* pH, homeostasis is maintained at 7.5 ± 0.5 while the environment ranges between pH 6-9. At lower or higher external pH values the maintenance of the internal pH is not achieved and leads to reduced growth rates. At neutral and low pH values oxidative stress occurs, the mechanisms for reactive oxygen species defense are impaired, and methionine and cysteine synthesis, as well as iron starvation response, are induced. All these effects indicate a functional link between pH maintenance, oxidative stress, iron homeostasis, and metabolic alterations [154].

The Hydrogen Peroxide Stimulon

C. glutamicum shows high natural resistance to H_2O_2 due to the production of catalase [155, 156], a fact used for the marketing of *C. glutamicum* catalase. The characterization of the hydrogen peroxide stimulon showed that only 2 minutes after the exposure to the agent, the genes *katA* (catalase) and *dps* (DNA protection protein) were the highest upregulated. Other genes overexpressed encode transcriptional regulators (*farR*, *cysR*, *ripA*, *oxyR*, *znr*, and *sufR*). These regulators control the second phase of the response, 3 minutes later, in which many genes for SOS and stress responses, as well as macro element and metal homeostasis, are induced. In contrast, many genes for carbohydrate metabolism and other specific biosynthesis pathways are repressed [156]. The hydrogen peroxide sensor-transcriptional regulator OxyR is the best-characterized regulator for hydrogen peroxide detoxification in bacteria. Two cysteine residues (206 and 215) in the C-terminus of the OxyR protein are supposed to be involved in the sensing of hydrogen peroxide [156]. Deletion of *oxyR* increased the resistance to hydrogen peroxide [155, 156], which suggests that OxyR has a repressor role in *C. glutamicum* under non-stress conditions. The members of the OxyR regulon were studied by global expression and OxyR binding sites were characterized, establishing a model for the OxyR-mediated redox-responsive regulatory mechanism. The OxyR regulon comprises 23 genes related to oxidative stress response, sulphur metabolism, and iron homeostasis. Some of these genes are part of the H_2O_2 stress experiment, e.g, the induced *dps*, *hemH*, *katA*, *oxyR*, genes of the *suf* cluster, *cg0311*, and *cg1746* [156].

Strain Development for Amino Acid Production Based on Global Expression Profile Studies

The production of amino acids is in the multi-million-ton-scale, with *C. glutamicum* playing a leading role in the industrial amino acid fermentation. The classical method for developing amino acid overproducing strains was the repeated random mutation and selection. Once the genome sequence was available [1, 2], the microarray technology provided detailed information on the metabolic pathways and their regulation to further increase the productivity of new designed strains.

Strain Development for Lysine Production

The transcriptome comparison between a lysine *C. glutamicum* industrial producer, B-6, and its parental strain revealed the higher expression levels of the pentose phosphate pathway and amino acid biosynthetic genes, especially *leuCD*, accompanied by a lower expression of the TCA cycle genes in the former strain [157]. The global amino acid upregulation in the producer mutant was suggested to be due to the induction of a general regulatory mechanism such as the stringent response [157]. The coding sequences of the amino acid biosynthetic genes in the lysine producer B-6 were searched for point mutations, and, as a result, a substitution in *leuC* gene (*leuC456*) was defined as a useful mutation. The *leuC456* allelic replacement increased the lysine production and led to a partial leucine requirement in the constructed strain ADL-3. Global expression profile due to the *leuC456* mutation in *rel*-negative and -positive backgrounds, showed that the upregulation of many amino acid biosynthetic genes is Rel-independent [158]. *NCgl0855* gene (coding a methyltransferase) and the *amtA-ocd-soxA* cluster (for an ammonium uptake system, an ornithine cyclodeaminase and an uncharacterized protein) were upregulated in a global expression comparison using the lysine producer MH20-22B and other *C. glutamicum* strains, and were suggested to be targets for the lysine production improvement [159].

Strain Development for Arginine Production

Global transcriptome analysis has been used to reveal points for arginine production optimization. The genetic comparison of three *C. glutamicum* classical producers obtained by multiple rounds of mutagenesis concluded with the need of the modification of *argB* (*argB26*), for deregulating the encoded N-acetyl-glutamate kinase biosynthetic key enzyme, and the deletion of *argR*, the *arg* operon repressor, for increasing *arg* genes expression. However, transcriptome analysis of such constructed strain, named RB, revealed that the *arg* operon of the classical producer was much more overexpressed than that of RB [160]. ChIP assays demonstrated the binding of ArgR to the upstream regions of *argC*, *argF*, *argG*, or *argB*, as well as those of *gdh* and *gltBD* genes (for glutamate dehydrogenase and glutamate synthase, respectively), which links the pathways of the direct precursor, glutamate, and the derived compound, arginine [161, 162]. Introduction of *leuC456* [157] into strain RB increased arginine production [160]. Besides the repression exerted by ArgR, FarR is also involved in the *arg* cluster regulation, as indicated by a global transcription profile of a FarR mutant. However, the overexpression of *arg* genes in a FarR mutant is not accompanied by an increased production of arginine, presumably due to the feedback inhibition of ArgB by arginine [163]. FarR also binds upstream regions of *arg* genes and *gdh* promoter as indicated by ChIP assays, and both regulators, ArgR and FarR, formed a complex with the *argB* promoter (ArgR-FarR-DNA) or formed protein-DNA complexes individually [164].

Metabolic engineering strategies have recently been applied to overproduce arginine and its precursor ornithine in *C. glutamicum*. For the production of arginine, these include the removal of the arginine operon repressor, the redirection of carbon flux to enrich the NADPH pool, and the disruption of the glutamate exporter to increase the arginine precursor availability [165]. The blocking of competitive pathways conducting to proline and arginine has been used for the production of ornithine [166].

STRAIN DEVELOPMENT FOR VALINE PRODUCTION

Acetohydroxy acid synthase (AHAS) is the key enzyme for the synthesis of the branched-chain amino acids (BCAAs) valine, isoleucine, and leucine, which exert feedback inhibition to AHAS. The *C. glutamicum* AHAS possesses two catalytic and two regulatory subunits encoded by *ilvB* and *ilvN*, respectively. The deletion of the C-terminal domain of IlvN rendered a feedback-resistant AHAS, insensitive to the BCAAs, with decreased AHAS activity, but increased lysine formation. Deletion of the *ilvB* gene, encoding the catalytic subunit of AHAS, led to BCAAs auxotrophy and further-improved lysine production. Comparative transcriptome analysis of a lysine producer and its *ilvB* derivative mutant revealed the differential expression of approximately 50 genes. However, none of them encode for enzymes involved in the lysine biosynthetic pathway, indicating that the beneficial effect of the *ilvB* deletion on lysine production is indirect [167]. Other studies in *C. glutamicum* showed a link between increased lysine formation and leucine auxotrophy [159] or with a limited leucine supply [158].

The strategies developed to improve valine production have been recently reviewed [168]. These include the deletion of pyruvate dehydrogenase complex and the overexpression of the *ilvBNCE* genes encoding the valine biosynthetic enzymes [169]; the overexpression of *ilvBNCE* and the deletion of the genes encoding pyruvate:quinone oxidoreductase, phosphoglucose isomerase, and pyruvate carboxylase [170]; the inactivation of the *sugR* gene and the replacement of acetate by ethanol in the media [171]; the introduction of a point mutation in *atpG* for a H (+) -ATPase [172]; the modulation of the promoter activity of the genes involved in the biosynthesis of branched-chain amino acids [173]; or the increase of NADPH availability [174].

STRAIN DEVELOPMENT FOR LEUCINE AND TRYPTOPHAN PRODUCTION

A defined deletion created in *ltbR* gene (for leucine and tryptophan biosynthesis regulator) led to the mutant strain *C. glutamicum* IB1486. According to whole-genome expression analysis the *trpEGDCFBA* operon, involved in tryptophan biosynthesis, and the *leuB* and *leuCD* genes, encoding enzymes of the leucine biosynthesis pathway, showed enhanced transcription in *C. glutamicum* IB1486 compared to the wild type strain. The regulatory role of LtbR in the expression of the leucine and tryptophan biosynthesis pathways was demonstrated by binding retardation experiments using DNA fragments localized in front of *leuB*, *leuC*, and *trpE*, which contain a specific 12-bp motif. The regulatory connection between the leucine and tryptophan biosynthesis pathways exerted by the LtbR repressor is new in bacteria [175]. Expression of the *leuCD* genes of *C. glutamicum* was also repressed under iron limitation by RipA [111]. It was suggested that the *trp* operon is regulated by classical attenuation [176].

CONCLUSION AND PERSPECTIVES

The availability of the *C. glutamicum* genome information has enabled the development of 'omics techniques and has caused a breakthrough in the understanding of the physiology of this microorganism. This chapter proves that microarrays constitute a useful tool in tracing changes in the gene expression level, facilitating the recognition of global expression patterns at the whole genome scale. In the mentioned studies, the microarrays have been employed for the elucidation of experimental hypothesis—revealing functional areas under regulation because of its transcriptional response to various stimuli—, for discovering new regulon members or identifying candidate target genes, for defining the sign—activation or repression—exerted by a transcriptional regulator over the controlled genes, or for building hierarchical cascades in the regulatory network. Moreover, the combination of Chromatin Immunoprecipitation (ChiP) with microarray technology (ChIP-on-chip) allows the testing of in vivo global protein-DNA interactions to identify functional elements in the genome.

The advances in the technology of next generation sequence—applied in techniques such as RNA-seq or ChIP-seq—are gradually replacing microarrays, especially for studies in human and higher eukaryotes. In *C. glutamicum* these approaches have achieved the characterization of regulon members by identifying in vivo targets of the tested regulator protein [50]; the discovery of a widespread antisense transcription and the mapping of promoter regions, gene starts and stops, 5′-UTRs, RBSs, transcript starts and ends [177]; or the detection and characterization of small RNA with potential regulatory properties [148, 178].

The strength of microarray technology rests on its accessibility with several available and proven platforms, the ease of use, the shorter turnaround time, and the lower cost. On the other hand, the storage and analysis of RNA-seq data are more challenging. Anyway, the results from RNA-seq and microarray are in good agreement [179]. Moreover, the two technologies are complementary [179-182] and microarrays and RNA-seq can be used for the orthogonal validation of the results of each other by simultaneously measuring thousands of transcripts.

REFERENCES

[1] Ikeda M, Nakagawa S. The *Corynebacterium glutamicum* genome: features and impacts on biotechnological processes. *Appl Microbiol Biotechnol.* 2003 Aug;62(2-3):99–109.

[2] Kalinowski J, Bathe B, Bartels D, Bischoff N, Bott M, Burkovski A, et al. The complete *Corynebacterium glutamicum* ATCC 13032 genome sequence and its impact

on the production of L-aspartate-derived amino acids and vitamins. *J Biotechnol.* 2003 Sep;104(1-3):5–25.

[3] Yukawa H, Omumasaba CA, Nonaka H, Kós P, Okai N, Suzuki N, et al. Comparative analysis of the *Corynebacterium glutamicum* group and complete genome sequence of strain R. Microbiology-SGM. 2007 Apr;153(Pt 4):1042–1058.

[4] Lv Y, Wu Z, Han S, Lin Y, Zheng S. Genome sequence of *Corynebacterium glutamicum* S9114, a strain for industrial production of glutamate. *J Bacteriol.* 2011 Nov;193(21):6096–6097.

[5] Lv Y, Liao J, Wu Z, Han S, Lin Y, Zheng S. Genome sequence of *Corynebacterium glutamicum* ATCC 14067, which provides insight into amino acid biosynthesis in coryneform bacteria. J Bacteriol. 2012 Feb;194(3):742–743.

[6] Brune I, Brinkrolf K, Kalinowski J, Pühler A, Tauch A. The individual and common repertoire of DNA-binding transcriptional regulators of *Corynebacterium glutamicum, Corynebacterium efficiens, Corynebacterium diphtheriae* and *Corynebacterium jeikeium* deduced from the complete genome sequences. BMC Genomics. 2005;6(1):86.

[7] Brinkrolf K, Brune I, Tauch A. The transcriptional regulatory network of the amino acid producer *Corynebacterium glutamicum*. J Biotechnol. 2007 Apr;129(2):191–211.

[8] Kohl TA, Tauch A. The GlxR regulon of the amino acid producer *Corynebacterium glutamicum*: Detection of the corynebacterial core regulon and integration into the transcriptional regulatory network model. *J Biotechnol.* 2009 Sep;143(4):239–246.

[9] Schröder J, Jochmann N, Rodionov DA, Tauch A. The Zur regulon of *Corynebacterium glutamicum* ATCC 13032. *BMC Genomics.* 2010;11:12.

[10] Pátek M, Nešvera J. Sigma factors and promoters in *Corynebacterium glutamicum. J Biotechnol.* 2011 Jul;154(2-3):101–113.

[11] Bott M. Offering surprises: TCA cycle regulation in *Corynebacterium glutamicum. Trends Microbiology.* 2007 Sep;15(9):417–425.

[12] Schröder J, Tauch A. Transcriptional regulation of gene expression in *Corynebacterium glutamicum*: the role of global, master and local regulators in the modular and hierarchical gene regulatory network. *FEMS Microbiol Rev.* 2010 Apr;34(5):685–737.

[13] Hong EJ, Park JS, Kim Y, Lee HS. Role of *Corynebacterium glutamicum sprA* encoding a serine protease in *glxR*-mediated global gene regulation. *Plos One.* 2014 Apr;9(4):e93587.

[14] Cocaign M, Monnet C, Lindley ND. Batch kinetics of *Corynebacterium glutamicum* during growth on various carbon substrates: use of substrate mixtures to localize metabolic bottlenecks. *Appl Microbiol Biotechnol.* 1993 Dec;40(4):526–530.

[15] Dominguez H, Cocaign-Bousquet M, Lindley N. Simultaneous consumption of glucose and fructose from sugar mixtures during batch growth of *Corynebacterium glutamicum. Appl Microbiol Biotechnol.* 1997;47(5):600–603.

[16] Kiefer P, Heinzle E, Wittmann C. Influence of glucose, fructose and sucrose as carbon sources on kinetics and stoichiometry of lysine production by *Corynebacterium glutamicum. J Ind Microbiol Biotechnol.* 2002 Jun;28(6):338–343.

[17] Liebl W. *Corynebacterium*–nonmedical. In: Dworkin M, Falkow S, Rosenberg E, Schleifer KH, Stackebrandt E, editors. *The Prokaryotes. Springer New York;* 2006. p. 796–818.

[18] Brinkrolf K, Plöger S, Solle S, Brune I, Nentwich SS, Hüser AT, et al. The LacI/GalR family transcriptional regulator UriR negatively controls uridine utilization of

Corynebacterium glutamicum by binding to catabolite-responsive element (*cre*)-like sequences. *Microbiology-SGM*. 2008 Apr;154(Pt 4):1068–1081.

[19] Krämer R, Lambert C, Hoischen C, Ebbighausen H. Uptake of glutamate in *Corynebacterium glutamicum*. Kinetic properties and regulation by internal pH and potassium. *Eur J Biochem*. 1990 Dec;194(3):929–935.

[20] Wendisch VF, de Graaf AA, Sahm H, Eikmanns BJ. Quantitative determination of metabolic fluxes during coutilization of two carbon sources: comparative analyses with *Corynebacterium glutamicum* during growth on acetate and/or glucose. *J Bacteriol*. 2000 Jun;182(11):3088–3096.

[21] Hüser AT, Becker A, Brune I, Dondrup M, Kalinowski J, Plassmeier J, et al. Development of a *Corynebacterium glutamicum* DNA microarray and validation by genome-wide expression profiling during growth with propionate as carbon source. *J Biotechnol*. 2003 Dec;106(2-3):269–286.

[22] Stansen C, Uy D, Delaunay S, Eggeling L, Goergen JL, Wendisch VF. Characterization of a *Corynebacterium glutamicum* lactate utilization operon induced during temperature-triggered glutamate production. *Appl Environ Microbiol*. 2005 Oct;71(10):5920–5928.

[23] Polen T, Schluesener D, Poetsch A, Bott M, Wendisch VF. Characterization of citrate utilization in *Corynebacterium glutamicum* by transcriptome and proteome analysis. *FEMS Microbiol Lett*. 2007 Aug;273(1):109–119.

[24] Shen XH, Zhou NY, Liu SJ. Degradation and assimilation of aromatic compounds by *Corynebacterium glutamicum*: another potential for applications for this bacterium? *Appl Microbiol Biotechnol*. 2012 Jul;95(1):77–89.

[25] Krings E, Krumbach K, Bathe B, Kelle R, Wendisch VF, Sahm H, et al. Characterization of myo-inositol utilization by *Corynebacterium glutamicum*: the stimulon, identification of transporters, and influence on L-lysine formation. *J Bacteriol*. 2006 Dec;188(23):8054–8061.

[26] Chaudhry MT, Huang Y, Shen XH, Poetsch A, Jiang CY, Liu SJ. Genome-wide investigation of aromatic acid transporters in *Corynebacterium glutamicum*. *Microbiology-SGM*. 2007 Mar;153:857–865.

[27] Arndt A, Auchter M, Ishige T, Wendisch VF, Eikmanns BJ. Ethanol catabolism in *Corynebacterium glutamicum*. J Mol Microbiol Biotechnol. 2008;15(4):222–233.

[28] Frunzke J, Engels V, Hasenbein S, Gaetgens C, Bott M. Co-ordinated regulation of gluconate catabolism and glucose uptake in *Corynebacterium glutamicum* by two functionally equivalent transcriptional regulators, GntR1 and GntR2. *Mol Microbiol*. 2008 Jan;67(2):305–322.

[29] Witthoff S, Mühlroth A, Marienhagen J, Bott M. C1 metabolism in *Corynebacterium glutamicum*: an endogenous pathway for oxidation of methanol to carbon dioxide. *Appl Environ Microbiol*. 2013 Nov;79(22):6974–6983.

[30] Parche S, Burkovski A, Sprenger GA, Weil B, Krämer R, Titgemeyer F. *Corynebacterium glutamicum*: a dissection of the PTS. *J Mol Microbiol Biotechnol*. 2001 Jul;3(3):423–428.

[31] Arndt A, Eikmanns BJ. The alcohol dehydrogenase gene *adhA* in *Corynebacterium glutamicum* is subject to carbon catabolite repression. *J Bacteriol*. 2007 Oct;189(20):7408–7416.

[32] Maeda T, Wachi M. 3' Untranslated region-dependent degradation of the *aceA* mRNA, encoding the glyoxylate cycle enzyme isocitrate lyase, by RNase E/G in *Corynebacterium glutamicum*. *Appl Environ Microbiol.* 2012 Dec;78(24):8753–8761.

[33] Tanaka Y, Okai N, Teramoto H, Inui M, Yukawa H. Regulation of the expression of phosphoenolpyruvate: carbohydrate phosphotransferase system (PTS) genes in *Corynebacterium glutamicum* R. *Microbiology-SGM.* 2008 Jan;154:264–274.

[34] Georgi T, Engels V, Wendisch VF. Regulation of L-lactate utilization by the FadR-type regulator LldR of *Corynebacterium glutamicum*. *J Bacteriol.* 2008 Feb;190(3):963–971.

[35] Radmacher E, Eggeling L. The three tricarboxylate synthase activities of *Corynebacterium glutamicum* and increase of L-lysine synthesis. *Appl Microbiol Biotechnol.* 2007 Sep;76(3):587–595.

[36] Brocker M, Schaffer S, Mack C, Bott M. Citrate utilization by *Corynebacterium glutamicum* is controlled by the CitAB two-component system through positive regulation of the citrate transport genes *citH* and *tctCBA*. *J Bacteriol.* 2009 Jun;191(12):3869–3880.

[37] Nentwich SS, Brinkrolf K, Gaigalat L, Hueser AT, Rey DA, Mohrbach T, et al. Characterization of the Lacl-type transcriptional repressor RbsR controlling ribose transport in *Corynebacterium glutamicum* ATCC 13032. *Microbiology-SGM.* 2009 Jan;155:150–164.

[38] Klaffl S, Brocker M, Kalinowski J, Eikmanns BJ, Bott M. Complex regulation of the phosphoenolpyruvate carboxykinase gene *pck* and characterization of its GntR-type regulator IolR as a repressor of myo-inositol utilization genes in *Corynebacterium glutamicum*. *J Bacteriol.* 2013 Sep;195(18):4283–4296.

[39] Chao H, Zhou NY. Involvement of the Global Regulator GlxR in 3-Hydroxybenzoate and Gentisate Utilization by *Corynebacterium glutamicum*. *Appl Environ Microbiol.* 2014 Jul;80(14):4215–4225.

[40] Kim HJ, Kim TH, Kim Y, Lee HS. Identification and characterization of *glxR*, a gene involved in regulation of glyoxylate bypass in *Corynebacterium glutamicum*. *J Bacteriol.* 2004 Jun;186(11):3453–3460.

[41] Kohl TA, Baumbach J, Jungwirth B, Pühler A, Tauch A. The GlxR regulon of the amino acid producer *Corynebacterium glutamicum*: In silico and in vitro detection of DNA binding sites of a global transcription regulator. *J Biotechnol.* 2008 Jul;135(4):340–350.

[42] Jungwirth B, Emer D, Brune I, Hansmeier N, Pühler A, Eikmanns BJ, et al. Triple transcriptional control of the resuscitation promoting factor 2 (*rpf2*) gene of *Corynebacterium glutamicum* by the regulators of acetate metabolism RamA and RamB and the cAMP-dependent regulator GlxR. *FEMS Microbiol Lett.* 2008 Apr;281(2):190–197.

[43] Bussmann M, Emer D, Hasenbein S, Degraf S, Eikmanns BJ, Bott M. Transcriptional control of the succinate dehydrogenase operon *sdhCAB* of *Corynebacterium glutamicum* by the cAMP-dependent regulator GlxR and the LuxR-type regulator RamA. *J Biotechnol.* 2009 Sep;143(3):173–182.

[44] Park SY, Moon MW, Subhadra B, Lee JK. Functional characterization of the *glxR* deletion mutant of *Corynebacterium glutamicum* ATCC 13032: involvement of GlxR in

acetate metabolism and carbon catabolite repression. *FEMS Microbiol Lett.* 2010 Mar;304(2):107–115.

[45] Toyoda K, Teramoto H, Inui M, Yukawa H. Genome-wide identification of in vivo binding sites of GlxR, a cyclic AMP receptor protein-type regulator in *Corynebacterium glutamicum. J Bacteriol.* 2011 Aug;193(16):4123–4133.

[46] Teramoto H, Inui M, Yukawa H. Transcriptional regulators of multiple genes involved in carbon metabolism in *Corynebacterium glutamicum. J Biotechnol.* 2011 Jul;154(2-3):114–125.

[47] van Ooyen J, Emer D, Bussmann M, Bott M, Eikmanns BJ, Eggeling L. Citrate synthase in *Corynebacterium glutamicum* is encoded by two *gltA* transcripts which are controlled by RamA, RamB, and GlxR. *J Biotechnol.* 2011 Jul;154(2-3):140–148.

[48] Toyoda K, Teramoto H, Gunji W, Inui M, Yukawa H. Involvement of regulatory interactions among global regulators GlxR, SugR, and RamA in expression of *ramA* in *Corynebacterium glutamicum. J Bacteriol.* 2013 Apr;195(8):1718–1726.

[49] Subhadra B, Lee JK. Elucidation of the regulation of ethanol catabolic genes and *ptsG* using a *glxR* and adenylate cyclase gene (*cyaB*) deletion mutants of *Corynebacterium glutamicum* ATCC 13032. *J Microbiol Biotechnol.* 2013 Dec;23(12):1683–1690.

[50] Jungwirth B, Sala C, Kohl TA, Uplekar S, Baumbach J, Cole ST, et al. High-resolution detection of DNA binding sites of the global transcriptional regulator GlxR in *Corynebacterium glutamicum. Microbiology-SGM.* 2013 Jan;159:12–22.

[51] Barzantny H, Schröder J, Strotmeier J, Fredrich E, Brune I, Tauch A. The transcriptional regulatory network of *Corynebacterium jeikeium* K411 and its interaction with metabolic routes contributing to human body odor formation. *J Biotechnol.* 2012 Jun;159(3):235–248.

[52] Nishimura T, Teramoto H, Inui M, Yukawa H. Gene expression profiling of *Corynebacterium glutamicum* during anaerobic nitrate respiration: induction of the SOS response for cell survival. *J Bacteriol.* 2011 Mar;193(6):1327–1333.

[53] Panhorst M, Sorger-Herrmann U, Wendisch VF. The *pstSCAB* operon for phosphate uptake is regulated by the global regulator GlxR in *Corynebacterium glutamicum. J Biotechnol.* 2011 Jul;154(2-3):149–155.

[54] Gerstmeir R, Cramer A, Dangel P, Schaffer S, Eikmanns BJ. RamB, a novel transcriptional regulator of genes involved in acetate metabolism of *Corynebacterium glutamicum. J Bacteriol.* 2004 May;186(9):2798–2809.

[55] Cramer A, Gerstmeir R, Schaffer S, Bott M, Eikmanns BJ. Identification of RamA, a novel LuxR-type transcriptional regulator of genes involved in acetate metabolism of *Corynebacterium glutamicum. J Bacteriol.* 2006 Apr;188(7):2554–2567.

[56] Engels V, Wendisch VF. The DeoR-type regulator SugR represses expression of *ptsG* in *Corynebacterium glutamicum. J Bacteriol.* 2007 Apr;189(8):2955–2966.

[57] Gaigalat L, Schlüter JP, Hartmann M, Mormann S, Tauch A, Pühler A, et al. The DeoR-type transcriptional regulator SugR acts as a repressor for genes encoding the phosphoenolpyruvate:sugar phosphotransferase system (PTS) in *Corynebacterium glutamicum. BMC Mol Biol.* 2007;8:104.

[58] Brinkrolf K, Schroeder J, Pühler A, Tauch A. The transcriptional regulatory repertoire of *Corynebacterium glutamicum*: reconstruction of the network controlling pathways involved in lysine and glutamate production. *J Biotechnol.* 2010 Sep;149(3):173–182.

[59] Toyoda K, Teramoto H, Inui M, Yukawa H. Expression of the *gapA* gene encoding glyceraldehyde-3-phosphate dehydrogenase of *Corynebacterium glutamicum* is regulated by the global regulator SugR. *Appl Microbiol Biotechnol.* 2008 Nov;81(2):291–301.

[60] Tanaka Y, Teramoto H, Inui M, Yukawa H. Regulation of expression of general components of the phosphoenolpyruvate: carbohydrate phosphotransferase system (PTS) by the global regulator SugR in *Corynebacterium glutamicum. Appl Microbiol Biotechnol.* 2008 Feb;78(2):309–318.

[61] Engels V, Lindner SN, Wendisch VF. The global repressor SugR controls expression of genes of glycolysis and of the L-lactate dehydrogenase LdhA in *Corynebacterium glutamicum. J Bacteriol.* 2008 Dec;190(24):8033–8044.

[62] Toyoda K, Teramoto H, Inui M, Yukawa H. The *ldhA* Gene, encoding fermentative L-lactate dehydrogenase of *Corynebacterium glutamicum*, is under the control of positive feedback regulation mediated by LldR. *J Bacteriol.* 2009 Jul;191(13):4251–4258.

[63] Dietrich C, Nato A, Bost B, Le Maréchal P, Guyonvarch A. Regulation of *ldh* expression during biotin-limited growth of *Corynebacterium glutamicum. Microbiology-SGM.* 2009 Apr;155(Pt 4):1360–1375.

[64] Cramer A, Eikmanns BJ. RamA, the transcriptional regulator of acetate metabolism in *Corynebacterium glutamicum*, is subject to negative autoregulation. *J Mol Microbiol Biotechnol.* 2007;12(1-2):51–59.

[65] Toyoda K, Teramoto H, Inui M, Yukawa H. Involvement of the LuxR-Type transcriptional regulator RamA in regulation of expression of the *gapA* gene, encoding glyceraldehyde-3-phosphate dehydrogenase of *Corynebacterium glutamicum. J Bacteriol.* 2009 Feb;191(3):968–977.

[66] Cramer A, Auchter M, Frunzke J, Bott M, Eikmanns BJ. RamB, the transcriptional regulator of acetate metabolism in *Corynebacterium glutamicum*, is subject to regulation by RamA and RamB. *J Bacteriol.* 2007 Feb;189(3):1145–1149.

[67] Jolkver E, Emer D, Ballan S, Krämer R, Eikmanns BJ, Marin K. Identification and characterization of a bacterial transport system for the uptake of pyruvate, propionate, and acetate in *Corynebacterium glutamicum. J Bacteriol.* 2009 Feb;191(3):940–948.

[68] Emer D, Krug A, Eikmanns BJ, Bott M. Complex expression control of the *Corynebacterium glutamicum* aconitase gene: Identification of RamA as a third transcriptional regulator besides AcnR and RipA. *J Biotechnol.* 2009 Mar;140(1-2):92–98.

[69] Blombach B, Cramer A, Eikmanns BJ, Schreiner M. RamB is an activator of the pyruvate dehydrogenase complex subunit E1p gene in *Corynebacterium glutamicum. J Mol Microbiol Biotechnol.* 2009;16(3-4):236–239.

[70] Auchter M, Arndt A, Eikmanns BJ. Dual transcriptional control of the acetaldehyde dehydrogenase gene *ald* of *Corynebacterium glutamicum* by RamA and RamB. *J Biotechnol.* 2009 Mar;140(1-2):84–91.

[71] Seibold GM, Hagmann CT, Schietzel M, Emer D, Auchter M, Schreiner J, et al. The transcriptional regulators RamA and RamB are involved in the regulation of glycogen synthesis in *Corynebacterium glutamicum. Microbiology-SGM.* 2010 Apr;156:1256–1263.

[72] Auchter M, Cramer A, Hüser A, Rückert C, Emer D, Schwarz P, et al. RamA and RamB are global transcriptional regulators in *Corynebacterium glutamicum* and control genes for enzymes of the central metabolism. *J Biotechnol.* 2011 Jul;154(2-3):126–139.

[73] Tanaka Y, Ehira S, Teramoto H, Inui M, Yukawa H. Coordinated Regulation of *gnd*, which encodes 6-phosphogluconate dehydrogenase, by the two transcriptional regulators GntR1 and RamA in *Corynebacterium glutamicum. J Bacteriol.* 2012 Dec;194(23):6527–6536.

[74] Gerstmeir R, Wendisch VF, Schnicke S, Ruan H, Farwick M, Reinscheid D, et al. Acetate metabolism and its regulation in *Corynebacterium glutamicum. J Biotechnol.* 2003 Sep;104(1-3):99–122.

[75] Veit A, Rittmann D, Georgi T, Youn JW, Eikmanns BJ, Wendisch VF. Pathway identification combining metabolic flux and functional genomics analyses: acetate and propionate activation by *Corynebacterium glutamicum. J Biotechnol.* 2009 Mar;140(1-2):75–83.

[76] Sidders B, Withers M, Kendall SL, Bacon J, Waddell SJ, Hinds J, et al. Quantification of global transcription patterns in prokaryotes using spotted microarrays. *Genome Biol.* 2007;8(12):R265.

[77] Youn JW, Jolkver E, Krämer R, Marin K, Wendisch VF. Identification and characterization of the dicarboxylate uptake system DccT in *Corynebacterium glutamicum. J Bacteriol.* 2008 Oct;190(19):6458–6466.

[78] Tanaka Y, Takemoto N, Ito T, Teramoto H, Yukawa H, Inui M. Genome-wide analysis of the role of global transcriptional regulator GntR1 in *Corynebacterium glutamicum. J Bacteriol.* 2014 Jun;196:3249–3258.

[79] Hyeon JE, Kang DH, Kim YI, You SK, Han SO. GntR-type transcriptional regulator PckR negatively regulates the expression of phosphoenolpyruvate carboxykinase in *Corynebacterium glutamicum. J Bacteriol.* 2012 May;194(9):2181–2188.

[80] Okibe N, Suzuki N, Inui M, Yukawa H. Isolation, evaluation and use of two strong, carbon source-inducible promoters from *Corynebacterium glutamicum. Lett Appl Microbiol.* 2010 Feb;50(2):173–180.

[81] Koch-Koerfges A, Kabus A, Ochrombel I, Marin K, Bott M. Physiology and global gene expression of a *Corynebacterium glutamicum* Delta F1FO-ATP synthase mutant devoid of oxidative phosphorylation. *Biochim Biophys Acta.* 2012 Feb;1817(2):370–380.

[82] Amon J, Titgemeyer F, Burkovski A. Common patterns - unique features: nitrogen metabolism and regulation in Gram-positive bacteria. *FEMS Microbiol Rev.* 2010 Mar;34:588–605.

[83] Silberbach M, Hüser A, Kalinowski J, Pühler A, Walter B, Krämer R, et al. DNA microarray analysis of the nitrogen starvation response of *Corynebacterium glutamicum. J Biotechnol.* 2005 Oct;119(4):357–367.

[84] Silberbach M, Schafer M, Huser AT, Kalinowski J, Pühler A, Krämer R, et al. Adaptation of *Corynebacterium glutamicum* to ammonium limitation: a global analysis using transcriptome and proteome techniques. *Appl Environ Microbiol.* 2005 May;71(5):2391–2402.

[85] Silberbach M, Burkovski A. Application of global analysis techniques to *Corynebacterium glutamicum*: new insights into nitrogen regulation. *J Biotechnol.* 2006 Oct;126(1):101–110.

[86] Beckers G, Strösser J, Hildebrandt U, Kalinowski J, Farwick M, Krämer R, et al. Regulation of AmtR-controlled gene expression in *Corynebacterium glutamicum*: mechanism and characterization of the AmtR regulon. *Mol Microbiol.* 2005 Oct;58(2):580–595.

[87] Nolden L, Farwick M, Krämer R, Burkovski A. Glutamine synthetases of *Corynebacterium glutamicum*: transcriptional control and regulation of activity. *FEMS Microbiol Lett.* 2001 Jul;201(1):91–98.

[88] Beckers G, Nolden L, Burkovski A. Glutamate synthase of *Corynebacterium glutamicum* is not essential for glutamate synthesis and is regulated by the nitrogen status. *Microbiology-SGM.* 2001 Nov;147(Pt 11):2961–2970.

[89] Beckers G, Bendt AK, Krämer R, Burkovski A. Molecular identification of the urea uptake system and transcriptional analysis of urea transporter- and urease-encoding genes in *Corynebacterium glutamicum. J Bacteriol.* 2004 Nov;186(22):7645–7652.

[90] Bendt AK, Beckers G, Silberbach M, Wittmann A, Burkovski A. Utilization of creatinine as an alternative nitrogen source in *Corynebacterium glutamicum. Arch Microbiol.* 2004 Jun;181(6):443–450.

[91] Jakoby M, Nolden L, Meier-Wagner J, Krämer R, Burkovski A. AmtR, a global repressor in the nitrogen regulation system of *Corynebacterium glutamicum. Mol Microbiol.* 2000 Aug;37(4):964–977.

[92] Hänssler E, Müller T, Palumbo K, Pátek M, Brocker M, Krämer R, et al. A game with many players: control of *gdh* transcription in *Corynebacterium glutamicum. J Biotechnol.* 2009 Jun;142(2):114–122.

[93] Buchinger S, Strösser J, Rehm N, Hänssler E, Hans S, Bathe B, et al. A combination of metabolome and transcriptome analyses reveals new targets of the *Corynebacterium glutamicum* nitrogen regulator AmtR. *J Biotechnol.* 2009 Mar;140(1-2):68–74.

[94] Muhl D, Jessberger N, Hasselt K, Jardin C, Sticht H, Burkovski A. DNA binding by *Corynebacterium glutamicum* TetR-type transcription regulator AmtR. *BMC Mol Biol.* 2009;10(1):73.

[95] Hasselt K, Rankl S, Worsch S, Burkovski A. Adaptation of AmtR-controlled gene expression by modulation of AmtR binding activity in *Corynebacterium glutamicum. J Biotechnol.* 2011 Jul;154(2-3):156–162.

[96] Strösser J, Ludke A, Schaffer S, Krämer R, Burkovski A. Regulation of GlnK activity: modification, membrane sequestration and proteolysis as regulatory principles in the network of nitrogen control in *Corynebacterium glutamicum. Mol Microbiol.* 2004 Jul;54(1):132–147.

[97] Nolden L, Ngouoto-Nkili CE, Bendt AK, Krämer R, Burkovski A. Sensing nitrogen limitation in *Corynebacterium glutamicum*: the role of *glnK* and *glnD*. *Mol Microbiol.* 2001 Dec;42(5):1281–1295.

[98] Torriani-Gorini A. Introduction: the Pho regulon of *Escherichia coli*. In: Torriani-Gorini A, Yagil E, Silver S, editors. Phosphate in Microoganisms: Cellular and Molecular Biology. *American Society for Microbiology Press;* 1994. p. 1–4.

[99] Hulett FM. The signal-transduction network for Pho regulation in *Bacillus subtilis. Mol Microbiol.* 1996 Mar;19(5):933–939.

[100] Sola-Landa A, Moura RS, Martín JF. The two-component PhoR-PhoP system controls both primary metabolism and secondary metabolite biosynthesis in *Streptomyces lividans. Proc Natl Acad Sci U S A.* 2003 May;100(10):6133–6138.

[101] Wendisch VF, Bott M. Phosphorus Metabolism and its Regulation. In: Burkovski A, editor. Corynebacteria: Genomics and Molecular Biology. *Caister Academic Press;* 2008. p. 203–216.
[102] Ishige T, Krause M, Bott M, Wendisch VF, Sahm H. The phosphate starvation stimulon of *Corynebacterium glutamicum* determined by DNA microarray analyses. *J Bacteriol.* 2003 Aug;185(15):4519–4529.
[103] Rittmann D, Sorger-Herrmann U, Wendisch VF. Phosphate starvation-inducible gene *ushA* encodes a 5' nucleotidase required for growth of *Corynebacterium glutamicum* on media with nucleotides as the phosphorus source. *Appl Environ Microbiol.* 2005 Aug;71(8):4339–4344.
[104] Kocan M, Schaffer S, Ishige T, Sorger-Herrmann U, Wendisch VF, Bott M. Two-component systems of *Corynebacterium glutamicum*: deletion analysis and involvement of the PhoS-PhoR system in the phosphate starvation response. *J Bacteriol.* 2006 Jan;188(2):724–732.
[105] Schaaf S, Bott M. Target genes and DNA-binding sites of the response regulator PhoR from *Corynebacterium glutamicum*. *J Bacteriol.* 2007 Jul;189(14):5002–5011.
[106] Lill R. Function and biogenesis of iron-sulphur proteins. *Nature.* 2009 Aug;460(7257):831–838.
[107] Andrews SC, Robinson AK, Rodríguez-Quiñones F. Bacterial iron homeostasis. *FEMS Microbiol Rev.* 2003;27(2-3):215–237.
[108] Boyd J, Oza MN, Murphy JR. Molecular cloning and DNA sequence analysis of a diphtheria *tox* iron-dependent regulatory element (*dtxR*) from *Corynebacterium diphtheriae*. *Proc Natl Acad Sci U S A.* 1990 Aug;87(15):5968–5972.
[109] Brune I, Werner H, Hüser AT, Kalinowski J, Pühler A, Tauch A. The DtxR protein acting as dual transcriptional regulator directs a global regulatory network involved in iron metabolism of *Corynebacterium glutamicum*. *BMC Genomics.* 2006;7:21.
[110] Wennerhold J, Bott M. The DtxR regulon of *Corynebacterium glutamicum*. *J Bacteriol.* 2006 Apr;188(8):2907–2918.
[111] Wennerhold J, Krug A, Bott M. The AraC-type regulator RipA represses aconitase and other iron proteins from *Corynebacterium* under iron limitation and is itself repressed by DtxR. *J Biol Chem.* 2005 Dec;280(49):40500–40508.
[112] Blombach B, Buchholz J, Busche T, Kalinowski J, Takors R. Impact of different CO2/HCO3- levels on metabolism and regulation in *Corynebacterium glutamicum*. *J Biotechnol.* 2013 Dec;168(4):331–340.
[113] Frunzke J, Gätgens C, Brocker M, Bott M. Control of heme homeostasis in *Corynebacterium glutamicum* by the two-component system HrrSA. *J Bacteriol.* 2011 Mar;193(5):1212–1221.
[114] Heyer A, Gätgens C, Hentschel E, Kalinowski J, Bott M, Frunzke J. The two-component system ChrSA is crucial for heme tolerance and interferes with HrrSA in heme-dependent gene regulation in *Corynebacterium glutamicum*. *Microbiology-SGM.* 2012 Dec;158(Pt 12):3020–3031.
[115] Schweitzer JE, Stolz M, Diesveld R, Etterich H, Eggeling L. The serine hydroxymethyltransferase gene *glyA* in *Corynebacterium glutamicum* is controlled by GlyR. *J Biotechnol.* 2009 Feb;139(3):214–221.
[116] Rey DA, Pühler A, Kalinowski J. The putative transcriptional repressor McbR, member of the TetR-family, is involved in the regulation of the metabolic network directing the

synthesis of sulfur containing amino acids in *Corynebacterium glutamicum*. *J Biotechnol.* 2003 Jun;103(1):51–65.

[117] Rey DA, Nentwich SS, Koch DJ, Rückert C, Pühler A, Tauch A, et al. The McbR repressor modulated by the effector substance S-adenosylhomocysteine controls directly the transcription of a regulon involved in sulphur metabolism of *Corynebacterium glutamicum* ATCC 13032. *Mol Microbiol.* 2005 May;56(4):871–887.

[118] Krömer JO, Bolten CJ, Heinzle E, Schröder H, Wittmann C. Physiological response of *Corynebacterium glutamicum* to oxidative stress induced by deletion of the transcriptional repressor McbR. *Microbiology-SGM.* 2008 Dec;154(Pt 12):3917–3930.

[119] Rückert C, Milse J, Albersmeier A, Koch DJ, Pühler A, Kalinowski J. The dual transcriptional regulator CysR in *Corynebacterium glutamicum* ATCC 13032 controls a subset of genes of the McbR regulon in response to the availability of sulphide acceptor molecules. BMC Genomics. 2008;9:483.

[120] Lee DS, Park JS, Kim Y, Lee HS. *Corynebacterium glutamicum sdhA* encoding succinate dehydrogenase subunit A plays a role in *cysR*-mediated sulfur metabolism. *Appl Microbiol Biotechnol.* 2014 Aug;98(15):6751–6759.

[121] Rückert C, Koch DJ, Rey DA, Albersmeier A, Mormann S, Pühler A, et al. Functional genomics and expression analysis of the *Corynebacterium glutamicum fpr2-cysIXHDNYZ* gene cluster involved in assimilatory sulphate reduction. *BMC Genomics.* 2005;6:121.

[122] Koch DJ, Rückert C, Albersmeier A, Hüser AT, Tauch A, Pühler A, et al. The transcriptional regulator SsuR activates expression of the *Corynebacterium glutamicum* sulphonate utilization genes in the absence of sulphate. *Mol Microbiol.* 2005 Oct;58(2):480–494.

[123] Nakunst D, Larisch C, Hüser AT, Tauch A, Pühler A, Kalinowski J. The extracytoplasmic function-type sigma factor SigM of *Corynebacterium glutamicum* ATCC 13032 is involved in transcription of disulfide stress-related genes. *J Bacteriol.* 2007 Jul;189(13):4696–4707.

[124] Teramoto H, Inui M, Yukawa H. *Corynebacterium glutamicum* Zur acts as a zinc-sensing transcriptional repressor of both zinc-inducible and zinc-repressible genes involved in zinc homeostasis. *FEBS J.* 2012 Dec;279(23):4385–4397.

[125] Nishimura T, Vertès AA, Shinoda Y, Inui M, Yukawa H. Anaerobic growth of *Corynebacterium glutamicum* using nitrate as a terminal electron acceptor. *Appl Microbiol Biotechnol.* 2007 Jun;75(4):889–897.

[126] Nishimura T, Teramoto H, Vertès AA, Inui M, Yukawa H. ArnR, a novel transcriptional regulator, represses expression of the *narKGHJI* operon in *Corynebacterium glutamicum*. *J Bacteriol.* 2008 May;190(9):3264–3273.

[127] Poole RK. Nitric oxide and nitrosative stress tolerance in bacteria. *Biochem Soc Trans.* 2005 Feb;33(Pt 1):176–180.

[128] Platzen L, Koch-Koerfges A, Weil B, Brocker M, Bott M. Role of flavohaemoprotein Hmp and nitrate reductase NarGHJI of *Corynebacterium glutamicum* for coping with nitrite and nitrosative stress. *FEMS Microbiol Lett.* 2014 Jan;350(2):239–248.

[129] Oguiza JA, Marcos AT, Martín JF. Transcriptional analysis of the *sigA* and *sigB* genes of *Brevibacterium lactofermentum*. *FEMS Microbiol Lett.* 1997 Aug;153(1):111–117.

[130] Larisch C, Nakunst D, Hüser AT, Tauch A, Kalinowski J. The alternative sigma factor SigB of *Corynebacterium glutamicum* modulates global gene expression during transition from exponential growth to stationary phase. *BMC Genomics*. 2007;8:4.

[131] Ehira S, Shirai T, Teramoto H, Inui M, Yukawa H. Group 2 sigma factor SigB of *Corynebacterium glutamicum* positively regulates glucose metabolism under conditions of oxygen deprivation. *Appl Environ Microbiol.* 2008 Aug;74(16):5146–5152.

[132] Halgasova N, Bukovska G, Ugorcakova J, Timko J, Kormanec J. The *Brevibacterium flavum* sigma factor SigB has a role in the environmental stress response. *FEMS Microbiol Lett*. 2002 Oct;216(1):77–84.

[133] Jakob K, Satorhelyi P, Lange C, Wendisch VF, Silakowski B, Scherer S, et al. Gene expression analysis of *Corynebacterium glutamicum* subjected to long-term lactic acid adaptation. *J Bacteriol*. 2007 Aug;189(15):5582–5590.

[134] Brockmann-Gretza O, Kalinowski J. Global gene expression during stringent response in *Corynebacterium glutamicum* in presence and absence of the *rel* gene encoding (p)ppGpp synthase. *BMC Genomics*. 2006;7:230.

[135] Ikeda M, Baba M, Tsukamoto N, Komatsu T, Mitsuhashi S, Takeno S. Elucidation of genes relevant to the microaerobic growth of *Corynebacterium glutamicum*. *Biosci Biotechnol Biochem*. 2009 Dec;73(12):2806–2808.

[136] Park SD, Youn JW, Kim YJ, Lee SM, Kim Y, Lee HS. *Corynebacterium glutamicum* σ^E is involved in responses to cell surface stresses and its activity is controlled by the anti-σ factor CseE. *Microbiology-SGM*. 2008 Mar;154(Pt 3):915–923.

[137] Ehira S, Teramoto H, Inui M, Yukawa H. Regulation of *Corynebacterium glutamicum* heat shock response by the extracytoplasmic-function sigma factor SigH and transcriptional regulators HspR and HrcA. *J Bacteriol.* 2009 May;191(9):2964–2972.

[138] Chi BK, Busche T, Van Laer K, Bäsell K, Becher D, Clermont L, et al. Protein S-mycothiolation functions as redox-switch and thiol protection mechanism in *Corynebacterium glutamicum* under hypochlorite stress. *Antioxid Redox Signal.* 2014 Feb;20(4):589–605.

[139] Busche T, Silar R, Pičmanová M, Pátek M, Kalinowski J. Transcriptional regulation of the operon encoding stress-responsive ECF sigma factor SigH and its anti-sigma factor RshA, and control of its regulatory network in *Corynebacterium glutamicum*. *BMC Genomics*. 2012;13:445.

[140] Ehira S, Teramoto H, Inui M, Yukawa H. A novel redox-sensing transcriptional regulator CyeR controls expression of an Old Yellow Enzyme family protein in *Corynebacterium glutamicum*. *Microbiology-SGM.* 2010 May;156(Pt 5):1335–1341.

[141] Bussmann M, Baumgart M, Bott M. RosR (Cg1324), a hydrogen peroxide-sensitive MarR-type transcriptional regulator of *Corynebacterium glutamicum*. *J Biol Chem.* 2010 Sep;285(38):29305–29318.

[142] Liu YB, Chen C, Chaudhry MT, Si MR, Zhang L, Wang Y, et al. Enhancing *Corynebacterium glutamicum* robustness by over-expressing a gene, *mshA*, for mycothiol glycosyltransferase. *Biotechnol Lett.* 2014 Jul;36(7):1453–1459.

[143] Chi BK, Roberts AA, Huyen TTT, Bäsell K, Becher D, Albrecht D, et al. S-bacillithiolation protects conserved and essential proteins against hypochlorite stress in firmicutes bacteria. *Antioxid Redox Signal.* 2013 Apr;18(11):1273–1295.

[144] Dalle-Donne I, Rossi R, Colombo G, Giustarini D, Milzani A. Protein S-glutathionylation: a regulatory device from bacteria to humans. *Trends Biochem Sci.* 2009 Feb;34(2):85–96.

[145] Engels S, Schweitzer JE, Ludwig C, Bott M, Schaffer S. *clpC* and *clpP1P2* gene expression in *Corynebacterium glutamicum* is controlled by a regulatory network involving the transcriptional regulators ClgR and HspR as well as the ECF sigma factor σ^H. *Mol Microbiol.* 2004 Apr;52(1):285–302.

[146] Jochmann N, Kurze AK, Czaja LF, Brinkrolf K, Brune I, Hüser AT, et al. Genetic makeup of the *Corynebacterium glutamicum* LexA regulon deduced from comparative transcriptomics and in vitro DNA band shift assays. *Microbiology-SGM.* 2009 May;155(Pt 5):1459–1477.

[147] Zemanová M, Kaderábková P, Pátek M, Knoppová M, Silar R, Nesvera J. Chromosomally encoded small antisense RNA in *Corynebacterium glutamicum. FEMS Microbiol Lett.* 2008 Feb;279(2):195–201.

[148] Mentz A, Neshat A, Pfeifer-Sancar K, Pühler A, Rückert C, Kalinowski J. Comprehensive discovery and characterization of small RNAs in *Corynebacterium glutamicum* ATCC 13032. *BMC Genomics.* 2013;14:714.

[149] Erill I, Campoy S, Barbé J. Aeons of distress: an evolutionary perspective on the bacterial SOS response. *FEMS Microbiol Rev.* 2007 Nov;31(6):637–656.

[150] Weinand M, Krämer R, Morbach S. Characterization of compatible solute transporter multiplicity in *Corynebacterium glutamicum. Appl Microbiol Biotechnol.* 2007 Sep;76(3):701–708.

[151] Möker N, Brocker M, Schaffer S, Krämer R, Morbach S, Bott M. Deletion of the genes encoding the MtrA-MtrB two-component system of *Corynebacterium glutamicum* has a strong influence on cell morphology, antibiotics susceptibility and expression of genes involved in osmoprotection. *Mol Microbiol.* 2004 Oct;54(2):420–438.

[152] Brocker M, Bott M. Evidence for activator and repressor functions of the response regulator MtrA from *Corynebacterium glutamicum. FEMS Microbiol Lett.* 2006 Nov;264(2):205–212.

[153] Fränzel B, Trötschel C, Rückert C, Kalinowski J, Poetsch A, Wolters DA. Adaptation of *Corynebacterium glutamicum* to salt-stress conditions. *Proteomics.* 2010 Feb;10(3):445–457.

[154] Follmann M, Ochrombel I, Krämer R, Trötschel C, Poetsch A, Rückert C, et al. Functional genomics of pH homeostasis in *Corynebacterium glutamicum* revealed novel links between pH response, oxidative stress, iron homeostasis and methionine synthesis. *BMC Genomics.* 2009;10:621.

[155] Teramoto H, Inui M, Yukawa H. OxyR acts as a transcriptional repressor of hydrogen peroxide-inducible antioxidant genes in *Corynebacterium glutamicum* R. *FEBS J.* 2013 Jul;280(14):3298–3312.

[156] Milse J, Petri K, Rückert C, Kalinowski J. Transcriptional response of *Corynebacterium glutamicum* ATCC 13032 to hydrogen peroxide stress and characterization of the OxyR regulon. *J Biotechnol.* 2014 Aug;190:40-54.

[157] Hayashi M, Ohnishi J, Mitsuhashi S, Yonetani Y, Hashimoto SI, Ikeda M. Transcriptome analysis reveals global expression changes in an industrial L-lysine producer of *Corynebacterium glutamicum. Biosci Biotechnol Biochem.* 2006 Feb;70(2):546–550.

[158] Hayashi M, Mizoguchi H, Ohnishi J, Mitsuhashi S, Yonetani Y, Hashimoto Si, et al. A *leuC* mutation leading to increased L-lysine production and *rel*-independent global expression changes in *Corynebacterium glutamicum*. *Appl Microbiol Biotechnol.* 2006 Oct;72(4):783–789.

[159] Sindelar G, Wendisch VF. Improving lysine production by *Corynebacterium glutamicum* through DNA microarray-based identification of novel target genes. *Appl Microbiol Biotechnol.* 2007 Sep;76(3):677–689.

[160] Ikeda M, Mitsuhashi S, Tanaka K, Hayashi M. Reengineering of a *Corynebacterium glutamicum* L-arginine and L-citrulline producer. *Appl Environ Microbiol.* 2009 Mar;75(6):1635–1641.

[161] Lee SY, Shin HS, Park JS, Kim YH, Min J. Proline reduces the binding of transcriptional regulator ArgR to upstream of *argB* in *Corynebacterium glutamicum*. *Appl Microbiol Biotechnol.* 2010 Mar;86(1):235–242.

[162] Lee SY, Kim YH, Min J. Conversion of phenol to glutamate and proline in *Corynebacterium glutamicum* is regulated by transcriptional regulator ArgR. *Appl Microbiol Biotechnol.* 2010 Jan;85(3):713–720.

[163] Hänssler E, Müller T, Jessberger N, Völzke A, Plassmeier J, Kalinowski J, et al. FarR, a putative regulator of amino acid metabolism in *Corynebacterium glutamicum*. *Appl Microbiol Biotechnol.* 2007 Sep;76(3):625–632.

[164] Lee SY, Park JM, Lee JH, Chang ST, Park JS, Kim YH, et al. Interaction of transcriptional repressor ArgR with transcriptional regulator FarR at the *argB* promoter region in *Corynebacterium glutamicum*. *Appl Environ Microbiol.* 2011 Feb;77(3):711–718.

[165] Park SH, Kim HU, Kim TY, Park JS, Kim SS, Lee SY. Metabolic engineering of *Corynebacterium glutamicum* for L-arginine production. *Nat Commun.* 2014;5:4618.

[166] Kim SY, Lee J, Lee SY. Metabolic engineering of *Corynebacterium glutamicum* for the production of L-ornithine. *Biotechnol Bioeng.* 2014. Advanced on line.

[167] Blombach B, Hans S, Bathe B, Eikmanns BJ. Acetohydroxyacid synthase, a novel target for improvement of L-lysine production by *Corynebacterium glutamicum*. *Appl Environ Microbiol.* 2009 Jan;75(2):419–427.

[168] Oldiges M, Eikmanns BJ, Blombach B. Application of metabolic engineering for the biotechnological production of L-valine. *Appl Microbiol Biotechnol.* 2014 Jul;98(13):5859–5870.

[169] Blombach B, Schreiner ME, Holátko J, Bartek T, Oldiges M, Eikmanns BJ. L-valine production with pyruvate dehydrogenase complex-deficient *Corynebacterium glutamicum*. *Appl Environ Microbiol.* 2007 Apr;73(7):2079–2084.

[170] Blombach B, Schreiner ME, Bartek T, Oldiges M, Eikmanns BJ. *Corynebacterium glutamicum* tailored for high-yield L-valine production. *Appl Microbiol Biotechnol.* 2008 Jun;79(3):471–479.

[171] Blombach B, Arndt A, Auchter M, Eikmanns BJ. L-valine production during growth of pyruvate dehydrogenase complex-deficient *Corynebacterium glutamicum* in the presence of ethanol or by inactivation of the transcriptional regulator SugR. *Appl Environ Microbiol.* 2009 Feb;75(4):1197–1200.

[172] Wada M, Hijikata N, Aoki R, Takesue N, Yokota A. Enhanced valine production in *Corynebacterium glutamicum* with defective H+-ATPase and C-terminal truncated acetohydroxyacid synthase. *Biosci Biotechnol Biochem.* 2008 Nov;72(11):2959–2965.

[173] Holátko J, Elisáková V, Prouza M, Sobotka M, Nesvera J, Pátek M. Metabolic engineering of the L-valine biosynthesis pathway in *Corynebacterium glutamicum* using promoter activity modulation. *J Biotechnol.* 2009 Feb;139(3):203–210.

[174] Bartek T, Blombach B, Lang S, Eikmanns BJ, Wiechert W, Oldiges M, et al. Comparative ^{13}C metabolic flux analysis of pyruvate dehydrogenase complex-deficient, L-valine-producing *Corynebacterium glutamicum. Appl Environ Microbiol.* 2011 Sep;77(18):6644–6652.

[175] Brune I, Jochmann N, Brinkrolf K, Hüser AT, Gerstmeir R, Eikmanns BJ, et al. The IclR-type transcriptional repressor LtbR regulates the expression of leucine and tryptophan biosynthesis genes in the amino acid producer *Corynebacterium glutamicum. J Bacteriol.* 2007 Apr;189(7):2720–2733.

[176] Seliverstov AV, Putzer H, Gelfand MS, Lyubetsky VA. Comparative analysis of RNA regulatory elements of amino acid metabolism genes in Actinobacteria. *BMC Microbiol.* 2005;5:54.

[177] Pfeifer-Sancar K, Mentz A, Rückert C, Kalinowski J. Comprehensive analysis of the *Corynebacterium glutamicum* transcriptome using an improved RNAseq technique. *BMC Genomics.* 2013;14:888.

[178]]Neshat A, Mentz A, Rückert C, Kalinowski J. Transcriptome sequencing revealed the transcriptional organization at ribosome-mediated attenuation sites in *Corynebacterium glutamicum* and identified a novel attenuator involved in aromatic amino acid biosynthesis. *J Biotechnol.* 2014 Nov;190:55-63.

[179] SEQC/MAQC-III Consortium. A comprehensive assessment of RNA-seq accuracy, reproducibility and information content by the Sequencing Quality Control Consortium. Nat Biotechnol. 2014 Sep;32(9):903–914.

[180] Kogenaru S, Qing Y, Guo Y, Wang N. RNA-seq and microarray complement each other in transcriptome profiling. *BMC Genomics.* 2012;13:629.

[181] Swindell WR, Xing X, Voorhees JJ, Elder JT, Johnston A, Gudjonsson JE. Integrative RNA-seq and microarray data analysis reveals GC content and gene length biases in the psoriasis transcriptome. *Physiol Genomics.* 2014 Aug;46(15):533–546.

[182] Miller JA, Menon V, Goldy J, Kaykas A, Lee CK, Smith KA, et al. Improving reliability and absolute quantification of human brain microarray data by filtering and scaling probes using RNA-Seq. *BMC Genomics.* 2014;15:154.

In: New Trends in *Corynebacterium glutamicum*
Editor: Carlos Barreiro
ISBN: 978-1-63483-280-9

Chapter 3

METHODS IN PROTEOMICS APPLIED TO *CORYNEBACTERIUM GLUTAMICUM*

Carlos Barreiro[1,2,*]
[1]INBIOTEC (Instituto de Biotecnología de León)
Proteomics Unit, Spain
[2]Área de Bioquímica y Biología Molecular
Departamento de Biología Molecular
Universidad de León, Spain

ABSTRACT

Nowadays, more than ten genome sequences of *Corynebacterium glutamicum* can be found in the databases. This fact has boosted the application of the 'omics' technologies to this bacterium. Accordingly, from the initial analysis attempt in 1998 to date, Proteomics has been focused on three different *C. glutamicum* sub-proteomes: intra-cellular, membrane and secretome. This chapter reviews and summarizes the procedures used in the study of those three proteome levels. A range of buffers, disruption methods, visualization and analysis procedures are straightened out in an understandable way to allow for a proper selection of the required method.

INTRODUCTION

The concept of proteome was coined by Valerie C. Wasinger, Marc R. Wilkins and co-workers in the late 1990s [1–3] to define the protein-based gene expression analysis. Thus, the term proteome is a portmanteau defined as 'the entire protein complement expressed by a genome, or by a cell or tissue' [2]. Due to the fact that this definition was slightly inaccurate, as it implied a static behaviour in the proteome similar to the lack of dynamism found in the genome, these authors delineated then the main differences between the genome and the proteome. Whereas there is only one genome per (micro)organism, or two if the

* Corresponding Author: Carlos Barreiro; c.barreiro@unileon.es

mitochondrial one is taken into account separately in eukaryotes; the proteome is a variable entity conditioned by different circumstances (e.g.: growth phase, external stimuli, cell function or location, etc.) [2].

Therefore, the genome has the whole planning of cellular capabilities, whereas the proteome is the tool that carries out such functions. As intermediaries between the genome and the proteome act the mRNAs, whose abundance is not usually proper correlated with the quantity of proteins. Thus, studying the mRNA amount and the resultant protein abundance in human liver samples, Anderson and Seilhamer (1999) defined the correlation coefficient as only of 0.48 [4]. The lack of this expected coherent correlation has also been detected in other microorganisms such as *Streptomyces* or *Bacillus* species [5, 6], which suggests that post-transcriptional regulation is a frequent phenomenon.

The definition of the concept of proteome was the starting point for the core definition of a new analytic science, Proteomics. In an exhaustive review from 1997 regarding the state of the protein identification in the so-called post-genome era, Peter James [7] was the first to define the 'rapid rise of Proteomics'. His manuscript, focused on the analysis of proteins as an ensemble, baptised the new science. The term was an analogy with Genomics, the study of the genome. The conclusion of the Human Genome Project encouraged the scientific community to adopt a new view of biology trying to analyse the relationships of all of the members involved in a particular biological system while it is functioning (Systems Biology). This fact enhanced the 'omics studies (derived from the Greek word –oma), which includes Proteomics as the analysis of the proteins as a whole [8–10].

Corynebacterium glutamicum is a well-known amino acid producer bacterium. Thus, L-glutamic acid and L-lysine have become bulk products at present (more than 2.93 and 1.95 million tons/year in 2012, respectively) within an opened worldwide market to threonine or tryptophan (230 and 4.8 thousand tons/year in 2010, respectively) due to their consumption from feed/food to pharmaceutical products [11–14]. The 21st century has opened the industrial point of view for the coryneform bacteria, in addition to other bacteria and fungi. Thus, their product portfolio is being increased considerably. The production of organic acids, pyruvate, isobutanol, D-lactic acid, carotenoids or alanine by *C. glutamicum* is a promising reality [15–20]. Probably, the main agent for this focal change has been the obtainment of the first genome sequence of *C. glutamicum* [21, 22]. Nowadays, several *C. glutamicum* strains have been fully sequenced and published (AR1 [23]; ATCC 13032 [21, 22]; ATCC 14067 [24]; ATCC 21300 [25]; K051 [26] S9114 [27]; R [28]), whereas other strains have just been directly deposited in the NCBI database [ATCC 21831 (acc. no.: CP007722); SCgG1 (acc.no: NC_021351); SCgG2 (acc. no: NC_021352); Z188 (acc. no: NZ_AKXP00000000, 35 contigs); (CT009589, 11 contigs)]. Besides, a prophage free version of the *C. glutamicum* ATCC13032 strain genome has been published recently [29]. All this huge information about genome sequences of *C. glutamicum* opened the Proteomics analysis of this industrial relevant bacteria in 1998 [30].

That new science (Proteomics) applied to *C. glutamicum* is the main topic of this chapter Thus, the methodologies used for the analysis of the intra-cellular, membrane and extra-cellular proteomes of *C. glutamicum* are detailed thoroughly in the following sections (Figure 1).

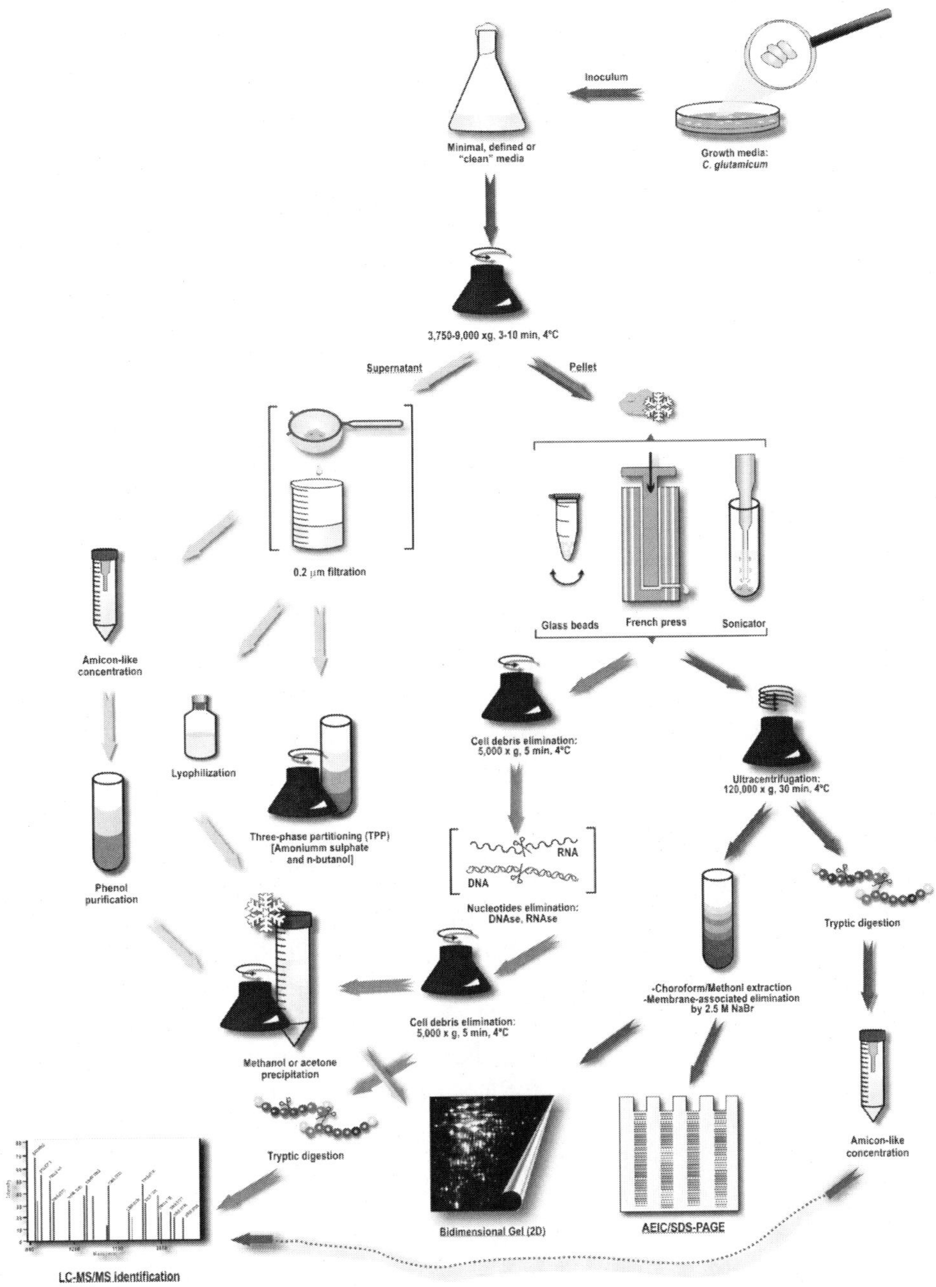

Figure 1. General overview of the protein extraction methods applied to the analysis of the intra-cellular (purple arrows), extra-cellular (green arrows) and membrane (red arrows) proteomes of *C. glutamicum*. Square brackets present those steps not used as standard in the extraction methods.

INTRA-CELLULAR PROTEOME: THE INITIAL ANALYSIS

Protein distribution based on their physicochemical properties as isoelectrical point (pI) or molecular weight is characteristic of the different taxonomical groups, which are directly conditioned by the genetic information. Thus, in contrast to eukaryotes, the prokaryotes present a positive correlation between the % G+C content and the total protein number and protein size [31]. In accordance, the protein pI is also characteristic. In 1999, VanBogelen and co-workers correlated the protein expression of the cellular state and the two-dimensional polyacrylamide gel electrophoresis (2D-PAGE) in *Escherichia coli* [32]. This study described a bimodal pattern with maxima fixed between pI 5.5 and pI 9.0, but the causes of this bimodality were not established at that time. The intra-cellular pH values were suggested as responsible for this binary distribution, since protein solubility is normally lower when close to their pI and the intra-cellular pH is ±7.0. This feature could avoid protein aggregation in prokaryotic proteomes [32]. Afterwards, this analysis was extended to three domains of Life, which confirms the bimodal character for the bacteria and archaea proteomes and a trimodal pattern in eukaryotes [33]. *C. glutamicum* is a reliable corroboration of this bacterial bimodal proteome pattern [30, 34–38]. Hence, the predicted pI of *C. glutamicum* varies from 3.3 (Cg1286: hypothetical protein) to 13.12 (Cg3432, RpmH: 50S ribosomal protein L34), where the pI gap ranges from about 7.27 to 7.67. From the whole predicted proteome, 2,363 proteins (77%) present pI values in the acidic pH region, whereas 695 proteins (23%) are located in the basic pH zone. Their pH medians are 4.98 and 9.5 per each group. Regarding the molecular weights, they go from 2.4 kDa (22 amino acids; Cg0491: hypothetical protein) to 316.7 kDa (2,996 amino acids; Cg2743: fatty acid synthase IA) [35,39].

Basically, there are two ways to tackle the proteome analysis: i) Bottom-Up: the digestion (e.g.: trypsin) of the proteins of interest results in "tryptic peptides" that are analysed by matrix-assisted laser desorption/ionization (MALDI) or electrospray ionization (ESI); or ii) Top-Down: it is an intact-protein identification method based on an ion trapping mass spectrometer to measure and analyse by tandem mass spectrometry protein ions and the fragment ions [40–42]. Protein identification is easily carried out by the bottom-up approach, since tryptic peptides are readily solubilized and separated in contrast with parent proteins. Unluckily, only a fraction of those tryptic peptides is detected. Due to this fact, it is considered suboptimal when defining alternative splice variants and modifications. In contrast, the top-down approach, with the detection of enough number of fragment ions, can show a complete description of the primary structure of the protein, its modifications and the connections among these modifications [42]. The Bottom-Up method can be subdivided in i) classical or gel based approach (bidimensional gels) and ii) shotgun approach or gel-free. Both methodological subdivisions have been used for the analysis of the intra-cellular proteome of *C. glutamicum* from the first study of Hermann and co-workers in 1998 to date [30] (Table 1A and B).

Poetsch and co-workers presented in 2011 [43] a complete review of the proteome analysis done in *C. glutamicum* and pathogenic *Corynebacterium* species, up to that date. These authors describe how a substantial number of proteome chromatographic separations have been done in *Corynebacterium* species. However, in most cases 2D-PAGE was the technology of choice [43] (Table 1A and B). Thus, as Vasco-Cárdenas and co-workers (2013) described [37], the proteome analysis of *C. glutamicum* has allowed obtaining the reference map [44, 45], the phosphoproteome [46]; several analysis focused on environmental stresses as: herbicides [47], heavy metals (e.g.: Hg, Ag or Co) [48, 49], antibiotics [50], heat shock [36], osmotic [51]; and also in nutritional stress as benzoic [52] or citric acids [53] and L-valine [54]. Thus, the present section gathers the media, collection, disruption and solubilisation methods used, up to date, for the intra-cellular proteome analysis of *C. glutamicum* (Table 1A and B). More than 30 papers have been published about this proteome study, having as culture media typically 'clean media' (e.g.: minimal media), which do not present insoluble particles or dark solutions (e.g.: soya flour, molasses, corn steep liquor). In contrast with the low pellet stability after centrifugation in *Streptomyces* species or fungi (e.g.: some *Penicillium* species), *C. glutamicum* yields stable pellets allowing their harvesting by centrifugation. The resuspension buffer used for disruption ranges from 7.0 to 8.0 pH with different salt combinations. Regarding the breaking systems, French press, glass beads and sonication disruption methods have been used routinely. Precipitation, if used, was mainly by acetone addition followed by a few washing steps or the dissolution of the disrupted supernatant directly in the isoelectric focusing (IEF) buffer, when 2D-PAGE approach was selected. Traditional IEF buffers for Proteomics were used, which range from 6-9 M of urea, 0-2 M of thiourea, 2-4% of CHAPS (3-[(3-cholamidopropyl)-dimethylammonio]-1-propanesulfonate) and 0.5-2% of Pharmalytes. The visualization (in 2D analysis) was carried out by means of visible stains (e.g.: colloidal Coomassie) and fluorescent dyes (e.g.: DIGE) (Table 1A and B).

MEMBRANE PROTEOME

The order Corynebacteriales contains corynebacteria, nocardia, mycobacteria, rhodococci and other related microorganisms (see chapter 1 [55]). A characteristic trademark of these microorganisms is the presence in the cellular envelope of 'mycolic acids' composed by α-branched and β-hydroxylated long-chain fatty acids. Those mycolic acids lengths are genus specific and vary from the shortest chains of *Corynebacterium* (C16 to C36, excluding some mycolateless species as *Corynebacterium amycolatum*) to the extremely-long ones of the *Segniliparus* genus (up to C100) or the most complex such as *Mycobacterium* (C60–C89) [55, 56]. Thus, the *C. glutamicum* cell envelope presents: plasma membrane, peptidoglycan covalently linked to arabinogalactan (esterified with mycolic acid), free mycolic acids and the S-layer (a PS2 protein crystalline layer) [56–58]. This complex cellular envelope strongly conditioned the methodological development for *C. glutamicum* membrane proteome analyses.

Initially, Marienfeld and co-workers (1997) [59] and later Lichtinger and co-workers (1998) [60] described the ultrastructure of the *C. glutamicum* cell wall, as well as the first protein purification method from the membrane. From this initial purification attempt to date, several methodological optimizations have been carried out to tackle the isolation and identification of the membrane proteins, as well as the membrane-associated proteins (Table 2). Hermann and co-workers in 1998 [30] carried out the first successful attempt at protein purification and visualization by means of bidimensional electrophoresis in *C. glutamicum.* Thus, 700 silver-stained protein spots were observed in the membrane fraction (and membrane-associated fraction) [30]. Besides, they detected the need for cytoplasm and membrane fractions separation as a prerequisite for proteome analyses. Two years later, the same group improved the analysis of the cell surface protein fraction. Thus, the modification of the lysis buffer (e.g.: thiourea, CHAPS concentration), the in-gel application of proteins or a milder sodium lauroyl sarcosinate extraction previous to acetone extraction yielded 100-120 silver-stained spots [34].

Table 1. A. Summary of the conditions (media, sample origin, disruption system, precipitation option, etc) developed or used for the analysis of the *C. glutamicum* intra-cellular proteome by Bidimensional Electrophoresis (2D-PAGE)

Media	Collection	Resuspension buffer	Disruption	Precipitation	IEF buffer	Visualization method	Reference
-Luria Bertani -Brain Heart Infusion (BHI) -MOPS-buffered minimal medium [101] -Modified MM1 minimal medium [102] -2x Ty + 2% glucose (TYG) -Modified CGXII minimal medium (mCGXII) [15]	Initial sample: -5-50 ml of bacterial culture -1 gr wet weight cells -20 ml culture + 10 ml pre-chilled buffer (20 mM Tris, 5 mM $MgCl_2$, 20 mM NaN_3, pH 7.5). Harvest: -Centrifugation: 3,750-9,000 ×g, 3-10 min, 4ºC. Washing step: -2x wash 0.9% NaC1	-*Desintegration buffer*: 2.92 g/L $NaH_2PO_4 \cdot H_2O$; 5.12 g/L $Na_2HPO_4 \cdot 2\ H_2O$; 0.247 g/L $MgSO_4 \cdot 7H_2O$ -TE buffer (10 mM Tris-HCl, pH 8.0; 1 mM EDTA) with 1 mM PMSF -Froze cells at -80ºC after harvesting or resuspended in 800 µl *washing buffer* + 1 pill of protease inhibitor mix (COMPLETE, Roche) per 20 ml -5 ml ice-cold TES buffer	Breaking system: -French Press: 3x 18,000 psi/172 Mpa -FastPrep [FastProtein BLUE tube, glass beads]: 3x, speed ratio: 6.5, 30 s -Sonication: i) 30 s activity / 1 min pause in an ice-cooled tube using a Branson sonicator (model B-12) at 75–100 W;	-Acetone precipitation (4-9 vol.), O/N. -14,000 ×g, 15 min, 4°C -Frozen (-78ºC) protein extract diluted with thiourea lysis buffer [(2 M thiourea, 7 M urea, 1% CHAPS (w/v), 0.4% DTT (w/v), and 0.5% pharmalyte 3–10 (v/v)] to a final concentration of less than 0.25% SDS	-Lysis buffer (9 M urea, 2-4% w/v CHAPS, 1% w/v DTT, 0.8% w/v 2-D Pharmalytes, pH range 3-10 and 8 mM PMSF [106]) -6 M urea, 2 M thiourea, 4% CHAPS, 0.5% Pharmalyte (pH 3 to 10), 0.4% DTT. Inhibitor mix: COMPLETE (Roche) -Rehydration buffer: 8 M urea; 2% (w/v) CHAPS; 0.01%	-Silver stain -Coomassie brilliant blue -Colloidal Comassie -DIGE (GE Healthcare) -BODIPY fluorescence (Life Technologies) -SYPRO red protein gel stain (Invitrogen Co.)	[30, 34–37,44–46, 48, 49, 53, 64, 65, 87, 103, 107–113]

Media	Collection	Resuspension buffer	Disruption	Precipitation	IEF buffer	Visualization method	Reference
-Medium F4 (per L: 100 g glucose, 1 g KH_2PO_4, 1 g $MgSO_4 \cdot 7H_2O$, 0.01 g $FeSO_4 \cdot 7H_2O$, 0.01 g $MnSO_4 \cdot 4\text{-}5\ H_2O$, 200 mg thiamine·HCl, 60 mg biotin, 0.02 mL antiforming agent, Adekanol LG-109, 13.8 mL soybean-meal hydrolysate (total nitrogen, 35 g/L)][103] -Salt medium containing 1% glucose [104]	-Wash 2x 10 ml of cold-water (4ºC). 1x *Washing buffer* (50 mM Tris–HCl, pH 7.2) -Preheated phosphate buffered saline (PBS) -20 mM Tris, 5 mM $MgCl_2$, 20 mM NaN_3, pH 7.5 -Continue the disruption step or shock-frozen in cold acetone (-60ºC) and stored at -70ºC	-Sodium dodecyl sulfate (SDS) buffer (100 mM Tris/HCl pH 9.5, 1% SDS w/v) -Lysis buffer (20 mM Tris, 5 mM EDTA, pH 7.5) -50 mM sodium phosphate buffer (pH 7.0) containing 12.5% (v/v) protease inhibitor cocktail for fungi (Sigma) -Urea/CHAPS alkylation buffer [8 M urea; 1% CHAPS; 1 mM EDTA; 200 mM Tris-HCl pH 8,0; 100 mM iodoacetamide][105]	ii) two times, 30 pulses; interval: 1 Hz, pulse duration: 0.3 s, 20 kHz homogeneous sound, power output: 60 W; iii) 220 W, 3 s vs. 5 s break, 50 cycles/time, 2 cycles using a JY92-II instrument; iv) ultrasonic homogenizer (UH-50, SMT Co.) at 50 W for 15 s with 15 s intervals for six times on ice	-100 mg of RNAse A and 25 mg of DNAse -1 mg/mL DNaseI and 0.25 mg/mL RNase A. -none Cell debris and unbroken cells elimination: -Centrifugation: 5,000-25,000 x g, 10-30 min, 4-10ºC. - Ultracentrifugation at 120,000-150,000 xg for 30-90 min. 4ºC.	-Mix 300 mg of proteins with rehydration solution (8 M urea, 2% CHAPS, 65 mM DTT, 2% Biolyte ampholyte) -Directly used for IEF separation in urea buffer Washing step post-precipitation: -2x wash *desintegration buffer* -80% acetone -none	Bromophenol blue, 20-60 mM DTT and 1% of IPGbuffer 3-11NL -6.4 M urea, 2% CHAPS, 2% protease inhibitor cocktail, 2% IPG buffer (GE Healthcare), 0.96% Destreak reagent (GE Healthcare), 0.01% bromophenol blue	-Radio-labelling (phosphoproteome)
Collection conditions: OD_{600nm} range: 0.6-7.0			Nucleotides elimination: -2 µl of Benzonase (DNAse+ RNAse; Merck) per sample 30 min (≅600–700 µl),				

Note: 3-[(3-cholamidopropyl)-dimethylammonio]-1-propanesulfonate (CHAPS); Isoelectrofocussing (IEF); Dithiothreitol (DTT); phosphate-buffered saline (PBS).

Table 1.B. Summary of the conditions (media, sample collection, digestion protocol and analyses method) developed or used for the analysis of the *C. glutamicum* intra-cellular proteome by LC-MS

Media	Collection	Disruption	Digestion	Analysis method	Reference
-Minimal medium -MME-SN modified MMES [114] -CGXII minimal medium [101]. $^{15}NH_4SO_4$ as the sole nitrogen Collection Conditions: OD_{600nm} range: 6.0-14.0	Collection: -4 mg biomass (cell dry weight) stored at -80°C Harvest: -Centrifugation: 8,500 rpm, 4°C, 5 min Washing step: -2x ice cold 0.1 M PBS solution (137 mM NaCl, 2.7 mM KCl, 10 mM Na_2HPO_4, and 1.8 mM KH_2PO_4; pH 7.4) -Sterile 0.9% sodium chloride Resuspension: -Lysis buffer (20 mM $MgCl_2$, 10 mM $MnCl_2$ and 200 units/mL DNase-I; in 0.1 M PBS) Incubate:1 h, 37°C and disrupt -PBS lysis buffer [PBS, 1 ml protease inhibitors (SIGMA) per 6 g cell, 250U DNase per mL buffer, 10mM $MgCl_2$ and 5mM $MnCl_2$] -Lysis buffer [50 mM potassium phosphate buffer, pH 8.0, 2 mM EDTA, 2 mM DTT, supplemented with complete protease inhibitor cocktail (Roche)]	Breaking system: -Sonication: CP130 Ultrasonic Processor (Cole Parmer Instrument Co.) a pulse mode with a 0.5 s interval under 15 W. -Mechanically: i) mixer mill; ii) beadmill (Retsch GmbH), 0.1mm glass beads, 5x, 30 s, highest settings, 4°C. -French press: 40K cell, volume of 35 mL, 4x at 20 000 psi. Cell debris elimination: -2x centrifugation: 4,500-5,000-16,000 xg, 15-60 min Dilute supernatant in PBS at 1 μg/μL -When protease inhibitors were used, 60°C for 1 h to inactivate	e.g.: -Denatured 200 μL of cell lysate in 120 μL of 8 M urea in 500 mM ammonium hydrogencarbonate. Mix 1 h at 37°C. -Disulfide bridges reduction by 12 μL of 300 mM DTT. Mix: 2 h, 37°C, 600 rpm. -Cysteine carboxymethylation by 8 μL of 2 M iodoacetic acid. 30 min. R/T. 16 μL of 1 M 2-mercaptoethanol. 20 min, R/T. -Dialysis: 16 h against 1 L distilled water (3,500 MW cut-off membranes) -Trypsin (1 μg for 50 μg protein) 24 h. -Stop by TFA (1.0% v/v end concentration)	-Hollow fibre flow field-flow fractionation (HF FlFFF) / (2D-LC-ESI-MS/MS) -Shotgun proteome analyses employing strong cation-exchange-ion-pair RP HPLC hyphenated to ESI MS/MS -MALDI-TOF/TOF and MALDI-LTQ-Orbitrap combined analysis -^{15}N-labelling MudPIT -Isotope dilution mass spectrometry coupled to high performance liquid chromatography (IDMS–LC–MS/MS)	[67, 69, 115–121]

Note: Dithiothreitol (DTT); Ethylenediaminetetraacetic acid (EDTA); phosphate-buffered saline (PBS); room temperature (R/T); trifluoroacetic acid (TFA).

Just one year later, Schaffer and co-workers (2001) published the first cytoplasmic (970 protein spots) and membrane-associated protein (660 spots) reference map [44]. Certainly, the most relevant improvement in the membrane proteome analysis was carried out by Schluesener and co-workers in 2005 [61]. They replaced the IEF (isoelectric focusing) step by AIEC (anion-exchange chromatography) followed by SDS-PAGE as the second dimension, in addition to washing with 2.5 M NaBr (elimination of membrane-associated proteins) and the membrane protein solubilisation improvement by detergent modification (ASB-14: amidosulfobetaine 14). As a result, 170 different proteins were identified, being 50 membrane integrants. The AIEC/SDSPAGE method was described subsequently as superior to the 6BAC-PAGE/SDS-PAGE (benzyldimethyl-n-hexadecylammonium chloride [62]) system [63]. This method permitted to define the effect of the membrane-bound AAA+ protease FtsH deletion or the pH stress in *C. glutamicum* by means of gel comparison between different conditions [64,65] (Table 2).

Next step in the membrane proteome analysis of *C. glutamicum* was the use of the two-dimensional chromatography. Hence, the combination of trypsin/cyanogen bromide or trypsin/chymotrypsin digestion and the multidimensional protein identification technology (MudPIT) analysis allowed for the identification of almost 50% of the predicted membrane proteins. More than 300 integral membrane proteins carrying small and large numbers of transmembrane helices were identified [66]. This method was tested also in Gram negative bacteria as *Escherichia coli* [67]. Subsequently, the Ansgar Poetsch's group used this methodology to analyse: i) the benzoate utilization as alternative carbon source using an internal standard metabolically labelled with ^{15}N; ii) the pH homeostasis and iii) the consumption of monocyclic aromatic carbon sources as protocatechuate by *C. glutamicum* [52, 68, 69].

The peculiar cellular membrane of *C. glutamicum* analysis has been tackled from the initial single protein isolation to the whole proteome analysis. Therefore, the procedures have been improved to increase the isolation of cell membranes, membrane-associated proteins elimination and the subsequent analysis of real membrane proteins (Table 2). Although the membrane proteins are sometimes recalcitrant for the isolation, the obtained results stand out in the analysis of transporters and membrane integrant proteins.

Secreted and Extra-Cellular Proteins (Secretome)

The secretome was defined in 2001 by Greenbaum and co-workers [70] as the population of gene products that are secreted from the cell. Later, it was specified by Makridakis and Vlahou [71] as the rich complex set of molecules secreted from living cells, which in a less canonical meaning also includes molecules shed from the living cells surface. Thus, the secretome, extra-cellular proteome (or less commonly named as exoproteome) involves the totality of the proteins secreted from a cell, entity or organism. These proteins can be observed as spontaneously released proteins and associated to the outer cell wall. Besides, the cellular machinery involved in the protein secretion has been proposed as part of the secretome [72]. The most crucial step for the secreted protein analysis is to distinguish secretome from 'degradome'. This means differentiating those proteins really secreted from those by-products coming from cell lysis events.

Table 2. Summary of the methodologies used for the analysis of *C. glutamicum* membrane proteome

Lysis	Disruption method	Purification	Resuspension method	Analysis method	Reference
-Centrifugation -Wash twice 0.9 % NaC1 -Resuspended in *desintegration buffer*: 2.92 g/L NaH_2PO_4 x H_20, 5.12 g/L $Na_2HPO_4 \cdot 2\ H_2O$, 0.247 g/L $MgSO_4 \cdot 7\ H_2O$	-French Press 3x 18,000 psi -Discard cell debris and unbroken cells by centrifugation 5,000 x g 10 min	-Ultracentrifugation 120,000 x g; 30 min. -Wash twice with *desintegration buffer*	-Resuspend in *lysis buffer* (9 M urea, 2% w/v CHAPS, 1% w/v DTT, 0.8% w/v 2-D Pharmalytes, pH range 3-10 and 8 mM PMSF [106]) -Inhibitor mix: COMPLETE (Roche)	-2D gel -Silver stain	[30]
-Centrifugation. -Wash twice 0.9 % NaCl. -Resuspended in *desintegration buffer*: 7.8 g/L NaH_2PO_4, 7.1 g/L Na_2HPO_4, 0.247 g/L $MgSO_4 \cdot 7\ H_2O$	-French Press 3x 18,000 psi -Discard cell debris and unbroken cells by centrifugation 5,000 x g 10 min	-Ultracentrifugation 120,000 x g; 30 min. -Wash twice with desintegration buffer	-Resuspend in *lysis buffer* (see above) with 2-4% CHAPS -Inhibitor mix: COMPLETE (Roche) -Extraction of cell surface proteins chloroform/methanol or sodium lauroyl sarcosinate -Diethylether precipitation -Resuspend *lysis buffer* containing 2% CHAPS	-2D gel -Silver stain	[34, 35, 122]
-Centrifugation: 4,500 xg, 15 min -Wash: PBS (137 mM NaCl, 2.7 mM KCl, 10 mM Na_2HPO_4, 1.8 mM KH_2PO_4, pH 7.4) -Resuspend 4 mL buffer/g wet cells in disintegration buffer (PBS containing additional 20 mM $MgCl_2$, 10 mM $MnCl_2$, 200 U/mL DNaseI, protease inhibitor (Sigma)	-French Pressure Cell (40K cell with a volume of 35 mL) with 4 passages at 20 000 psi. -Discard cell debris and unbroken cells by 2x centrifugation 5,000 xg 4ºC	-Ultracentrifugation at 100,000 xg, 4ºC, 30 min. -Resuspend pellet with ice-cold PBS buffer and ultracentrifugation again	-Remove membrane-associated by incubation with 20 mM Tris-HCl, pH 8.0; 100 mM sodium carbonate; 6 M urea; 4 M guanidine thiocyanate; 2.5 M NaBr; 4ºC; 15 rpm 30 min -Centrifugation at 100,000 xg, 30 min -Incubate membranes 1 h, R/T in solubilization buffer (20 mM Tris-HCl, pH 8.0, 150 mM NaCl, 1 mM DTT, 2% w/v ASB-14 at a ratio 10:1 to protein w/w); 15 rpm. -Sediment unsolubilized proteins: 100,000 xg; 4ºC; 30 min.	-Anion-exchange chromatography (AIEC) followed by sodium dodecyl sulfate (SDS)-PAGE. -Colloidal Coomassie	[53, 61, 63–65]
-Centrifugation: 6,000 xg, 15 min -Washing and resuspension steps were as previous row describes	-French Pressure Cell (volume of 35 mL) at 40,000 psi. -Discard cell debris and unbroken cells by centrifugation 6,000 xg 4ºC. -Ultracentrifugation at 100,000 xg, 4ºC, 30 min.	Membranes treatment: a) -Wash with ammonium carbonate (0.1 M, pH 11) -Dissolve in 1 ml of 90% formic acid -Digest with CNBr, 24 h R/T. -Decrease volume to 200 µl, adjusted to pH 8.6 with ammonium carbonate and 2 M urea -Digest with trypsin, O/N	-Remove undigested proteins and membrane fragments by 3-kDa cut-off filters (Microcon YM-3, Millipore) -Adjust pH 2.0 with acetic acid	-Multidimensional protein identification technology (MudPIT)	[52,66,68, 69]

Lysis	Disruption method	Purification	Resuspension method	Analysis method	Reference
		b) -Trypsin digestion of isolated membranes O/N, 37ºC, 25 mM ammonium bicarbonate buffer, pH 8.6. -Centrifugate at 100,000 xg. -Wash twice with ice-cold deionized water -Resuspend in 60% methanol and 40% 25 mM ammonium bicarbonate, sonication for 2x 10 min. Adjust pH to 8.6 with ammonium hydroxide -Trypsin and chymotrypsin digestion O/N, 37ºC	-Dialysis: 12 kDa cut-off dialysis tubing against 25 mM ammonium hydrogencarbonate buffer, 12 h	-Multidimensional protein identification technology, (MudPIT)	
-Centrifugation: 4,000 xg, 15 min. -Resuspension: PBS lysis buffer (8 g/L NaCl, 0.2 g/L KCL, 1.81 g/L $Na_2HPO_4 \cdot 2H_2O$, 0.24 g/L KH_2PO4 pH 7.4 with HCl), 1 mL protease inhibitors (Sigma) per cell gram, 250 U DNase per millilitre buffer, 10 mM $MgCl_2$ and 5 mM $MnCl_2$	-French Pressure Cell (40K cell with a volume of 35 mL) at 20,000 psi. -Discard cell debris and unbroken cells by 2x centrifugation 4,000 xg -ultracentrifugation at 100,000 xg and stored at -70ºC	-Wash and ultracentrifuge (100,000 xg; 35 min) 2 mg membranes: 3x 2.5 M sodium bromide 1x water -Incubate: 1% formic acid, 37ºC, 2 h. Neutralise with ammonium hydroxide -Dilute pretreated membranes in 500 µL of a biphasic system. Centrifuge: 1000 xg, 10 min. -3x 250 µL fresh PEG extraction. -Wash: 3x 100 µL fresh dextran			[58,67]

Note: 3-[(3-cholamidopropyl)-dimethylammonio]-1-propanesulfonate (CHAPS); Dithiothreitol (DTT); overnight (O/N); phosphate-buffered saline (PBS); room temperature (R/T).

Table 3. Summary of the procedures developed for the analysis of *C. glutamicum* secretome (extra-cellular proteome)

Microorganism	Extraction method	Reference
C. glutamicum ATCC 13032 *Corynebacterium efficiens* YS-314	-1,000 ml Luria-Broth (Invitrogen) -$4.5x10^{11}$ cells. Centrifugation: 5,000 xg, 10 min. Centrifugation: 10,000 xg, 90 min. 3 tablets of COMPLETE (protease inhibitor; Roche) -Freezing -80ºC and vacuum drying. Rehydratation in 25 mL H_2O -Discard small peptides and proteins [<5 kDa; Amicon-Ultra-15 filter units (Millipore)]. Centrifugation: 3,500 xg, 4ºC, 4 h. -Dissolve supernatant in solution A (20.8 mL H_2O, 4 mL 0.5 M Tris-HCl buffer pH 6.8 and 200 mL 1 M DTT). 15 min R/T. -Add water-saturated phenol (10 mL). Mix: 4ºC, 2 h. Centrifugation: 3,000 xg, 4ºC, 30 min -Precipitate proteins from phenolic phase by adding 200 mL 1 M DTT, 300 mL 8 M ammonium acetate and 30 mL of pre-cooled methanol (–20ºC). Incubation: -20ºC, 12 h -Centrifugation: 3,500 xg, 4ºC, 30 min. Wash twice: 5 mL 70% (v/v) ethanol (-20ºC) and once with 5 mL acetone (-20ºC); 2 h. -Dry pellet (R/T). Solubilizate in 450 μL rehydration buffer (8 M urea, 2% w/v CHAPS and 0.01% v/v bromophenol blue)	[87]
C. glutamicum	Similar to the Hansmeier and co-workers (2006) [87] protocol, but: -800 mL of MM1 broth (MMYE without yeast extract [123]). -Centrifugation: 3800 xg, 4ºC, 15 min and 3800 xg, 4ºC, 45 min. 1 tablets of COMPLETE mini (protease inhibitor; Roche) per 400 mL of broth -Amicon step was discarded -Lyophilisate of 400 mL of culture was rehydrated in 25 mL H_2O, supplemented with 4 mL 0.5 M Tris-HCl (pH 6.8) and 200 mL 1M DTT. R/T, 30 min -After methanol precipitation the precipitated proteins were washed with cold (-20 ºC) ethanol (70%) and acetone -Dry pellet were rehydrated in 400 mL 0.1M Tris-HCl, 1% (w/v) SDS; R/T	[90]

Microorganism	Extraction method	Reference
C. glutamicum	-100 ml minimal medium containing 4% glucose -Centrifugation: 10,000 xg, 10 min, 4°C -Concentration: up to 1 ml [Centriprep YM-30 (Millipore)] -Precipitation: acetone. Resuspended: 1 ml solubilization buffer (9 M urea, 4% CHAPS, and 1% DTT)	[124]
Corynebacterium jeikeium K411	Similar to the Hansmeier and co-workers (2006) [87] protocol, but: -500 mL culture (BYT complex medium [125]) -After the initial centrifugation (5000 x*g* for 90 min) a filtration step was included [0.2 µm pore-size filters (Schleicher & Schuell)]	[88]
Corynebacterium pseudotuberculosis	<u>Three-phase partitioning (TPP) technique:</u> 500 mL of CDM (chemically defined medium [126]). -Centrifugation: 4,000 rpm, 20 min, R/T -Collect 400 mL, add 20 µL Protease Inhibitor Cocktail P8465 (Sigma-Aldrich). Filter through 0.22 µm -Add 30% ammonium sulfate to the culture supernatants. R/T. Mix -Adjust pH by HCl addition at 4.0 -Add n-butanol (vol:vol). Vortex 1 min and rest 1 h (until the mixtures separated into three phases) -Centrifugation: 1350 xg, 10 min. Collect the interfacial precipitation and dissolved in 1 ml of Tris 20 mM, pH 7.4 buffer + 10 µL protease inhibitor (500 µl buffer/5 ml culture supernatant) [-Diafiltration and buffer exchange with NH_4HCO_3 (100 mM) and 5kDa cut off spin columns (Millipore)]	[91–93]

Note: 3-[(3-cholamidopropyl)-dimethylammonio]-1-propanesulfonate (CHAPS); dithiothreitol (DTT); min: minutes; room temperature (R/T); sodium dodecyl sulphate (SDS).

This key process for secretome analysis can be faced up sequentially from three different aspects: i) sampling; ii) post-identification protein analysis and iii) unconventional secreted protein or 'moonlighting proteins'. Thus, a careful bacteria elimination process (e.g.: rigorous centrifugation, nylon filters and/or 0.45 μm filtration) and the use of low temperatures to avoid cell lysis are highly recommended. Sample collection is capital to circumvent the non-natural secreted protein collection. So, the correlation between growth phase and secreted protein gathering is a useful tip to discard the cell lysis. Therefore, it is recommended to perform the sampling at the earliest time point compatible with the optimal moment of extracellular protein secretion.

Additionally, bioinformatics tools have been developed to characterize the secreted proteins either in a classical or non-classical way [e.g.: SignalP (for classical secretion) or SecretomeP (for non-classical)] [73–75]. These pieces of software help to confirm the reliability of the putative secreted proteins. However, some unconventional proteins, named 'moonlighting proteins' [76–78], can be detected in secretomes even when the above-mentioned cautions are taken into account. These proteins have unconventional secretion systems [79,80], which complicates their authenticity as secreted proteins [81, 82].

Extra-cellular proteins for specific analyses like western blot or GFP detection can be isolated in a simple manner: i) acetone precipitation and PBS resuspension [83] or ii) centrifugation to obtain cell-free supernatants [84]. Yet, the analysis of whole secretomes complicates the protein extraction due to the required large culture broth volumes. Thus, lyophilisation, precipitation (e.g.: acetone, trichloroacetic acid) or concentration (e.g.: Amicon-type filtering) steps are mandatory.

Even *C. glutamicum* is a soil isolated bacterium [85,86] and these inhabitants frequently present a huge array of secreted proteins (e.g.: *Streptomyces* species or fungi) just a few number of papers can be found focused on its secretome analysis. Thus, some other *Corynebacterium* species analyses have been included in this section. Table 3 summarizes several methods applied to *Corynebacterium* species for extra-cellular proteome study. Basically, the methods are based on phenol extraction including Amicon-like concentration steps [87–90] or *n*-butanol precipitation [91–93]. In a first approach, those methods include several steps that always mean a protein lost along the process. In addition, these large extraction processes contrast with the secretomes from other soil microorganism like Actinobacteria: i) *Streptomyces coelicolor* [94], ii) *Streptomyces griseus* [95], iii) the close-related bacteria *Brevibacterium linens* [96] or fungi as *Penicillium chrysogenum* [97], where it is enough with a trichloroacetic acid precipitation step and the subsequent acetone washing one. This difference in protocols could be due to a low presence of secreted proteins in the culture broth of *Corynebacterium* species.

CONCLUSION

From the initial manuscript of Hermann and co-workers in 1998 [30] about the analysis of the proteome of *C. glutamicum* to date, more than sixty papers have been focused specifically on the optimization, update and use of Proteomics methodologies (Figure 1). Initially, the reference map as an example of technical capabilities paved the way for more complex analyses, sometimes combined with other omics, in order to improve the skills of *C.*

glutamicum. Yet, several Proteomics analyses remain incomplete such as the global scale phosphorylation sites location by gel-free methodologies [39] or the putative introduction of novel methodologies similar to SWATH™ 2.0 data independent acquisition (ABSciex). But, in parallel, the development of bioinformatics tools such as CoryneRegNet [98], BioCyc [99] or those that allow integrating all the current routine omics as ProMeTra system [100] were also needed and very welcomed. Hence, Proteomics is becoming a routinely methodology used for *C. glutamicum* research.

ACKNOWLEDGMENTS

The author would like to thank the European Union program ERA-IB [SCILS project (EIB.12.057)] through the Spanish Ministry of Economy and Competitiveness (MINECO. Madrid, Spain) (PCIN-2013-063). Special thanks go to A. Mulero for her excellent critical reading and to the degree and master students E. Sabugo, E. Fernández-Alegre, A. Ibañez and A. Morales for their positive cooperation.

REFERENCES

[1] Wasinger VC, Cordwell SJ, Cerpa-Poljak A, Yan JX, Gooley AA, Wilkins MR, et al., Progress with gene-product mapping of the Mollicutes: *Mycoplasma genitalium*. Electrophoresis 1995;16:1090–4.

[2] Wilkins MR, Sanchez JC, Gooley A a., Appel RD, Humphery-Smith I, Hochstrasser DF, et al., Progress with proteome projects: why all proteins expressed by a genome should be identified and how to do it. *Biotechnol Genet. Eng. Rev*. 1996;13:19–50.

[3] Wilkins MR, Pasquali C, Appel RD, Ou K, Golaz O, Sanchez JC, et al., From proteins to proteomes: large scale protein identification by two-dimensional electrophoresis and amino acid analysis. *Biotechnology* (N Y) 1996;14:61–5.

[4] Anderson L, Seilhamer J. A comparison of selected mRNA and protein abundances in human liver. Electrophoresis n.d.;18:533–7.

[5] Rodríguez-García A, Barreiro C, Santos-Beneit F, Sola-Landa A, Martín JF. Genome-wide transcriptomic and proteomic analysis of the primary response to phosphate limitation in *Streptomyces coelicolor* M145 and in a Delta*phoP* mutant. *Proteomics* 2007;7:2410–29.

[6] Mäder U, Antelmann H, Buder T, Dahl MK, Hecker M, Homuth G. *Bacillus subtilis* functional genomics: genome-wide analysis of the DegS-DegU regulon by transcriptomics and proteomics. *Mol. Genet. Genomics* 2002;268:455–67.

[7] James P. Protein identification in the post-genome era: the rapid rise of proteomics. *Q. Rev. Biophys.* 1997;30:279–331.

[8] Carrera Mouriño M. Proteómica bottom-up y top-down de organismos poco secuenciados identificación de especies de la familia Merlicciidae. PhD Thesis. Vigo (Spain), 2008.

[9] Ideker T, Galitski T, Hood L. A new approach to decoding life: systems biology. *Annu Rev Genomics Hum. Genet*. 2001;2:343–72.

[10] Oltvai ZN, Barabási A. Systems biology. Life's complexity pyramid. *Science* 2002;298:763–4.
[11] Woo HM, Park J-B. Recent progress in development of synthetic biology platforms and metabolic engineering of *Corynebacterium glutamicum*. *J. Biotechnol.* 2014;180:43–51.
[12] Hermann T. Industrial production of amino acids by coryneform bacteria. *J. Biotechnol.* 2003;104:155–72.
[13] Ajinomoto. Food products business 2010. http://www.ajinomoto.com/ir/pdf/Food-Oct2010.pdf
[14] Ajinomoto. Feed-Use Amino Acids Business. 2011. http://www.ajinomoto.com/ir/pdf/FeeduseAA-Oct2011.
[15] Wieschalka S, Blombach B, Eikmanns BJ. Engineering *Corynebacterium glutamicum* for the production of pyruvate. *Appl. Microbiol. Biotechnol.* 2012;94:449–59.
[16] Wieschalka S, Blombach B, Eikmanns BJ. Bio-based production of organic acids with *Corynebacterium glutamicum* 2012:1–16.
[17] Production I, Blombach B, Riester T, Wieschalka S, Ziert C, Youn J, et al., *Corynebacterium glutamicum* Tailored for Efficient 2011;77:3300–10.
[18] Okino S, Suda M, Fujikura K, Inui M, Yukawa H. Production of D-lactic acid by *Corynebacterium glutamicum* under oxygen deprivation. *Appl. Microbiol. Biotechnol.* 2008;78:449–54.
[19] Heider SAE, Wolf N, Hofemeier A, Peters-Wendisch P, Wendisch VF. Optimization of the IPP Precursor Supply for the Production of Lycopene, Decaprenoxanthin and Astaxanthin by *Corynebacterium glutamicum*. *Front Bioeng Biotechnol* 2014;2:28.
[20] Yamamoto S, Gunji W, Suzuki H, Toda H, Suda M, Jojima T, et al., Overexpression of Genes Encoding Glycolytic Enzymes in *Corynebacterium glutamicum* Enhances Glucose Metabolism and Alanine Production under Oxygen Deprivation Conditions 2012.
[21] Ikeda M, Nakagawa S. The *Corynebacterium glutamicum* genome: features and impacts on biotechnological processes. *Appl. Microbiol. Biotechnol.* 2003;62:99–109.
[22] Kalinowski J, Bathe B, Bartels D, Bischoff N, Bott M, Burkovski A, et al., The complete *Corynebacterium glutamicum* ATCC 13032 genome sequence and its impact on the production of L-aspartate-derived amino acids and vitamins. *J. Biotechnol.* 2003;104:5–25.
[23] Park SH, Kim HU, Kim TY, Park JS, Kim S-S, Lee SY. Metabolic engineering of *Corynebacterium glutamicum* for L-arginine production. *Nat. Commun.* 2014;5:4618.
[24] Lv Y, Liao J, Wu Z, Han S, Lin Y, Zheng S. Genome sequence of *Corynebacterium glutamicum* ATCC 14067, which provides insight into amino acid biosynthesis in coryneform bacteria. *J. Bacteriol.* 2012;194:742–3.
[25] Lee C, Nam J, Son E, Kwon O, Han W, Cho J, et al., Next-generation sequencing-based genome-wide mutation analysis of L-lysine-producing *Corynebacterium glutamicum* ATCC 21300 strain. *J. Microbiol.* 2012;50:860–3.
[26] Binder S, Schendzielorz G, Stäbler N, Krumbach K, Hoffmann K, Bott M, et al., A high-throughput approach to identify genomic variants of bacterial metabolite producers at the single-cell level. *Genome. Biol.* 2012;13:R40.
[27] Lv Y, Wu Z, Han S, Lin Y, Zheng S. Genome sequence of *Corynebacterium glutamicum* S9114, a strain for industrial production of glutamate. *J. Bacteriol.* 2011;193:6096–7.

[28] Yukawa H, Omumasaba CA, Nonaka H, Watanabe J, Ikeda Y, Verte AA. Comparative analysis of the *Corynebacterium glutamicum* group and complete genome sequence of strain R 2007:1042–58.

[29] Baumgart M, Unthan S, Rückert C, Sivalingam J, Grünberger A, Kalinowski J, et al., Construction of a prophage-free variant of *Corynebacterium glutamicum* ATCC 13032 for use as a platform strain for basic research and industrial biotechnology. *Appl. Environ. Microbiol.* 2013;79:6006–15.

[30] Hermann T, Wersch G, Uhlemann EM, Schmid R, Burkovski A. Mapping and identification of *Corynebacterium glutamicum* proteins by two-dimensional gel electrophoresis and microsequencing. *Electrophoresis* 1998;19:3217–21.

[31] Tiessen A, Pérez-Rodríguez P, Delaye-Arredondo LJ. Mathematical modeling and comparison of protein size distribution in different plant, animal, fungal and microbial species reveals a negative correlation between protein size and protein number, thus providing insight into the evolution of proteomes. *BMC Res Notes* 2012;5:85.

[32] VanBogelen RA, Schiller EE, Thomas JD, Neidhardt FC. Diagnosis of cellular states of microbial organisms using proteomics. *Electrophoresis* 1999;20:2149–59.

[33] Schwartz R, Ting CS, King J. Whole proteome pI values correlate with subcellular localizations of proteins for organisms within the three domains of life. *Genome. Res.* 2001;11:703–9.

[34] Hermann T, Finkemeier M, Pfefferle W, Wersch G, Krämer R, Burkovski A. Two-dimensional electrophoretic analysis of *Corynebacterium glutamicum* membrane fraction and surface proteins. *Electrophoresis* 2000;21:654–9.

[35] Hermann T, Pfefferle W, Baumann C, Busker E, Schaffer S, Bott M, et al. Proteome analysis of *Corynebacterium glutamicum*. *Electrophoresis* 2001;22:1712–23.

[36] Barreiro C, González-Lavado E, Brand S, Tauch A, Martín JF. Heat Shock Proteome Analysis of Wild-Type *Corynebacterium glutamicum* ATCC 13032 and a Spontaneous Mutant Lacking GroEL1, a Dispensable Chaperone. *J. Bacteriol.* 2005;187:884–9.

[37] Vasco-Cárdenas MF, Baños S, Ramos A, Martín JF, Barreiro C. Proteome response of *Corynebacterium glutamicum* to high concentration of industrially relevant C4 and C5 dicarboxylic acids. *J. Proteomics* 2013;85:65–88.

[38] Schaffer S, Burkovski A. Proteomics. In: Eggeling L, Bott M, editors. Handb. *Corynebacterium glutamicum*, London: CRC Press (Taylor & Francis Group); 2005, p. 99–118.

[39] Wendisch VF, Polen T. Transcriptome/Proteome Analysis of *Corynebacterium glutamicum*. In: Yukawa H, Inui M, editors. *Corynebacterium glutamicum* Biol. Biotechnol., vol. 23, Berlin, Heidelberg: Springer Berlin Heidelberg; 2013, p. 173–216.

[40] Kelleher NL. Top-down proteomics. *Anal. Chem.* 2004;76:197A – 203A.

[41] Sze SK, Ge Y, Oh H, McLafferty FW. Top-down mass spectrometry of a 29-kDa protein for characterization of any posttranslational modification to within one residue. *Proc. Natl. Acad. Sci. U S A* 2002;99:1774–9.

[42] Chait BT. Chemistry. Mass spectrometry: bottom-up or top-down? *Science* 2006;314:65–6.

[43] Poetsch A, Haussmann U, Burkovski A. Proteomics of corynebacteria: From biotechnology workhorses to pathogens. *Proteomics* 2011;11:3244–55.

[44] Schaffer S, Weil B, Nguyen VD, Dongmann G, Günther K, Nickolaus M, et al. A high-resolution reference map for cytoplasmic and membrane-associated proteins of *Corynebacterium glutamicum*. *Electrophoresis* 2001;22:4404–22.

[45] Li L, Wada M, Yokota A. Cytoplasmic proteome reference map for a glutamic acid-producing *Corynebacterium glutamicum* ATCC 14067. *Proteomics* 2007;7:4317–22.

[46] Bendt AK, Burkovski A, Schaffer S, Bott M, Farwick M, Hermann T. Towards a phosphoproteome map of *Corynebacterium glutamicum*. *Proteomics* 2003;3:1637–46.

[47] Fanous A, Weiland F, Lück C, Görg A, Friess A, Parlar H. A proteome analysis of *Corynebacterium glutamicum* after exposure to the herbicide 2,4-dichlorophenoxy acetic acid (2,4-D). *Chemosphere* 2007;69:25–31.

[48] Fanous A, Weiss W, Görg A, Jacob F, Parlar H. A proteome analysis of the cadmium and mercury response in *Corynebacterium glutamicum*. *Proteomics* 2008;8:4976–86.

[49] Fanous A, Hecker M, Görg A, Parlar H, Jacob F. *Corynebacterium glutamicum* as an indicator for environmental cobalt and silver stress--a proteome analysis. *J. Environ. Sci. Health B* 2010;45:666–75.

[50] Fränzel B, Frese C, Penkova M, Metzler-Nolte N, Bandow JE, Wolters DA. *Corynebacterium glutamicum* exhibits a membrane-related response to a small ferrocene-conjugated antimicrobial peptide. *J. Biol. Inorg. Chem.* 2010;15:1293–303.

[51] Fränzel B, Trötschel C, Rückert C, Kalinowski J, Poetsch A, Wolters DA. Adaptation of *Corynebacterium glutamicum* to salt-stress conditions. *Proteomics* 2010;10:445–57.

[52] Haussmann U, Qi S-W, Wolters D, Rögner M, Liu S-J, Poetsch A. Physiological adaptation of *Corynebacterium glutamicum* to benzoate as alternative carbon source - a membrane proteome-centric view. *Proteomics* 2009;9:3635–51.

[53] Polen T, Schluesener D, Poetsch A, Bott M, Wendisch VF. Characterization of citrate utilization in *Corynebacterium glutamicum* by transcriptome and proteome analysis. *FEMS Microbiol. Lett.* 2007;273:109–19.

[54] Lange C, Rittmann D, Wendisch VF, Bott M, Sahm H. Global Expression Profiling and Physiological Characterization of *Corynebacterium glutamicum* Grown in the Presence of L-Valine Global Expression Profiling and Physiological Characterization of *Corynebacterium glutamicum* Grown in the Presence of L -Valine 2003.

[55] Goodfellow M, Kämpfer P, Busse H-J, Trujillo ME, Suzuki K, Ludwig W, et al., Bergey's Manual of Systematic Bacteriology. The Actinobacteria, part B. Volumen five. Springer; 2012.

[56] Lanéelle M, Tropis M, Daffé M. Current knowledge on mycolic acids in *Corynebacterium glutamicum* and their relevance for biotechnological processes. *Appl. Microbiol. Biotechnol.* 2013;97:9923–30.

[57] Bayan N, Houssin C, Chami M, Leblon G. Mycomembrane and S-layer: two important structures of *Corynebacterium glutamicum* cell envelope with promising biotechnology applications. *J. Biotechnol.* 2003;104:55–67.

[58] Fränzel B, Fischer F, Trötschel C, Poetsch A, Wolters D. The two-phase partitioning system-a powerful technique to purify integral membrane proteins of *Corynebacterium glutamicum* for quantitative shotgun analysis. *Proteomics* 2009;9:2263–72.

[59] Marienfeld S, Uhlemann EM, Schmid R, Krämer R, Burkovski A. Ultrastructure of the *Corynebacterium glutamicum* cell wall. Antonie Van Leeuwenhoek 1997;72:291–7.

[60] Lichtinger T, Burkovski A, Niederweis M, Krämer R, Benz R. Biochemical and biophysical characterization of the cell wall porin of *Corynebacterium glutamicum*: the

channel is formed by a low molecular mass polypeptide. *Biochemistry* 1998;37:15024–32.

[61] Schluesener D, Fischer F, Kruip J, Rögner M, Poetsch A. Mapping the membrane proteome of *Corynebacterium glutamicum*. *Proteomics* 2005;5:1317–30.

[62] Hartinger J, Stenius K, Högemann D, Jahn R. 16-BAC/SDS-PAGE: a two-dimensional gel electrophoresis system suitable for the separation of integral membrane proteins. *Anal. Biochem.* 1996;240:126–33.

[63] Schluesener D, Rögner M, Poetsch A. Evaluation of two proteomics technologies used to screen the membrane proteomes of wild-type *Corynebacterium glutamicum* and an L-lysine-producing strain. *Anal. Bioanal. Chem.* 2007;389:1055–64.

[64] Lüdke A, Krämer R, Burkovski A, Schluesener D, Poetsch A. A proteomic study of *Corynebacterium glutamicum* AAA+ protease FtsH. *BMC Microbiol.* 2007;7:6.

[65] Barriuso-Iglesias M, Schluesener D, Barreiro C, Poetsch A, Martín JF. Response of the cytoplasmic and membrane proteome of *Corynebacterium glutamicum* ATCC 13032 to pH changes. *BMC Microbiol.* 2008;8:225.

[66] Fischer F, Wolters D, Rögner M, Poetsch A. Toward the complete membrane proteome: high coverage of integral membrane proteins through transmembrane peptide detection. *Mol. Cell. Proteomics* 2006;5:444–53.

[67] Fränzel B, Wolters D a. Advanced MudPIT as a next step toward high proteome coverage. *Proteomics* 2011;11:3651–6.

[68] Follmann M, Ochrombel I, Krämer R, Trötschel C, Poetsch A, Rückert C, et al., Functional genomics of pH homeostasis in *Corynebacterium glutamicum* revealed novel links between pH response, oxidative stress, iron homeostasis and methionine synthesis. *BMC Genomics* 2009;10:621.

[69] Haussmann U, Poetsch A. Global proteome survey of protocatechuate- and glucose-grown *Corynebacterium glutamicum* reveals multiple physiological differences. *J. Proteomics* 2012;75:2649–59.

[70] Greenbaum D, Luscombe NM, Jansen R, Qian J, Gerstein M. Interrelating different types of genomic data, from proteome to secretome: 'oming in on function. *Genome. Res.* 2001;11:1463–8.

[71] Makridakis M, Roubelakis MG, Vlahou A. Stem cells: insights into the secretome. *Biochim. Biophys. Acta.* 2013;1834:2380–4.

[72] Tjalsma H, Bolhuis A, Jongbloed JD, Bron S, van Dijl JM. Signal peptide-dependent protein transport in *Bacillus subtilis*: a genome-based survey of the secretome. *Microbiol. Mol. Biol. Rev.* 2000;64:515–47.

[73] Bendtsen JD, Kiemer L, Fausbøll A, Brunak S. Non-classical protein secretion in bacteria 2005;13:1–13.

[74] Bendtsen JD, Nielsen H, von Heijne G, Brunak S. Improved prediction of signal peptides: SignalP 3.0. *J. Mol. Biol.* 2004;340:783–95.

[75] Bendtsen JD, Jensen LJ, Blom N, Von Heijne G, Brunak S. Feature-based prediction of non-classical and leaderless protein secretion. *Protein. Eng. Des. Sel.* 2004;17:349–56.

[76] Jeffery CJ. Moonlighting proteins. *Trends Biochem. Sci.* 1999;24:8–11.

[77] Jeffery CJ. Moonlighting proteins-an update. *Mol. Biosyst.* 2009;5:345–50.

[78] Jeffery CJ. An introduction to protein moonlighting. *Biochem. Soc. Trans* 2014;42:1679–83.

[79] Nickel W, Rabouille C. Mechanisms of regulated unconventional protein secretion 2009;10.

[80] Nickel W. Unconventional Secretory Routes : Direct Protein *Export Across the Plasma Membrane of Mammalian Cells* 2005:607–14.

[81] Barreiro C, Martín JF, García-Estrada C. Proteomics Shows New Faces for the Old Penicillin Producer *Penicillium chrysogenum. J. Biomed. Biotechnol.* 2012;2012:1–15.

[82] Barreiro C, Martín JF, García-Estrada C. Proteomics Methodology Applied to the Analysis of Filamentous Fungi - New Trends for an Impressive Diverse Group of Organisms. In: Prasain JK, editor. Tandem Mass Spectrom. - Appl. Princ., Rijeka, Croatia: InTech; 2012.

[83] Yim SS, An SJ, Choi JW, Ryu AJ, Jeong KJ. High-level secretory production of recombinant single-chain variable fragment (scFv) in *Corynebacterium glutamicum. Appl. Microbiol. Biotechnol.* 2014;98:273–84.

[84] Teramoto H, Watanabe K, Suzuki N, Inui M, Yukawa H. High yield secretion of heterologous proteins in *Corynebacterium glutamicum* using its own Tat-type signal sequence. *Appl. Microbiol. Biotechnol* 2011;91:677–87.

[85] Kinoshita S, Nakayama S, Akita S. Taxonomical study of glutamic acid accumulating bacteria, *Micrococcus glutamicus*, nov. sp. *Bull Agric Chem. Soc. Jpn* 1958;22:176–85.

[86] Abe S, Takayama K, Kinoshita S. Taxonomical studies on glutamic acid-producing bacteria. *J. Gen. Appl. Microbiol.* 1967;13:279–301.

[87] Hansmeier N, Chao T, Pühler A, Tauch A, Kalinowski J. The cytosolic, cell surface and extracellular proteomes of the biotechnologically important soil bacterium Corynebacterium efficiens YS-314 in comparison to those of *Corynebacterium glutamicum* ATCC 13032. Proteomics 2006;6:233–50.

[88] Hansmeier N, Chao T-C, Daschkey S, Müsken M, Kalinowski J, Pühler A, et al. A comprehensive proteome map of the lipid-requiring nosocomial pathogen *Corynebacterium jeikeium* K411. *Proteomics* 2007;7:1076–96.

[89] Hansmeier N, Chao T, Kalinowski J, Pühler A, Tauch A. Mapping and comprehensive analysis of the extracellular and cell surface proteome of the human pathogen *Corynebacterium diphtheriae* 2006:2465–76.

[90] Mahne M, Tauch A, Pühler A, Kalinowski J. The *Corynebacterium glutamicum* gene pmt encoding a glycosyltransferase related to eukaryotic protein-O-mannosyltransferases is essential for glycosylation of the resuscitation promoting factor (Rpf2) and other secreted proteins. *FEMS Microbiol. Lett* 2006;259:226–33.

[91] Paule BJA, Meyer R, Moura-Costa LF, Bahia RC, Carminati R, Regis LF, et al., Three-phase partitioning as an efficient method for extraction/concentration of immunoreactive excreted-secreted proteins of *Corynebacterium pseudotuberculosis. Protein Expr. Purif.* 2004;34:311–6.

[92] Pacheco LGC, Slade SE, Seyffert N, Santos AR, Castro TLP, Silva WM, et al. A combined approach for comparative exoproteome analysis of *Corynebacterium pseudotuberculosis. BMC Microbiol.* 2011;11:12.

[93] Silva WM, Seyffert N, Santos A V, Castro TLP, Pacheco LGC, Santos AR, et al. Identification of 11 new exoproteins in *Corynebacterium pseudotuberculosis* by comparative analysis of the exoproteome. *Microb. Pathog.* 2013;61-62:37–42.

[94] Kim D, Chater K, Lee K, Hesketh A. Changes in the extracellular proteome caused by the absence of the *bldA* gene product, a developmentally significant tRNA, reveal a new

target for the pleiotropic regulator AdpA in *Streptomyces coelicolor*. *J. Bacteriol.* 2005;187:2957–66.

[95] Akanuma G, Hara H, Ohnishi Y, Horinouchi S. Dynamic changes in the extracellular proteome caused by absence of a pleiotropic regulator AdpA in *Streptomyces griseus*. *Mol. Microbiol.* 2009;73:898–912.

[96] Shabbiri K, Botting CH, Adnan A, Fuszard M. Charting the cellular and extracellular proteome analysis of *Brevibacterium linens* DSM 20158 with unsequenced genome by mass spectrometry-driven sequence similarity searches. *J. Proteomics* 2013;83:99–118.

[97] Jami M-S, García-Estrada C, Barreiro C, Cuadrado A-A, Salehi-Najafabadi Z, Martín J-F. The *Penicillium chrysogenum* extracellular proteome. Conversion from a food-rotting strain to a versatile cell factory for white biotechnology. *Mol. Cell. Proteomics* 2010;9:2729–44.

[98] Baumbach J. CoryneRegNet 4.0 - A reference database for corynebacterial gene regulatory networks. *BMC Bioinformatics* 2007;8:429.

[99] Caspi R, Altman T, Billington R, Dreher K, Foerster H, Fulcher CA, et al., The MetaCyc database of metabolic pathways and enzymes and the BioCyc collection of Pathway/Genome Databases. *Nucleic Acids Res* 2014;42:D459–71.

[100] Neuweger H, Persicke M, Albaum SP, Bekel T, Dondrup M, Hüser AT, et al., Visualizing post genomics data-sets on customized pathway maps by ProMeTra-aeration-dependent gene expression and metabolism of *Corynebacterium glutamicum* as an example. *BMC Syst. Biol.* 2009;3:82.

[101] Keilhauer C, Eggeling L, Sahm H. Isoleucine synthesis in *Corynebacterium glutamicum*: molecular analysis of the *ilvB-ilvN-ilvC* operon. *J. Bacteriol.* 1993;175:5595–603.

[102] Kase H, Nakayama K. Production of L-threonine by analog-resistant mutants. *Agric. Biol. Chem.* 1972;36:1611–21.

[103] Li L, Wada M, Yokota A. A comparative proteomic approach to understand the adaptations of an H+ -ATPase-defective mutant of *Corynebacterium glutamicum* ATCC14067 to energy deficiencies. *Proteomics* 2007;7:3348–57.

[104] Eikmanns BJ, Thum-Schmitz N, Eggeling L, Lüdtke KU, Sahm H. Nucleotide sequence, expression and transcriptional analysis of the *Corynebacterium glutamicum gltA* gene encoding citrate synthase. *Microbiology* 1994;140 (Pt 8:1817–28.

[105] Chi BK, Gronau K, Mäder U, Hessling B, Becher D, Antelmann H. S-bacillithiolation protects against hypochlorite stress in *Bacillus subtilis* as revealed by transcriptomics and redox proteomics. *Mol. Cell Proteomics* 2011;10:M111.009506.

[106] Görg A, Boguth G, Obermaier C, Posch A, Weiss W. Two-dimensional polyacrylamide gel electrophoresis with immobilized pH gradients in the first dimension (IPG-Dalt): the state of the art and the controversy of vertical versus horizontal systems. *Electrophoresis* 1995;16:1079–86.

[107] Silberbach M, Schäfer M, Hüser AT, Kalinowski J, Pühler A, Krämer R, et al., Adaptation of *Corynebacterium glutamicum* to ammonium limitation: a global analysis using transcriptome and proteome techniques. *Appl. Environ. Microbiol.* 2005;71:2391–402.

[108] Beckers G, Strösser J, Hildebrandt U, Kalinowski J, Farwick M, Krämer R, et al., Regulation of AmtR-controlled gene expression in *Corynebacterium glutamicum*:

mechanism and characterization of the AmtR regulon. *Mol. Microbiol.* 2005;58:580–95.

[109] Valbuena N, Letek M, Ramos A, Ayala J, Nakunst D, Kalinowski J, et al., Morphological changes and proteome response of *Corynebacterium glutamicum* to a partial depletion of FtsI. *Microbiology* 2006;152:2491–503.

[110] Qi S-W, Chaudhry MT, Zhang Y, Meng B, Huang Y, Zhao K-X, et al. Comparative proteomes of *Corynebacterium glutamicum* grown on aromatic compounds revealed novel proteins involved in aromatic degradation and a clear link between aromatic catabolism and gluconeogenesis via fructose-1,6-bisphosphatase. *Proteomics* 2007;7:3775–87.

[111] Krömer JO, Bolten CJ, Heinzle E, Schröder H, Wittmann C. Physiological response of *Corynebacterium glutamicum* to oxidative stress induced by deletion of the transcriptional repressor McbR. *Microbiology* 2008;154:3917–30.

[112] Kim J, Fukuda H, Hirasawa T, Nagahisa K, Nagai K, Wachi M, et al. Requirement of de novo synthesis of the OdhI protein in penicillin-induced glutamate production by *Corynebacterium glutamicum. Appl. Microbiol. Biotechnol.* 2010;86:911–20.

[113] Chi BK, Busche T, Van Laer K, Bäsell K, Becher D, Clermont L, et al., Protein S-mycothiolation functions as redox-switch and thiol protection mechanism in *Corynebacterium glutamicum* under hypochlorite stress. *Antioxid. Redox. Signal* 2014;20:589–605.

[114] Koch DJ, Rückert C, Rey DA, Mix A, Pühler A, Kalinowski J. Role of the *ssu* and *seu* genes of *Corynebacterium glutamicum* ATCC 13032 in utilization of sulfonates and sulfonate esters as sulfur sources. *Appl. Environ. Microbiol.* 2005;71:6104–14.

[115] Kim KH, Kang D, Koo HM, Moon MH. Molecular mass sorting of proteome using hollow fiber flow field-flow fractionation for proteomics. *J. Proteomics* 2008;71:123–31.

[116] Lasaosa M, Delmotte N, Huber CG, Melchior K, Heinzle E, Tholey A. A 2D reversed-phase x ion-pair reversed-phase HPLC-MALDI TOF/TOF-MS approach for shotgun proteome analysis. *Anal. Bioanal. Chem.* 2009;393:1245–56.

[117] Delmotte N, Lasaosa M, Tholey A, Heinzle E, Huber CG. Two-dimensional reversed-phase x ion-pair reversed-phase HPLC: an alternative approach to high-resolution peptide separation for shotgun proteome analysis. *J. Proteome Res.* 2007;6:4363–73.

[118] Delmotte N, Lasaosa M, Tholey A, Heinzle E, van Dorsselaer A, Huber CG. Repeatability of peptide identifications in shotgun proteome analysis employing off-line two-dimensional chromatographic separations and ion-trap MS. *J. Sep. Sci.* 2009;32:1156–64.

[119] Baeumlisberger D, Rohmer M, Arrey TN, Mueller BF, Beckhaus T, Bahr U, et al., Simple dual-spotting procedure enhances nLC-MALDI MS/MS analysis of digests with less specific enzymes. *J. Proteome Res.* 2011;10:2889–94.

[120] Voges R, Noack S. Quantification of proteome dynamics in *Corynebacterium glutamicum* by (15)N-labeling and selected reaction monitoring. *J. Proteomics* 2012;75:2660–9.

[121] Rietschel B, Baeumlisberger D, Arrey TN, Bornemann S, Rohmer M, Schuerken M, et al., The benefit of combining nLC-MALDI-Orbitrap MS data with nLC-MALDI-TOF/TOF data for proteomic analyses employing elastase. *J. Proteome Res.* 2009;8:5317–24.

[122] Beckers G, Bendt AK, Krämer R, Burkovski A. Molecular identification of the urea uptake system and transcriptional analysis of urea transporter- and urease-encoding genes in *Corynebacterium glutamicum*. *J. Bacteriol.* 2004;186:7645–52.

[123] Katsumata R, Ozaki A, Oka T, Furuya A. Protoplast transformation of glutamate-producing bacteria with plasmid DNA. *J. Bacteriol* 1984;159:306–11.

[124] Suzuki N, Watanabe K, Okibe N, Tsuchida Y, Inui M, Yukawa H. Identification of new secreted proteins and secretion of heterologous amylase by *C. glutamicum*. *Appl. Microbiol. Biotechnol.* 2009;82:491–500.

[125] Hansmeier N, Bartels FW, Ros R, Anselmetti D, Tauch A, Pühler A, et al., Classification of hyper-variable *Corynebacterium glutamicum* surface-layer proteins by sequence analyses and atomic force microscopy 2004;112:177–93.

[126] Moura-Costa L, Paule B, Azevedo V, Freire S, Nascimento I, Schaer R, et al., Chemically defined synthetic medium for *Corynebacterium pseudotuberculosis* culture. *Rev Bras Saúde E Produção Anim.* 2002;3:1–9.

In: New Trends in *Corynebacterium glutamicum*
Editor: Carlos Barreiro

ISBN: 978-1-63483-280-9

Chapter 4

SCALE-DOWN STUDIES OF *CORYNEBACTERIUM GLUTAMICUM* CULTIVATIONS FOR A CONSISTENT BIOPROCESS DEVELOPMENT

Anja Lemoine, Stefan Junne** and Peter Neubauer
Technische Universität Berlin, Department of Biotechnology,
Chair of Bioprocess Engineering, Berlin, Germany

ABSTRACT

Due to the wide industrial application of *Corynebacterium glutamicum,* scale-down studies of this organism have gained more interest recently. Cells are steadily exposed to a changed environment in these studies, while the reactor concept aims to simulateheterogeneities, which appear in industrial scale fed-batch cultivations. Mixing times usually exceed the substrate uptake rates in these scales, thus leading to zones of excess substrate at the zone near to the feed addition. In contrast, substrate depletion occurs in zones far away from the feeding. Hence, cells are steadily exposed to an oscillating environment, which is mimicked with the combination of different reactor designs and cultivation conditions. In the present chapter, the overall response of *C. glutamicum* main carbon metabolite synthesis, amino acid accumulation and the energy charge as well as the viability and polarisability of cells are described, and comparisons with other microbes are made. The potential intensification of investigations of the cellular response to oscillating conditions for strain engineering and process development is discussed with respect to recently introduced novel *on line* monitoring tools and micro-bioreactors.

The response of *C. glutamicum* to the oscillating conditions varies significantly from those of other organisms: the cells show quite a robustness, as the growth rate is hardly affected. However, the lysine production is reduced. Since the oscillating substrate concentration appears along with different dissolved oxygen availability, also zones of oxygen depletion are occurring in the large scale. Oxygen limitation indicates a great influence on the metabolite fluxes, when corresponding zones are mimicked in scale-

* Corresponding author: Stefan Junne; stefan.junne@tu-berlin.de.

down cultivations. Lactate synthesis plays a pivotal role in the redirection of carbon when energy regeneration is reduced due to the absence of oxygen.

INTRODUCTION

Inhomogeneities in Industrial Nutrient-Limited Fed-Batch Processes

A step-wise scale-up from milliliter to cubic meter scale is usually applied to select the optimum parameters, operating modes and reactor types during process optimization. However, conditions in laboratory scale are usually not transferable to the production scale [1, 2]. This situation often leads to an unwanted accumulation of byproducts in the large scale, accompanied with a change of the biomass yield and product selectivity due to inhomogeneities [1, 3, 4].

Gradients are intensified when the typical nutrient limited fed-batch procedure is applied due to the addition of a highly concentrated and often viscous nutrient solution and its 10,000-fold or even greater dilution in the bioreactor's liquid phase [5]. When the limiting nutrient is added to the headspace or liquid core of the reactor, a zone of excess nutrient supply is created. Mixing times increase to the order of minutes from the lab to the industrial scale, thus substrate gradients appear above a certain scale. It has been reported that mixing times of a stirred tank reactor (STR) are 50 s at a scale of 20 m^3 instead of 5 s at laboratory scale [6]. At a scale of 150 m^3, the mixing time can rise to several minutes [7]. With respect to many catabolic reactions that expose turnover rates in the range of 1.5 - 2.0 s^{-1} [8, 9], significant impacts on the substrate uptake rate and concomitant intracellular fluxes in cells can be assumed.

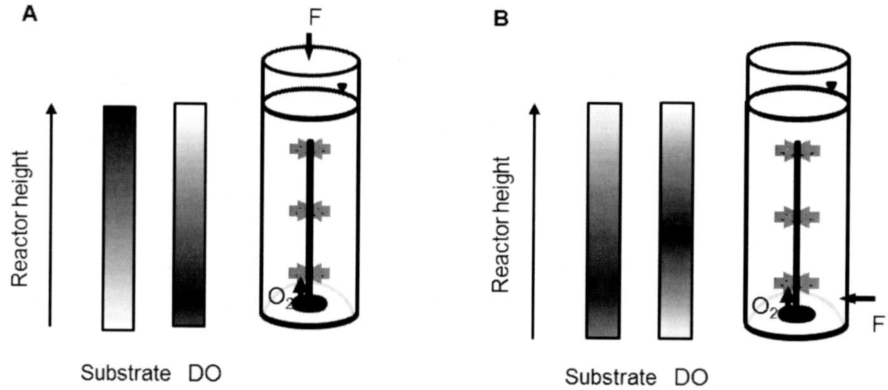

Figure 1. Distribution of substrate and dissolved oxygen (DO) in large scale bioprocesses and scale-down reactors for the investigation of responses caused by these inhomogeneities. Distribution of substrate and dissolved oxygen (DO) when fed from the top (A) or from the bottom (B).

Furthermore, increased mixing times lead to severe concentration gradients of dissolved gases, temperature, acids and bases. The existence of limitations of oxygen transfer rates at different locations in the fluid phase in industrial bioreactors are known for several decades [10, 11]. Insufficient mixing leads to the formation of axial and radial gradients in the oxygen distribution. Substrate and oxygen gradients are often opposed to each other along the height of the reactor (Figure 1), generated e.g. by feeding substrates from the top while the air is sparged from the bottom. The conduction of high-cell density cultivations increases the gradients. A larger volume leads to a lower volumetric power input, which is usually caused by longer distances from the stirrer blades to the wall. Hydrostatic pressure can also support the formation of gradients of dissolved gases [2, 12].

Due to the complex gradient formation, *on line* determination of representative concentrations of a limiting substrate or dissolved oxygen is challenging. Various effects on the process performance due to scale-up effects have been described in literature. It has been shown that the size of these zones as well as the residence time and the response of the bacteria in these zones influences the performance of a bioprocess [1, 13] and frequently yields in a heterogeneous cell population [6]. Cells experience these gradients as frequent stimulations while they are exposed steadily to the different zones of the reactor. The adaptation to this environmental stress provokes a large amount of cellular resources. It includes general stress adaptation to guarantee the survival of at least some parts of the population. However, a crucial part of this adaptation strategy is not necessarily beneficial for a successful bioprocess, where adaptation to all eventualities does not provide a benefit.

Spatial gradients of oxygen concentrations at *Escherichia coli* cultivations support acetic acid accumulation and lowers process productivity [14]. Investigations of cultivations at different scales revealed that temporal changes in glucose concentrations and dissolved oxygen (DO) extensively affected the cell physiology of *E. coli* W3110. Insufficient mixing at a scale of 20 m^3 led to a decrease of the biomass yield of 35.0%. Interestingly, the cell viability was not affected negatively when compared to a well-mixed 5.0×10^{-3} m^3 bench-scale cultivation [6]. When a kinetic model approach of the main carbon metabolism was connected to computational fluid dynamics, the impact of environmental changes could be predicted. Lapin et al. investigated the impact of glucose concentration gradients in a fluid volume of 900 L. It was shown that the intracellular ratio of phosphoenolpyruvic acid (PEP) to pyruvic acid (PYR) in *E. coli* cells depended on the location of the cells in the fluid phase of the bioreactor. This ratio plays a pivotal role in the sugar uptake mechanism driven by the phosphotransferase system (PTS) [15].

In order to resolve these issues, which occur during scale-up, attempts to improve the reactor design and power input have been made. An extensive study regarding the effects of stirrer types and their position in the liquid phase on mixing times were performed by Vrábel et al. [16]. They found that a considerable reduction of mixing time can be achieved if the upper impellers were axial instead of radial blade stirrers. In order to improve the substrate distribution at fed-batch cultivations, the feed input can be located near to an impeller in the fluid phase. However, own studies showed that gradients still exist when the pipe is not reaching very close to the stirrers. Zones far away from the glucose addition still exhibit gradients, since these regions are also poorly mixed in large scale.

Larsson et al. [17] suggested to apply multiple feeding points located in well-mixed zones to reduce mixing problems. The requirement of a very sophisticated regulation mechanism to ensure exact feeding volumes basically hindered any application. With regard to pipe blockage, contamination and mechanical stability, many industrial scale fed-batch cultivations apply surface feeding, which is typically used during scale-up as well [12]. Thus, the appearance of inhomogeneities is not easy to counteract by altered reactor designs. The examination of the cellular response to the oscillating environmental conditions and the conclusions of it for strain and bioprocess optimization remain the crucial part for reducing negative effects on yield and selectivity during scale-up.

Scale-Down Bioreactor Concepts and Analytical Methods

The investigation of regulatory responses of cells in a perturbed environment is a frequently applied experiment to identify metabolic bottlenecks and targets for strain optimization. A classical approach is the combination of a substrate pulse and sampling with a high frequency thereafter as applied in systems biology. However, these pulse experiments are performed with random settings of concentration steps and durations, as they focus on a better understanding of the biological system, but not an improved process optimization. In order to include the multiple tools for analyzing cell regulation and issues that arise from scale-up, suitable scale-down reactor and experimental designs have to be applied. Multi compartment scale-down reactors (Figure 2) have been widely used to study scaling effects. In this paragraph two-compartment and three-compartment reactor concepts are described.

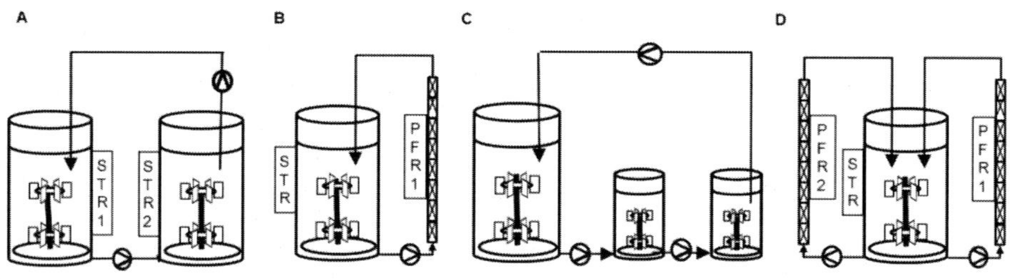

Figure 2. Two-compartment scale-down reactor concepts, comprising of two stirred-tank reactors (A) or one stirred-tank and one plug-flow reactor (B). Three-compartment reactor concepts, comprising of three stirred-tank reactors (C) or one stirred-tank and two plug-flow reactors (D).

Beside several approaches, which have been designed during the last 20 years, novel miniaturized approaches allow a higher throughput and a higher degree of optimization within process development in order to investigate regulatory responses to oscillating environments like in industrial scale fed-batch cultivations. The requirements for suitable scale-down systems are (i) an exact dosage of metabolites for achieving defined concentration gradients, (ii) a flexible adjustment of cultivation settings due to the various processes that exist (e.g. feed rate, aeration rate), (iii) the availability of a certain sampling volume (at least in the mL range) in order to perform a suitable *off line* analysis, and (iv) the achievement of reproducibility by a controlled environment (pH, temperature, etc.) as it is usually the case in bioreactors. Hence, the application of common shake flasks is critical for the achievement of suitable and precisely defined scale-down conditions. Therefore, several reactor concepts have been developed earlier, e.g. multi-compartment reactor systems. Additionally, in recent times, multi-bioreactor systems, which are connected to a liquid handling platform, allow the parallel investigation of multiple situations as they occur during the course of an industrial fed-batch cultivation (Figure 3). A parallel approach enables to mimic different situations e.g. caused by the increased cell density. The experimental conditions, which are crucial for the process, can be selected in the parallel approach and investigated in more detail in larger scale-down concepts, which are usually better equipped in terms of *on line* and *at line* monitoring tools.

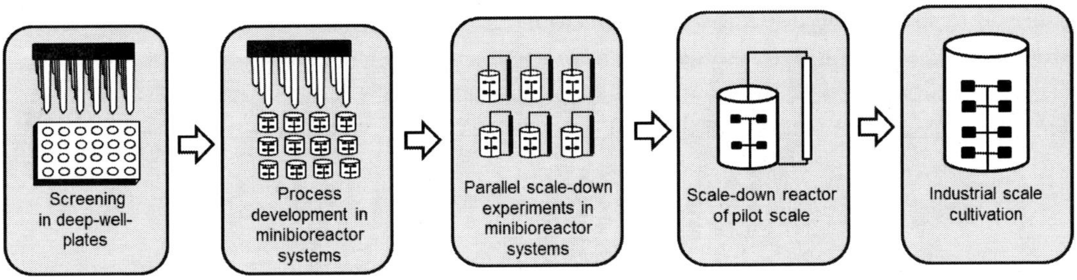

Figure 3. Consistent process development including the application of micro-bioreactor systems coupled to automated liquid handling platforms and suitable scale-down conditions, which differ depending on the cultivation vessel size and degree of parallelization. The final scale-down concept will be set up based on the data gained about sensitive parameters and crucial conditions in parallel and automated cultivation experiments.

Monitoring devices, which are able to qualitatively and quantitatively report the population heterogeneity based on a physiologic or morphologic analysis, can increase the relevance of scale-down experiments significantly. Firstly, the comparison of such parameters with the industrial scale allows a more precise scale-down, since the conditions of cells should be very similar in both scales applied. Secondly, these parameters provide a valuable picture of the impact of stress due to the environmental oscillations on the cells (e.g. osmolality and cell size, fluctuating substrate concentration and polarisability). In some cases, direct correlations between the monitored parameters and the process performance can be found, which helps to identify possible targets for strain improvements (e.g. in order to counteract deficiencies in membranes). Hence, a suitable concept consists of a first screening phase for the identification of crucial conditions, which might be performed in a parallel cultivation system, and a second experimental design phase, where as many parameters are measured as possible when the (relevant) crucial conditions are applied in a larger scale-down bioreactor concept. A statistical (and model-based) data analysis procedure is able to identify sensitive parameters for process-relevant features like yield and selectivity. In the forthcoming subchapters, several scale-down reactor concepts, partly applicable in parallel, and recently developed monitoring tools suitable for scale-down experiments are described.

Analytical Methods Applied in Scale-Down Experiments

Numerous studies of various organisms have been performed since the introduction of scale down reactor systems. Many of them used state-of-the-art metabolome analysis methods in order to describe the response of cells to environmental perturbation. Beyond this, many attempts have been made to gain a broader understanding of the cellular response by either *on line* or *off line* methods as summarized in Figure 4. In parallel to the progress in the -omics technologies, proteome and transcriptome studies were included in the analysis of scale-down experiments [18].

Since the overall impact of stress on the physiology, but also morphology, was a major question, flow cytometry studies were performed already a decade ago, accompanying scale-down experiments [1, 6]. The outcomes gave insight into the robustness of *E. coli* against a heterogeneous environment. In order to gain a more detailed knowledge about the cell status with a high time-resolution, novel *on line* tools are currently applied, which are sometimes referred to as *on line* flow cytometry. These methods have in common that they are able to track the particle size and distribution or the polarisability of the cell membrane directly in or at the bioreactor in an automated way.

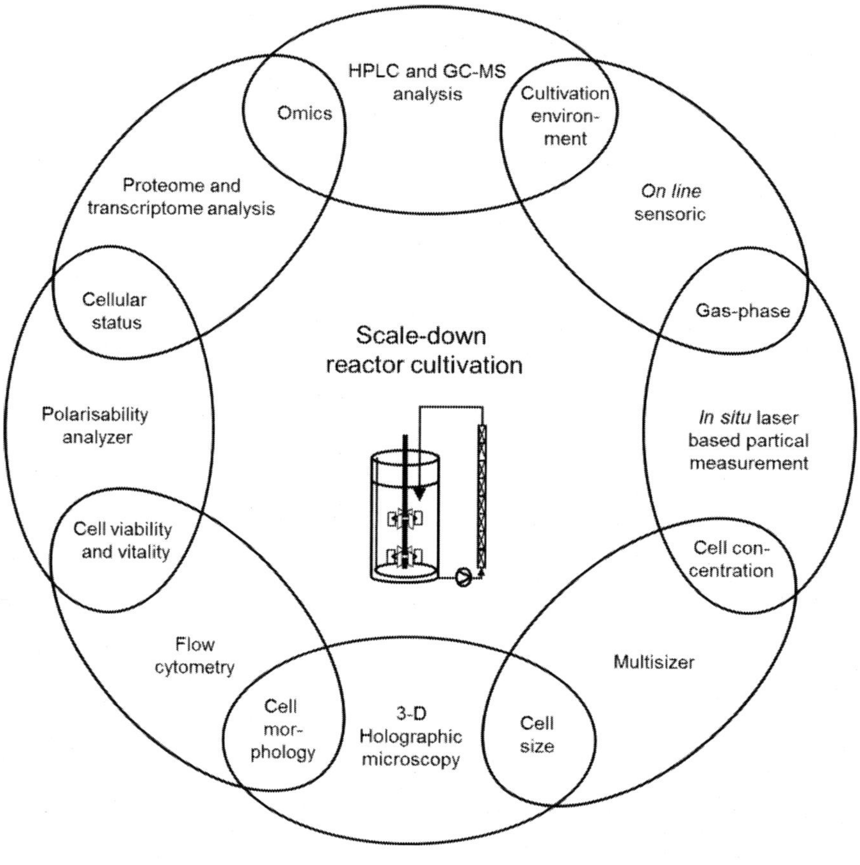

Figure 4. Overview of analytical methods, which have been applied to observe the response of cells to oscillating environmental conditions in scale-down cultivations.

Since the degree of homogeneity of a culture matters for the achievement of a high productivity in a production process, these methods are highly interesting for process improvement, if the data is suitable to obtain time-point specific histograms. This allows for a conclusion whether the formation of subpopulations is related to certain events in the cultivation, or at which certain conditions they appear under dynamic changes of the cultivation environment. The data also allows for more accurate (hybrid) models which are simulating a large scale environment under the consideration of cellular dynamics. One example among these process analytical tools is the biocell analyzer (Sequip GmbH, Germany), which is based on particle tracking in a field generated by a laser-light beam. The particles, which flow through this field, are reflecting light, which is measured. The shape of the reflectance allows to draw conclusions about the particle size and shape, and its intensity about the cell structure [19]. The particle detecting sensor of SOPAT (Germany) is based on photooptical detection and has an integrated software for the analysis of pictures shot *in situ* in bioreactor cultivations. These two examples show that the tracking of cellular particles based on their optical properties is feasible. Another promising approach is the tracking of fluorescent markers in the cells. This concept was successfully applied to investigate the effect of heterogeneities on the cell viability and protein leakage [20-22]. Thus, a real-time resolution of the morphological and physiological status can be used for a suitable scale-down approach if classical process parameters like the power input to volume ratio failed. Then, the achievement of the same cellular status and particle distribution in the large and lab scale would prove the correctness of the scale-down approach.

Two-Compartment Reactor Concepts

One established technique for simulating the heterogeneity at the large scale is based on the division of the liquid phase into different parts, each characterized by distinctly different conditions. For example, one zone reflects the bulk zone of the fluid phase and the other one the top zone when the feed is added to the headspace. The reactor systems based on this principle either consist of two coupled stirred tank reactors (STR-STR), or of a STR coupled with a plug flow reactor (STR-PFR) as shown in Figure 2. The advantage of the STR-STR system is the distinct separation of conditions in two cultivation vessels, the miniaturization is facilitated.

One of the first studies using a two-compartment system was performed by Oosterhuis et al. The authors used two STRs in order to study the impact of fluctuating oxygen concentrations in a gluconic acid fermentation [23]. Such systems with two coupled STRs were used in further studies, mostly to investigate the effect of oxygen fluctuations [24], but also the effect of oscillations of the pH-value [25]. Oxygen oscillations have been simulated in a single well-mixed STR, either by using time pulsed mixtures of nitrogen and oxygen [26, 27] or by the variation of the reactor head pressure.

An advantage of the PFR-STR system is the achievement of a gradient, to which cells are exposed in a defined residence time. From the early 1990s on, STR-PFR scale-down reactors have been used to study different zones in large scale bioreactors. The response of the cells to these gradients can be observed along the height of the PFR. Beside standard PFRs, the application of PFRs equipped with static mixers is common, as these provide efficient mixing of gas and liquid, while maintaining plug flow characteristics [12]. Thus, the static mixers enable a certain gas transfer in the PFR and the adjustment of a distinct volumetric oxygen transfer rate [28]. Furthermore, ports along the PFR module enable sampling and the application of sensors, so that the conditions in the PFR can be monitored and samples can be taken for analysis at various residence times in the same experiment [29].

The conditions in each compartment depend on the type of zone or mixing situation that shall be simulated. Typically, the feeding zone is represented by the PFR, while the STR mimics the bulk zone of the fluid phase in the reactor [25]. The volumetric ratio between the two compartments is equal to the ratio of the feeding zone to the bulk zone in the large scale reactor, which can be estimated using computational fluid dynamics [1].

Among the first studies was the investigation of the response of the yeast *Saccharomyces cerevisiae* when exposed to high glucose concentrations [5]. Scale-down studies beside *S. cerevisiae* were performed with *E. coli* [4, 30], applying an aerated PFR, in which the feed was introduced at the entrance. Many investigations have been performed with a non-aerated plug flow module with *E. coli* [3, 31], *Bacillus subtilis* [29] and *Corynebacterium glutamicum* [32]. The feeding zone in a large scale bioreactor, where substrate and oxygen limitation occur simultaneously, was mimicked in these studies in the PFR module. Furthermore, the STR-PFR scale-down reactor was suitable to study the effect of pH-inhomogeneities, e.g. on *E. coli* [33] and *B. subtilis* [34]. In each case differences compared to the well-mixed situation were found, depending on the size of the two zones and the circulation rate between them. Although the STR-PFR system is very useful to study the behaviour in large scale reactors in the lab scale, it is rarely applied, perhaps as it is not commercially available.

Three-Compartment Reactor Concept

Basically, two concepts for the design of a three-compartment reactor can be applied : (i) three stirred tank reactors connected as cascade, (ii) a stirred tank reactor with two coupled plug flow modules. Consequently three characteristic zones in a large scale cultivation can be simulated, and thus the cellular response can be monitored. In a top-fed cultivation, three different zones were identified as (i) the top zone, characterized by substrate excess and oxygen depletion, (ii) the bulk zone of the fermenter, characterized by substrate limitation and oxygen excess, and (iii) the bottom zone, characterized by substrate depletion and oxygen excess (Figures 1 and 2). A stirred tank cascade enables the adjustment of defined and controlled gradients, whereas the response of cells to gradients can be examinated with respect to a defined residence time in plug flow modules.

However, not only top-fed reactors are used in industry. Another location for the feed addition is the liquid core in the bottom of the reactor. In this case, three zones occur, too. The first is located near the feed addition in the bottom part of the reactor, which is characterized by substrate excess and oxygen depletion, a second one in the bulk zone, characterized by substrate limitation and oxygen excess, and a third one in the top part of the reactor, where oxygen and substrate is depleted. A scale-down reactor with a non-aerated plug flow reactor, where the feed is introduced at the entrance of the PFR is simulating the first zone, an aerated STR the second zone, and a non-aerated second PFR the third zone, respectively. This concept has been evolved by coupling an existing STR-PFR two-compartment reactor [29] with another PFR module, which is identical to the previously established one. It also consists of static mixers, five sampling and ten sensor ports. It is sterilizable and comprises of the same residence times as the first PFR module. It can be aerated and an addition of any compound can be realized at the entrance of it.

The concept has been used to conduct cultivations, in which the second PFR module represented a bottom zone of a top-fed nutrient-limited *E. coli* fed-batch cultivation. While substrate was added to the first PFR module, in which a concomitant oxygen limitation occurred, no substrate was added to the second PFR module. Due to the starvation, which occurred in this module, oxygen from the STR module was not consumed, leading to oxygen excess. The exposure of the cells to these perturbations led to a greater portion of non-viable cells in *E.coli* cultivations as revealed by flow cytometry (own data). The cultivation exhibited a distinctly different behavior than in the two-compartment reactor. The cell stress led to a diminished growth rate and biomass yield. Furthermore, *C. glutamicum* cultivations were performed as well in this reactor concept. Again, different response patterns were observed in comparison to the two-compartment reactor. Thus, the choice of the scale-down conditions is essential for the outcome of a scale-down experiment.

Miniaturization of Scale-Down Approaches

In order to miniaturize scale-down reactor concepts, as many features of a usual lab-scale or industrial-scale bioreactor should be maintained, as the correct setting of conditions is crucial for the experimental outcome. Miniaturization for the purpose of parallelization is considered to be of a scale of several milliliters of liquid volume.

Parallel cultivation is usually performed in microwell plates. However, the plates applied are usually not suited for the application in bioprocess optimization, since they are not equipped with any *on line* measurement tools. The achievable gas transfer rates are also much lower than the ones obtained in stirred tank bioreactors. When *in situ* fluorescence sensor technology is coupled to microwell plates, the pH-value, DO and dissolved CO_2 can be measured *on line* (PreSens Precision Sensing GmbH, Regensburg, Germany). The BioLector® (m2p-labs GmbH, Aachen, Germany) allows to follow the time courses of biomass concentration, metabolites and reporter proteins by measuring the intensities of fluorescence and scattered light *on line* in microwell plates [35]. The cavities of the wells have a specific flower shape, which provides higher oxygen transfer rates. The integration of a fed-batch procedure realized with hydrogel channels as bridge between a feed reservoir and each well. The so-called fed-batch microtiter plate (FB-MTP) enables 44 parallel experiments. The feed rate can be adjusted by the geometries of and the concentration gradient between the two ports of the hydrogel channel [36]. In the μ-24 MicroReactor (distributed by Pall Inc., Port Washington, NY), the pH-value and DO is controlled with *on line* measurements applying the MicroReactor® software [37]. A high well-to-well reproducibility was proven [38].

A micro-bioreactor system was developed in a polymer block, comprising of a cultivation volume of 150 μL at each cultivation chamber. It is mixed with a magnetic stirrer. The cultivation chamber is mounted on a platform equipped with sensors for the *on line* measurement of pH, DO and the optical density. A micro-valves driven dosage of the feed solution is applied [39]. BioProcessors' SimCell system (BioProcessor, Woburn MA) comprises of an array of 210 micro-bioreactors, segregated in 6 gas-permeable bioreactors of a working volumen of up to 800 μL. Lamping et al. [40] designed a bioreactor that comprises of a working volume of 10 mL. The STR is equipped with pH, temperature, DO and cell density monitoring devices. The BioReactor 48 system (2mag AG, München, Germany) enables the parallel cultivation in 48 stirred tanks, which comprise of a volume of 8 to 12 mL. Monitoring of pH and DO and defined aeration is feasible. The system is coupled to a liquid handling robot and allows pulse-wise feeding. Bower et al. [41] described a micro-bioreactor for a working volume of 1 mL. The growth chamber is connected to a sampling port and to three liquid reservoirs. Feeding and sampling are driven by miniaturized peristaltic pumps. The ambr 24 reactor system (distributed by Sartorius Stedim, Germany) consists of 24 or 48 single-use bioreactors, which are equipped with an impeller and a sparging device. The reactors are equipped with sensor spots for dissolved oxygen and the pH-value [42].

All the presented systems basically includethe necessary features to consider scale-up issues already in very small cultivation volumes at a very early screening and process optimization stage. However, up to now, the recently developed micro-bioreactor systems had been applied only rarely for scale-down experiments. Nevertheless, research is initiated to include parallel systems as initial step into scale-down strategies for the identification of crucial conditions and sensitive oscillating parameters.

Corynebacterium glutamicum Scale-Down Experiments

Various scale-down experiments have been performed in order to observe the response of *C. glutamicum* in the production scale. Schilling et al. published the first scale-down reactor cultivation in 1999 [43]. It described the effect of a prolonged mixing time from 10 to 130 sec on a lysine-producing *C. glutamicum* DSM 5715 strain (DSMZ GmbH, Braunschweig, Germany). Leucine was fed as the limiting substrate into the headspace of a 42 L STR. The biomass and lysine formation, as well as the specific oxygen uptake rate and carbon dioxide production rate, have been determined. Furthermore, the enzyme activity of aspartate kinase, citrate synthase, dihydrodipicolinate synthase, and phosphoenolpyruvate carboxylase were determined [43]. Käß and co-authors [32, 44] performed scale-down experiments in a two-compartment reactor system consisting of an aerated STR coupled to a non-aerated PFR module. Moreover, Käß et al. applied different residence times in the PFR-module up to 3 min and investigated two strains *C. glutamicum* DM 1933 [44] and ATCC13032 [32]. The strain *C. glutamicum* DM 1933 is a lysine overproducing strain, in which the *pep-carboxykinase (pck)* gene is deleted and which contains a point mutation of the aspartate kinase and a chromosomal integration of the lysine exporter lysE. *C. glutamicum* ATCC13032 is a wildtype strain. The product (lysine) yield, maximum glucose uptake capacity, specific oxygen uptake, nucleotide and main carbon metabolite concentration, mRNA transcription for genes of the glycolysis and the TCA-cycle and protein expression of *C. glutamicum* DM 1933 were studied.

Recently, a three-CR cultivation with *C. glutamicum* DM 1800 (*pyc*P458S and *lys*CT3111 point mutations [45], carrying plasmid pEKEx2cadA [46]) was conducted with an aerated STR and two non-aerated coupled PFRs, one with and the second without feed addition [47]. Furthermore, a two-CR experiment was performed and the measurements were compared to a single-CR cultivation in the aerated STR. Biomass concentration, main carbon metabolite, amino acid and nucleotide concentrations. Moreover, effects on the cell structure were observed, including morphology and cell polarizability. Bucholz et al. investigated the effect of oscillating CO_2/HCO_3^- levels in a three-compartment cascade bioreactor on the biomass yield, as well as the metabolome and transcriptional response of the wild type strain *C. glutamicum* ATCC13032 and the lysine producer *C. glutamicum* DM1933 [48]. The residence time in the CO_2/HCO_3^- elevated environment was adjusted to 3.6 min at a volume ratio of 8.4%, which is recognized as relevant in industrial scale. Pressure gradients of pCO_2 of 75-315 mbar were applied. No remarkable changes could be detected in the biomass and product concentration compared to the reference batch cultivation. However, 66 genes were either up- or down regulated under these conditions in dependence of the pressure gradient and residence time. These findings indicate the broad response to oscillating CO_2 availability, although detailed mechanisms remain to be resolved.

Experimental Setup of *C. glutamicum* Scale-Down Cultivations at Oscillating Oxygen and Substrate Availability

Recently, a study was presented, which describes a methodology to evaluate the importance of oxygen transfer rates for the biomass and product yield in *C. glutamicum* cultures by applying a parallel cultivation procedure [49]. The oxygen-dependency of cellular turnover rates was investigated on the microwell-based cultivation system BioLector. The threshold value of 14 mmol L^{-1} h^{-1} was identified for affecting the strains' productivity. The method is suitable to identify conditions of scale-down investigations, which lead to a sensitive response of the cell.

In other recent reports, a scale-down reactor concept was applied, which consisted of a STR and one or two PFR modules (Figure 2D). PFR module(s) equipped with several DO sensors for the adjustment of the dissolved oxygen concentration were applied for the cultivation indistinct areas of oxygen depletion [32, 44, 47]. At increased cell concentration, the oxygen uptake is increased, leading to greater areas of oxygen depletion.Thus, A similar development always occurs in the scale-down bioreactor: the residence time within the PFR, during which cells are confronted with oxygen depletion, becomes greater at an increased cell concentration. Oxygen depletion occurs only in a small part of the PFR at low cell densities. This is not only true for the PFR1, where the oxygen consumption depends mainly on the volumetric substrate consumption at the feed addition at the entrance of the PFR module, but also in the PFR2 of the three-CR cultivation, although no substrate is supplied. Nevertheless, the oxygen consumption is reduced compared to the PFR1, hence anaerobic zones appear during a later time point in the course of the cultivation.

Käß et al. [44] showed that the total oxygen uptake in the two-CR cultivation is similar as in the reference cultivation. The reduced oxygen uptake in the PFR part is compensated with an increased uptake in the aerated STR module.

The cellular growth for the scale-down experiments in the two-, and three-CR cultivation with *C. glutamicum* DM 1800 measured in the STRshowed biomass yields of 0.4 $g \cdot g_{DCW}^{-1}$ in the single-CR, 0.42 $g \cdot g_{DCW}^{-1}$ in the two-CR and 0.37 $g \cdot g_{DCW}^{-1}$ in the three-CR cultivation. Cultivations with *C. glutamicum* DM 1933 and DM ATCC13032 exhibited similar biomass yields in the scale-down as well as in the reference cultivations, too [32, 44]. Schilling et al. [43] observed a reduced biomass production of 7% after 53 hours of feeding, while the substrate uptake was diminished by 14% in *C. glutamicum* DSM 5715 cultivations.

Metabolite Concentrations

The concentration measured in cell-free supernatant samples of scale-down cultivations with *C. glutamicum* DM 1800 of pyruvate, lactate, succinate and malate are shown in Figure 5.

Figure 5. Main carbon metabolism and extracellular metabolite concentrations of a single-CR cultivation (black bar), a two-CR cultivation (grey bar) and a three-CR cultivation (white bar) of *C. glutamicum* DM 1800.

The concentration in cell suspension samples of the plug flow reactor modules in scale-down cultivations of *C. glutamicum* DM 1800 of pyruvate, lactate, succinate and malate are shown in Figure 6. Part A shows results of PFR 1 of the two-CR cultivation (fed PFR), part B and C the PFR1 and PFR2 (no feed addition) of the three-CR cultivation, respectively, and the black and grey bar the results measured at 9 and 12 hours after feed start. The pyruvate concentration decreased in the PFR1 of the two-CR cultivation as well as in the PFR1 and PFR2 of the three-CR cultivation under oxygen limitation. Most distinct is the decrease in the two-CR cultivation, where the concentration was reduced by 50%. Earlier, Käß et al. [44] showed an slight decrease of extracellular pyruvate concentration in the PFR module at a residence time of 3 min, too. In contrast, measurements of the extracellular pyruvate concentration in the aerated STR compartment showed an accumulation in the two-CR cultivation up to 0.1 mM 12.2 hours after feed start.

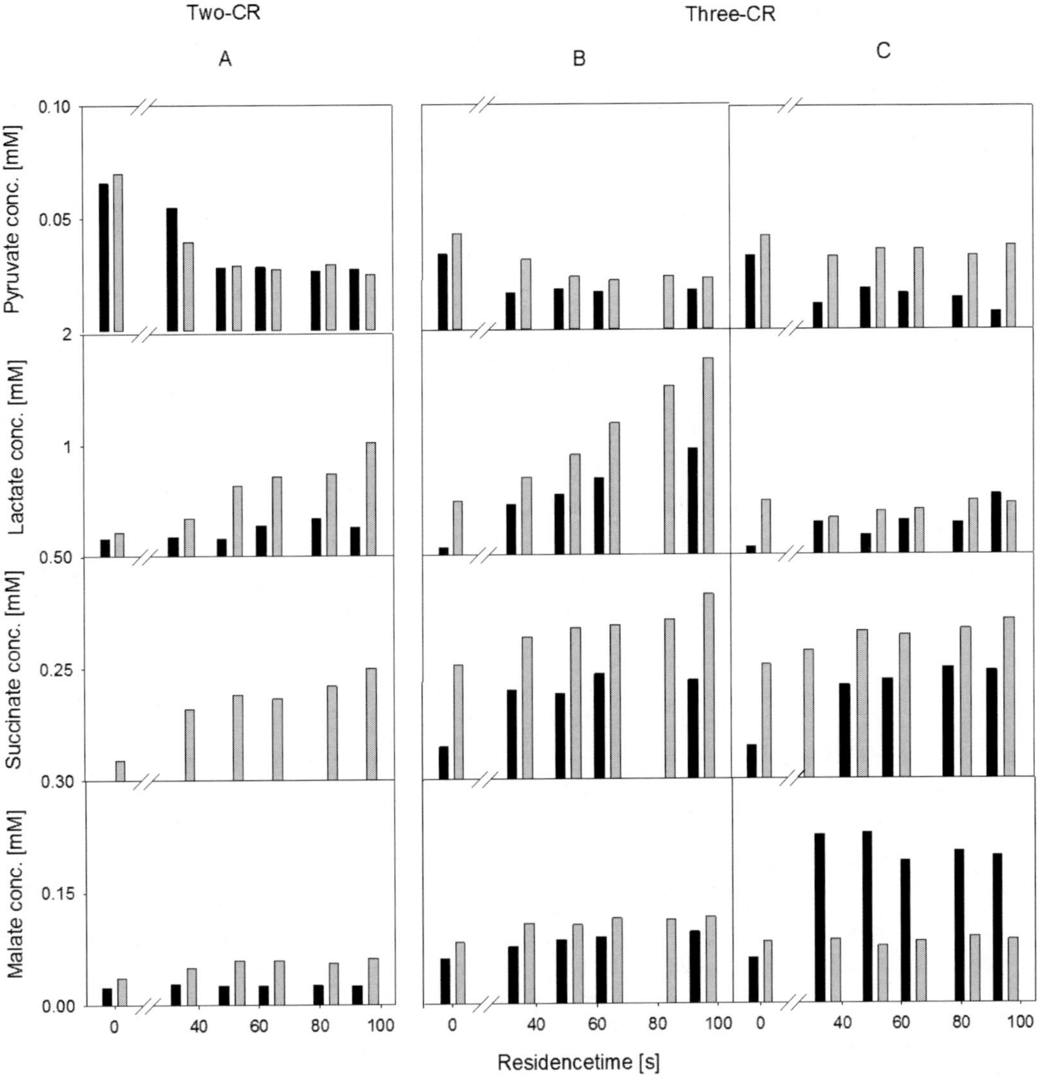

Figure 6. Metabolite concentration of the cell suspension measured in the PFR in scale-down experiments with *C. glutamicum* DM 1800 in two sampling times: samples taken 9 hours after feed start are shown with black bars, samples taken after 12 hours after feed start are shown with grey bars. Part A shows the PFR1 (with feed addition at the entrance of the PFR and without oxygen supply) of the two-CR cultivation, part B shows the PFR1 of the three-CR cultivation and part C shows the PFR2 (without substrate and oxygen supply) of the three-CR cultivation.

The lactate concentration is increasing in the PFR 1 of the two-CR cultivation. The accumulation of lactate is most significant in the PFR 1 of the three-CR. The concentration rises from 0.5 mM in the STR to 1.8 mM at the fifth port of PFR 1 with an applied residence time of 2 min 12 hours after feed start. Käß et al. [44] measured an extracellular lactate concentration of 2.3 mM 12 hours after feed start (residence time of 3 min) in a two-compartment cultivation.

The increased residence time for the cells in oxygen limited zones in the three-CR cultivation resulted in an increased lactate accumulation already in the earlier sample point where the cell concentration is 5 g L^{-1}. These findings can be explained with an up-regulation of the IdhA gene and therefore increased expression of lactate dehydrogenase (LDH) under oxygen limitation [1]. In the aerated STR compartment the extracellular lactate concentration is 0.18 mM in the three-CR and 0.05 mM in the single-, and two-CR 12.2 hours after feed start. Käß et al. [44] had the same findings in the two-CR cultivation for *C. glutamicum* DM 1933 [44] and DM ATCC13032 [32]. This indicates a re-uptake of extracellular lactate, following the conversion to pyruvate in the STR module. Lactate accumulation is increased due to the prolonged time in the non-aerated plug flow modules and cannot be completely consumed in the aerated STR in the three-CR experiment. Due to the oxygen limitation, the regeneration of energetic co-factors is inhibited and therefore reactions are preferred, which recover these co-factors. Hence, the reaction from pyruvate to lactate might be beneficial for the cell since menaquinone and nicotinamide adenine dinucleotide (NAD), co-factors involved in the energy household, are recovered in this reaction.

Succinate accumulation is detected in the second sample point at a cell concentration of 10 g L^{-1} in the PFR1 of the two-CR cultivation In the PFR2 of the three-CR experiment, succinate is accumulated to similar amounts as in the PFR1 module. This indicates that succinate accumulated mainly due to anaerobic conditions within the plug flow modules. Käß et al. [44] found an increase of extracellular succinate concentration in the two-CR cultivation up to 0.5 mM in the last port. In the STR compartment of the three-CR succinate was accumulated 2-fold higher compared to the single-CR cultivation.

In the two-, and three-CR cultivation malate is accumulated slightly in PFR1, whereas malate is increased 3-fold in the PFR2 at the first sample point 9 hours after feed-start. The increase is observable within the first seconds under substrate and oxygen depletion in the PFR2, which might indicate a retarded TCA-cycle. The carbon flux in the TCA is reduced at the later time point, so that the accumulation of succinate and malate was much less pronounced. Schilling et al. [43] found that the enzyme activity of citrate synthetase, and phosphenol pyruvate carboxylase is reduced in the scale-down reactor concept, which indicates a decreased turnover rate in the TCA-cycle and the carbon flux into the TCA-cycle, respectively. Extracellular malate accumulated two-fold higher in the two-CR and 3-fold higher in the three-CR compared to the single-CR experiment.

Free Amino Acid Synthesis

The extracellular amino acid concentration at 12 hours feed-time in single-, two-, and three-CR cultivations with *C. glutamicum* DM 1800 are shown in Table 1. The extracellular amino acid concentration of serine was increased under oscillatory conditions, whereas glycine only accumulated under conditions in the three-CR system. The total glycine content is increased in the three-CR cultivation, too. Alanine is accumulating in both cultivations, but to higher amounts in three-CR compared to both other cultivation conditions. The total alanine content is increased under oscillations. Glutamate and glutamine accumulate much more in the three-CR cultivation. Threonine is only accumulated extracellularly in the three-CR cultivation. Furthermore, aspartate is accumulating to the same amount in both scale-down cultivations. Most interestingly, lysine production is reduced under oscillatory conditions in the two-, and three-CR, in detail the lysine yield is 0.1 g· g_{DCW}^{-1} for the single-CR, 0.8 g· g_{DCW}^{-1} for the two-CR and 0.7 g· g_{DCW}^{-1} for the three-CR cultivation. Schilling et al. [43] observed a reduced lysine production of 12% in the scale-down reactor cultivation of *C. glutamicum* DSM 5715 after 53 hours of feeding compared to the reference. Howeversubstrate consumption was reduced, too.

No reduction in the lysine yield in the two-CR set-up compared to the single-CR reference was found at *C. glutamicum* DM 1933 cultivations [44]. In summary, early synthesized amino acids are accumulated in the scale-down cultivations. Several later synthesized amino acids are incorporated to a minor portion, which indicates bottlenecks of the conversion of the corresponding amino acids.

Table 1. Intracellular free and total amino acid concentration in *C. glutamicum* DM 1800 single-CR (homogeneous conditions), two-CR and three-CR fed-batch cultivations (both heterogenous conditions)

	Single-CR		Two-CR		Three-CR	
	free amino acid	total amino acid	free amino acid	total amino acid	free amino acid	total amino acid
Ala	0.018	0.214	0.016	0.228	0.025	0.345
Asp	0.013	0.137	0.012	0.135	0.012	0.091
Cys	n.d.	0.002	n.d.	0.003	n.d.	0.003
Glu	0.575	1.023	0.457	1.141	0.459	1.335
Gln	0.034	n.d.	0.030	n.d.	0.033	n.d.
Gly	0.023	0.101	0.020	0.132	0.089	0.113
Hom	n.d.	0.008	n.d.	0.008	n.d.	0.011
Iso	n.d.	0.053	n.d.	0.048	n.d.	0.051
Leu	n.d.	0.112	n.d.	0.115	n.d.	0.079
Lys	n.d.	1.974	n.d.	1.682	n.d.	1.617
Phe	n.d.	0.039	n.d.	0.043	n.d.	0.025
Pro	n.m.	0.038	n.m.	0.036	n.m.	0.042
Ser	0.009	0.078	0.008	0.075	0.009	0.08
Thr	n.m.	0.107	n.m.	0.107	n.m.	0.093
Val	n.m.	0.081	n.m.	0.084	n.m.	0.094

n.m.: not measured compounds; n.d.: not detected compounds

Cell Size and Viability

The cell size and granularity distribution of *C. glutamicum* DM 1800 population in scale-down cultivations as shown in Figure 7 did not indicate any change during the experiment. The cell size differed only marginal, whereas the results of cultivations in different reactor types (single-, two-, and three-CR) were similar. Propidium iodide (PI) staining was applied for the investigation of viability of the cell population. PI is widely used as an indicator for the cellular membrane integrity, as the dye is not detected in a cell with an intact cell membrane. Results of PI staining at homogeneous and inhomogeneous conditions obtained in scale-down cultivations are shown in Figure 8. In general, the percentage of PI stained cells remained beneath 1.5%. The viability of the cells was not affected by the applied oscillations.

These findings prove the general robustness of *C. glutamicum* to oscillations although the synthesis of several metabolites is altered as described in the previous sections.

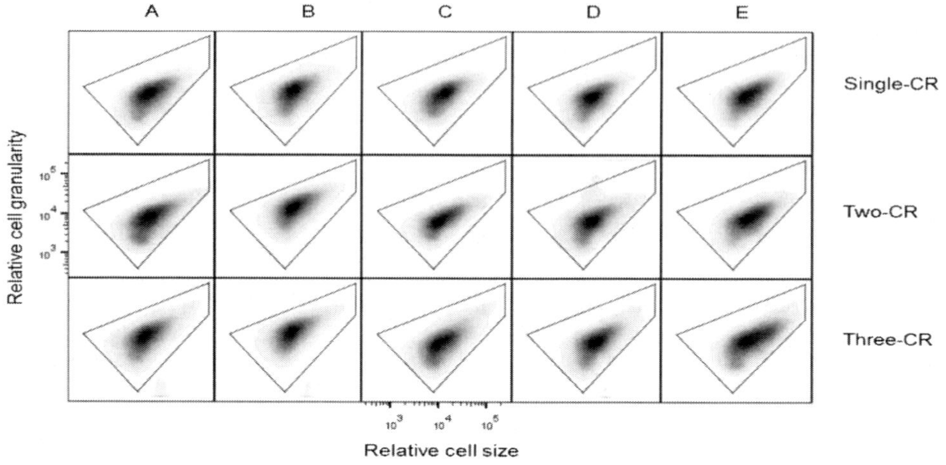

Figure 7. Comparison of unstained samples, taken from the STR compartment, in single-, two-, and three-CR cultivations of *C. glutamicum* DM 1800. Part A shows results after 14 hours of the batch-phase, part B shows results after 18 hours of the batch-phase (right before feed start), part C shows results after 4 hours, part D after 8 hours and part E after 12 hours of feeding.

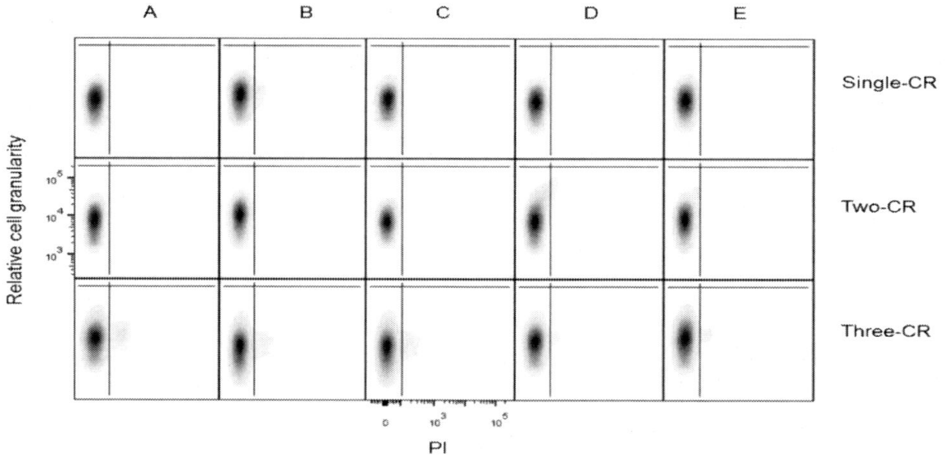

Figure 8. Comparison of samples taken from the STR compartment, stained by PI in single-, two-, and three-CR cultivations of *C. glutamicum* DM 1800. Part A shows results after 14 hours of the batch-phase, part B shows results after 18 hours of the batch-phase (right before feed start), part C shows results after 4 hours, part D after 8 hours and part E after 12 hours of feeding.

Bis-(1,3-dibutylbarbituric acid) trimethine oxonol (BOX) staining is used to evaluate the membrane potential. BOX enters the cell in unpolar regions, therefore a higher membrane potential decreases the percentage of stained cells. In the scale-down as well as in the reference cultivation, the percentage of BOX-stained cells decreased from the onset of feed start (Figure 9). The percentage of BOX-stained cells remained similar throughout the cultivation in a single-, and two-CR cultivation. In a three-CR cultivation, a decreased number of BOX-stained cells was detected compared to the two-, and single-CR cultivation.

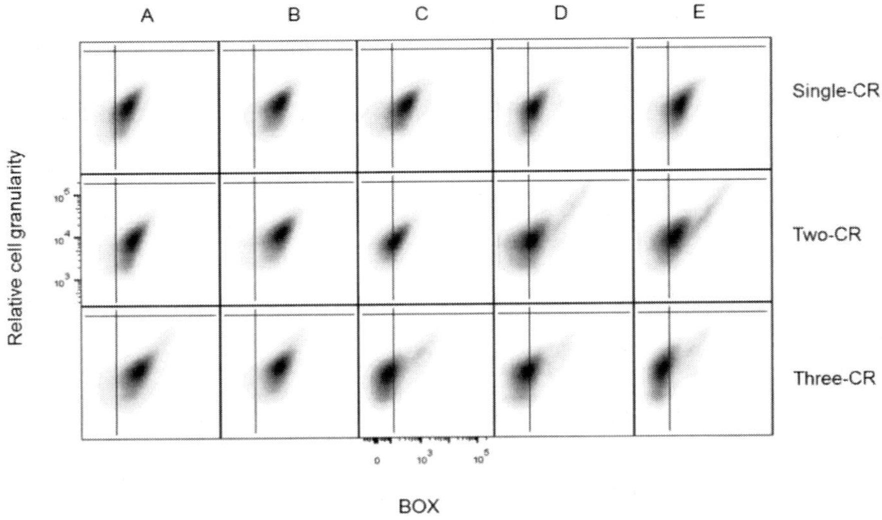

Figure 9. Comparison of samples, taken from the STR compartment, stained by BOX in single-, two-, and three-CR cultivations of *C. glutamicum* DM 1800. Part A shows results after 14 hours of the batch-phase, part B shows results after 18 hours of the batch-phase (right before feed start), part C shows results after 4 hours, part D after 8 hours and part E after 12 hours of feeding.

Responses to Heterogeneities of Other Microbiols

Gradients of dissolved oxygen were among the first parameters considered for scale-down experiments [27]. *Gluconobacter oxidans* ATCC 621H was cultivated in two equal stirred tank reactors (STR-STR), in which the DO control was set to 30% of saturation in one vessel. In the other one, dissolved oxygen was nearly depleted. The gluconic acid production rate was four times lower compared to experiments when the concentration of DO was maintained at 30.0% at both reactors. Oosterhuis et al. [23] concluded that the application of oscillating gas compositions is a less suitable method to scale-down a gluconic acid cultivation compared to a circulation of cells between two cultivation vessels, which are sparged with different rates. It was found that time intervals of at least 180 s, in which the cells are exposed to oxygen depletion, are required to determine a reduction of the specific oxygen uptake rate of *G. oxidans*. The utilization of pure oxygen was found to be critical during a scale-down experiment, because the occurrence of peak oxygen concentrations above 0.6 mmol L-1 resulted in a decreased gluconic acid production and finally cessation of growth. A Monte Carlo method approach was applied in order to simulate fluctuations of the aeration and feeding. A higher biomass yield compared to continuous feeding or periodic cycling (STR-PFR) was achieved with this approach in *S. cerevisiae* cultivations [26]. Similar fluctuations of the supply of dissolved oxygen led to a prolonged growth phase in a cultivation of a recombinant *E. coli* strain [50].

Among the earliest applications of STR-PFR scale-down reactors were investigations of *S. cerevisiae* cultivations [5, 50]. The volume of the PFR was 10% of the total volume and therefore equal to the volumetric part of the feeding zone at large scale reactors, estimated by CFD simulations of a 30.0 m^3 STR fed-batch cultivation [51]. These cultivations exhibited a biomass yield reduction of 6%, but also an increase of product selectivity under inhomogeneic conditions. *E.coli* K12 W3110 cultivations where characterized by an increase of glucose consumption and acetic acid formation as response to the exposure to repeated short-term glucose excess in combination with oxygen depletion [4, 30]. The biomass yield decreased by approx. 20%. Acetic acid formation, as it occurred in the anaerobic PFR module, is assumed to be a metabolic response to oxygen limitation, which finally results in metabolic overflow due to a reduction of the conversion rates in the TCR cycle [1, 52]. Formic acid accumulates significantly in the culture broth as the most preferred short cut of the cell to convert carbon to a stable and excretable metabolite. Acetic and formic acid are re-assimilated at oxygen sufficient conditions in the STR part – at least partially. Addition of the trace elements selenium, nickel and molybdenum, necessary for the function of formate hydrogen lyase (FHL), prevents the accumulation of formic acid, norvaline and norleucine [53, 54]. The investigation of the stress response of *E.coli* K12 W3110 revealed that different mRNA levels increased significantly after entering a high-glucose and low-oxygen zone in the PFR part [18]. The mRNA level of *pfl*, the oxygen starvation induced pyruvate formate lyase, increased up to threefold within 42 seconds. The mRNA level of the oxygen sensitive gene *ackA* revealed a similar increase after 27 seconds. Multi-parameter flow cytometric analysis revealed that *E. coli* cells show a higher viability at oscillating conditions [1]. These results confirm findings from large scale multi-parameter flow cytometric analysis reported by Hewitt et al. [6].

A different approach to investigate variations in the environment on the production of recombinant protein by *E. coli* was performed by Jazini and Herwig [55]. Post-induction oscillatory feeding led to a higher yield of protein per substrate (0.027 C-mol C-mol^{-1}) and higher activity per protein (0.79 U mg^{-1}) when compared to a constant feeding rate (0.011 C-mol C-mol^{-1} and 0.11 U mg^{-1}). However, the impact of oscillatory conditions on the recombinant protein production varied depending on the strain and target protein. Sandoval-Basurto et al. [24] showed that the productivity is reduced substantially in *E. coli* W3110 cultivations. Bylund et al. [3] found that oscillations can result in a higher product yield of recombinant protein. A ratio of 0.5 and 16 (v/v) between aerobic and anaerobic zones were applied in this study. Research was also focused on the intracellular accumulation of branched-chain amino acids in *E. coli.* Norvaline, norleucine, and β-methylnorleucine (βMNL) can accumulate and are secreted under certain cultivation conditions, especially during cultivation in mineral salt media. Several articles have been published in recent years, e.g. describing the incorporation of βMNL in a recombinant hirudine [56, 57], norleucine in recombinant human brain derived neurotrophic factor [58] and interleukin 2 [59, 60], and norvaline in recombinant hemoglobin [61]. Since the production of heterologous proteins is often desired, additional stress is imposed on microorganisms while foreign proteins are expressed in a host [62]. It was found that the detrimental effect of AP50 production on the cell was reduced in recombinant *E. coli* by the effect of heterogeneities. This can be related to the induced stress responses, which drive the cells into a dormant state [12]. It has also been revealed that rapid glucose oscillations favour the plasmid stability and recombinant protein production rate [63]. Thus, the stress imposed on the cells can also support the protein production process. Therefore, a case-sensitive observation of scale-up effects is necessary. Accumulation of short chain fatty acids can lead to an acidification of the surrounding. This could enforce the cells to spend energy on maintaining the intracellular pH [33]. In a study in a scale-down reactor, a changing pH-value was shown to induce changes in the metabolism of *B. subtilis*, probably linked to the activity of metabolic enzymes [25].

In addition to the stress responses to high substrate concentrations or temperatures, cells also react to low substrate or oxygen concentrations. Low oxygen concentrations in combination with a concomitant high substrate concentration force the cell to switch to different metabolic pathways. The conditions lead to an induction of fermentative pathways and byproduct synthesis. It was shown for *E. coli* in large scale fermentations [64] as well as in small scale experiments [31, 65]. In *B. subtilis* cultures, the growth rate is reduced and byproduct concentrations increase when DO oscillations are present [25, 29]. Similar results were obtained in *S. cerevisiae* cultivations [66].

Besides the oxygen starvation, the response to carbon source starvation is essential. One of the main reactions of a cell exposed to glucose starvation is the increased synthesis of guanosine tetraphosphate (ppGpp), an alarmone that induces the 'stringent control'. In a study using a scale-down reactor, an increase of the intracellular ppGpp concentration was observed in response to short-time glucose starvation in *E. coli* [30]. The expression of signaling molecules like ppGpp involve the expression of stress proteins. In consequence, the DNA replication is inhibited, the rRNA synthesis is decreased and the protein production is diminished. Furthermore, the glycolytic activity, DNA metabolism and the synthesis of structural components is reduced [12]. Hence the response of the cells to glucose starvation is a survival strategy, as the cells save energy by entering a dormant status, in which less proteins are expressed [67, 68]. As a consequence, the product formation is reduced. While lactate production under anaerobic conditions was observed at *C. glutamicum* cultivations as a response to oxygen depletion, no indication for a starvation response was noticed. In this case, *C. glutamicum* seems to be robust against short-term substrate depletion in comparison to other bacteria.

CONCLUSION

C. glutamicum is an organism widely used in large scale bioprocesses. Therefore, the response to an oscillating environment is of high interest. Scale-down studies should be performed for the systematic investigation of strain performance and productivity, but already in the screening stage. Thereby conditions, which are relevant in the industrial scale, can be considered in order to identify their sensitivity on the cellular response mechanisms. Automated parallel bioreactor systems, in which a screening for crucial cultivation conditions is performed, can be combined not only with strain selection and optimization, but multi-compartment scale-down approaches. The consideration of cell-wide physiological and morphological responses to oscillating conditions, including novel tools for *on line* measurements of higher resolution [69] provides data that allow the integration of systems biology approaches into bioprocess scaleup. The combination of scale-down cultivations and microfluidic devices will allow the identification and separation of more robust cells [70] and a general evaluation of the population heterogeneity caused by oscillations.

C. glutamicum indicated a comparably high robustness against oscillating conditions, which might be one reason for the quite successful industrial application. However, large scale effects should be considered as early as possible during strain development in order to ensure a successful application as production strain. This relies not only on single product synthesis features, but includes the way the organisms react to an oscillating environment.

Finally, the collection and consideration of data from computational fluid dynamics and higher resolved (multi-position) monitoring in industrial bioreactors will allow a case-wise adjustment of gradients in scale-down systems. Since conditions vary a lot among production sites due to different reactor and stirrer geometries as well as feed line positions, which are barely considered yet, a further improvement of the specific simulation of the large scale in the lab scale becomes feasible.

References

[1] Enfors SO, Jahic M, Rozkov A, Xu B, Hecker M, Jurgen B, et al. Physiological responses to mixing in large scale bioreactors. *J Biotechnol* 2001 Feb 13;85(2):175-85.

[2] Schmidt FR. Optimization and scale up of industrial fermentation processes. Appl Microbiol Biotechnol 2005 Sep;68(4):425-35.

[3] Bylund F, Castan A, Mikkola R, Veide A, Larsson G. Influence of scale-up on the quality of recombinant human growth hormone. *Biotechnol Bioeng* 2000;69(2):119-28.

[4] Neubauer P, Häggström L, Enfors SO. Influence of substrate oscillations on acetate formation and growth yield in *Escherichia coli* glucose limited fed-batch cultivations. *Biotechnol Bioeng* 1995;47(2):139-46.

[5] George S, Larsson G, Enfors SO. A scale-down two-compartment reactor with controlled substrate oscillations: Metabolic response of *Saccharomyces cerevisiae*. *Bioprocess Eng* 1993 1993/11/01;9(6):249-57.

[6] Hewitt CJ, Nebe-Von Caron G, Axelsson B, McFarlane CM, Nienow AW. Studies related to the scale-up of high-cell-density *E. coli* fed-batch fermentations using multiparameter flow cytometry: effect of a changing microenvironment with respect to glucose and dissolved oxygen concentration. *Biotechnol Bioeng* 2000 Nov 20;70(4):381-90.

[7] Namdev PK, Thompson BG, Gray MR. Effect of feed zone in fed-batch fermentations of *Saccharomyces cerevisiae*. *Biotechnol Bioeng* 1992 Jun 20;40(2):235-46.

[8] Buchholz A, Hurlebaus J, Wandrey C, Takors R. Metabolomics: quantification of intracellular metabolite dynamics. *Biomol Eng* 2002 Jun;19(1):5-15.

[9] Hoque MA, Ushiyama H, Tomita M, Shimizu K. Dynamic responses of the intracellular metabolite concentrations of the wild type and pykA mutant *Escherichia coli* against pulse addition of glucose or NH3 under those limiting continuous cultures. *Biochem Eng J* 2005 Nov 1;26(1):38-49.

[10] Bryant J. Characterisation of mixing in fermenters. *Advanced Biochemical Engineering* 1977;5:101-23.

[11] Vardar F, Lilly MD. The measurement of oxygen-transfer coefficients in fermentors by frequency response techniques. *Biotechnol Bioeng* 1982 Jul;24(7):1711-9.

[12] Hewitt CJ, Nienow AW. The scale-up of microbial batch and fed-batch fermentation processes. *Adv Appl Microbiol* 2007;62:105-35.

[13] Lin HY, Mathiszik B, Xu B, Enfors SO, Neubauer P. Determination of the maximum specific uptake capacities for glucose and oxygen in glucose-limited fed-batch cultivations of *Escherichia coli*. *Biotechnol Bioeng* 2001 Jun 5;73(5):347-57.

[14] Phue JN, Shiloach J. Impact of dissolved oxygen concentration on acetate accumulation and physiology of *E. coli* BL21, evaluating transcription levels of key genes at different dissolved oxygen conditions. *Metab Eng* 2005 Sep-Nov;7(5-6):353-63.

[15] Lapin A, Schmid J, Reuss M. Modeling the dynamics of *E. coli* populations in the three-dimensional turbulent field of a stirred-tank bioreactor—A structured–segregated approach. *Chem Eng Sci* 2006;61(14):4783-97.

[16] Vrabel P, van der Lans RGJM, Luyben KCAM, Boon L, Nienow AW. Mixing in large-scale vessels stirred with multiple radial or radial and axial up-pumping impellers: modelling and measurements. *Chem Eng Sci* 2000 Dec;55(23):5881-96.

[17] Larsson H, Eliasson AC. Phase separation of wheat flour dough studied by ultracentrifugation and stress relaxation .2. Influence of mixing time, ascorbic acid, and lipids. *Cereal Chem* 1996 Jan-Feb;73(1):25-31.

[18] Schweder T, Kruger E, Xu B, Jurgen B, Blomsten G, Enfors SO, et al. Monitoring of genes that respond to process-related stress in large-scale bioprocesses. *Biotechnol Bioeng* 1999 Oct 20;65(2):151-9.

[19] Brognaux A, Bugge J, Schwartz FH, Thonart P, Telek S, Delvigne F. Real-time monitoring of cell viability and cell density on the basis of a three dimensional optical reflectance method (3D-ORM): investigation of the effect of sub-lethal and lethal injuries. *J Ind Microbiol Biotechnol* 2013 Jul;40(7):679-86.

[20] Brognaux A, Han SS, Sorensen SJ, Lebeau F, Thonart P, Delvigne F. A low-cost, multiplexable, automated flow cytometry procedure for the characterization of microbial stress dynamics in bioreactors. *Microbial Cell Factories* 2013 Oct 31;12.

[21] Brognaux A, Bugge J, Schwartz FH, Thonart P, Telek S, Delvigne F. Real-time monitoring of cell viability and cell density on the basis of a three dimensional optical reflectance method (3D-ORM): investigation of the effect of sub-lethal and lethal injuries. *J Ind Microbiol Biot* 2013 Jul;40(7):679-86.

[22] Delvigne F, Brognaux A, Francis F, Twizere JC, Gorret N, Sorensen SJ, et al. Green fluorescent protein (GFP) leakage from microbial biosensors provides useful information for the evaluation of the scale-down effect. *Biotechnology Journal* 2011 Aug;6(8):968-78.

[23] Oosterhuis NM, Kossen NW, Olivier AP, Schenk ES. Scale-down and optimization studies of the gluconic acid fermentation by *Gluconobacter oxydans*. *Biotechnol Bioeng* 1985 May;27(5):711-20.

[24] Sandoval-Basurto EA, Gosset G, Bolivar F, Ramirez OT. Culture of *Escherichia coli* under dissolved oxygen gradients simulated in a two-compartment scale-down system: metabolic response and production of recombinant protein. *Biotechnol Bioeng* 2005 Feb 20;89(4):453-63.

[25] Amanullah A, Hewitt CJ, Nienow AW, Lee C, Chartrain M, Buckland BC, et al. Measurement of strain-dependent toxicity in the indene bioconversion using multiparameter flow cytometry. *Biotechnol Bioeng* 2003 Feb 20;81(4):405-20.

[26] Namdev PK, Yegneswaran PK, Thompson BG, Gray MR. Experimental Simulation of Large-Scale Bioreactor Environments Using a Monte-Carlo Method. *Can J Chem Eng* 1991 Apr;69(2):513-9.

[27] Oosterhuis NM, Kossen NW. Dissolved oxygen concentration profiles in a production-scale bioreactor. *Biotechnol Bioeng* 1984 May;26(5):546-50.

[28] Neubauer P, Junne S. Scale-down simulators for metabolic analysis of large-scale bioprocesses. *Curr Opin Biotech* 2010 Feb;21(1):114-21.

[29] Junne S, Klingner A, Kabisch J, Schweder T, Neubauer P. A two-compartment bioreactor system made of commercial parts for bioprocess scale-down studies: Impact of oscillations on *Bacillus subtilis* fed-batch cultivations. *Biotechnol J* 2011;6(8):1009-17.

[30] Neubauer P, Åhman M, Törnkvist M, Larsson G, Enfors SO. Response of guanosine tetraphosphate to glucose fluctuations in fed-batch cultivations of *Escherichia coli*. *J Biotechnol* 1995;43(3):195-204.

[31] Xu B, Jahic M, Blomsten G, Enfors SO. Glucose overflow metabolism and mixed-acid fermentation in aerobic large-scale fed-batch processes with *Escherichia coli*. *Appl Microbiol Biotechnol* 1999 1999/05/01;51(5):564-71.

[32] Käß F, Junne S, Neubauer P, Wiechert W, Oldiges M. Process inhomogeneity leads to rapid side product turnover in cultivation of *Corynebacterium glutamicum*. *Microbial Cell Factories* 2014;13(1).

[33] Onyeaka H, Nienow AW, Hewitt CJ. Further studies related to the scale-up of high cell density *Escherichia coli* fed-batch fermentations: the additional effect of a changing microenvironment when using aqueous ammonia to control pH. *Biotechnol Bioeng* 2003 Nov 20;84(4):474-84.

[34] Amanullah A, McFarlane CM, Emery AN, Nienow AW. Scale-down model to simulate spatial pH variations in large-scale bioreactors. *Biotechnol Bioeng* 2001 Jun 5;73(5):390-9.

[35] Kensy F, Zang E, Faulhammer C, Tan R-K, Buechs J. Validation of a high-throughput fermentation system based on online monitoring of biomass and fluorescence in continuously shaken microtiter plates. *Microbial Cell Factories* 2009 Jun 4;8.

[36] Wilming A, Bahr C, Kamerke C, Buchs J. Fed-batch operation in special microtiter plates: a new method for screening under production conditions. *J Ind Microbiol Biot* 2014 Mar;41(3):513-25.

[37] Tang YJ, Laidlaw D, Gani K, Keasling JD. Evaluation of the effects of various culture conditions on Cr(VI) reduction by *Shewanella oneidensis* MR-1 in a novel high-throughput mini-bioreactor. *Biotechnol Bioeng* 2006;95(1):176-84.

[38] Isett K, George H, Herber W, Amanullah A. Twenty-four-well plate miniature bioreactor high-throughput system: Assessment for microbial cultivations. *Biotechnol Bioeng* 2007;98(5):1017-28.

[39] Zhang Z, Perozziello G, Boccazzi P, Sinskey A, Geschke O, Jensen K. Microbioreactors for Bioprocess Development. *J Assoc Lab Autom* 2007;12(3):143-51.

[40] Lamping SR, Zhang H, Allen B, Shamlou PA. Design of a prototype miniature bioreactor for high throughput automated bioprocessing. *Chem Eng Sci* 2003 Feb-Mar;58(3-6):747-58.

[41] Bower DM, Lee KS, Ram RJ, Prather KLJ. Fed-Batch Microbioreactor Platform for Scale-down and Analysis of a Plasmid DNA Producing Process. *Biotechnol Bioeng* 2012;109(8):1976 - 86.

[42] Moses S, Manahan M, Ambrogelly A, Ling WLW. Assessment of AMBRTM as a model for high-throughput cell culture process development strategy. *Adv Biosci Biotechnol* 2012;3:918-27.

[43] Schilling BM, Pfefferle W, Bachmann B, Leuchtenberger W, Deckwer WD. A special reactor design for investigations of mixing time effects in a scaled-down industrial L-lysine fed-batch fermentation process. *Biotechnol Bioeng* 1999;64(5):599-606.

[44] Käß F, Hariskos I, Michel A, Brandt H-J, Spann R, Junne S, et al. Assessment of robustness against dissolved oxygen/substrate oscillations for *C. glutamicum* DM1933 in two-compartment bioreactor. *Bioprocess Biosyst Eng* 2013 2013/11/12:1-12.

[45] Georgi T, Rittmann D, Wendisch VF. Lysine and glutamate production by *Corynebacterium glutamicum* on glucose, fructose and sucrose: Roles of malic enzyme and fructose-1,6-bisphosphatase. *Metab Eng* 2005;7(4):291-301.

[46] Eikmanns BJ, Kleinertz E, Liebl W, Sahm H. A family of *Corynebacterium glutamicum/Escherichia coli* shuttle vectors for cloning, controlled gene expression, and promoter probing. *Gene* 1991 Jun 15;102(1):93-8.

[47] Lemoine A, Maya Martínez-Iturralde N, Spann R, Neubauer P, Junne S. Response of *Corynebacterium glutamicum* exposed to oscillations cultivation conditions in a two- and a novel three compartment scale down bioreactor. *Biotechnol Bioeng* 2015; 112(6):1220-31.

[48] Buchholz J, Graf M, Freund A, Busche T, Kalinowski J, Blombach B, et al. CO2/HCO3 (-) perturbations of simulated large scale gradients in a scale-down device cause fast transcriptional responses in *Corynebacterium glutamicum*. *Appl Microbiol Biotechnol* 2014 Oct;98(20):8563-72.

[49] Käß F, Prasad A, Tillack J, Moch M, Giese H, Buchs J, et al. Rapid assessment of oxygen transfer impact for *Corynebacterium glutamicum*. *Bioprocess Biosyst Eng* 2014 Jul 1.

[50] Namdev PK, Irwin N, Thompson BG, Gray MR. Effect of oxygen fluctuations on recombinant *Escherichia coli* fermentation. *Biotechnol Bioeng* 1993 Mar 15;41(6):666-70.

[51] George S, Larsson G, Olsson K, Enfors SO. Comparison of the Baker's yeast process performance in laboratory and production scale. *Bioprocess Eng* 1998 Feb;18(2):135-42.

[52] Soini J, Falschlehner C, Liedert C, Bernhardt J, Vuoristo J, Neubauer P. Norvaline is accumulated after a down-shift of oxygen in *Escherichia coli* W3110. *Microbial Cell Factories* 2008 Oct 21;7.

[53] Soini J, Ukkonen K, Neubauer P. High cell density media for *Escherichia coli* are generally designed for aerobic cultivations - consequences for large-scale bioprocesses and shake flask cultures. *Microbial Cell Factories* 2008 Aug 7;7.

[54] Biermann M, Linnemann J, Knüpfer U, Vollstädt S, Bardl B, Seidel G, et al. Trace element associated reduction of norleucine and norvaline accumulation during oxygen limitation in a recombinant *Escherichia coli* fermentation. *Microbial Cell Factories* 2013(12:116).

[55] Jazini M, Herwig C. Effect of post-induction substrate oscillation on recombinant alkaline phosphatase production expressed in Escherichia coli. *J Biosci Bioeng* 2011 Dec;112(6):606-10.

[56] Muramatsu R, Misawa S, Hayashi H. Finding of an isoleucine derivative of a recombinant protein for pharmaceutical use. *J Pharm Biomed Anal* 2003 Apr 1;31(5):979-87.

[57] Muramatsu R, Negishi T, Mimoto T, Miura A, Misawa S, Hayashi H. Existence of beta-methylnorleucine in recombinant hirudin produced by *Escherichia coli. J Biotechnol* 2002 Feb 14;93(2):131-42.

[58] Sunasara KM, Cramer SM, Hauer CR, Rupp RG, Shoup VA. Characterization of recombinant human brain-derived neurotrophic factor variants. *Arch Biochem Biophys* 1999 Dec 15;372(2):248-60.

[59] Lu HS, Tsai LB, Kenney WC, Lai PH. Identification of unusual replacement of methionine by norleucine in recombinant interleukin-2 produced by *E. coli. Biochem Biophys Res Commun* 1988 Oct 31;156(2):807-13.

[60] Tsai LB, Lu HS, Kenney WC, Curless CC, Klein ML, Lai PH, et al. Control of Misincorporation of Denovo Synthesized Norleucine into Recombinant Interleukin-2 in *Escherichia-Coli. Biochem Bioph Res Co* 1988 Oct 31;156(2):733-9.

[61] Apostol I, Levine J, Lippincott J, Leach J, Hess E, Glascock CB, et al. Incorporation of norvaline at leucine positions in recombinant human hemoglobin expressed in *Escherichia coli. J Biol Chem* 1997 Nov 14;272(46):28980-8.

[62] Borth N, Mitterbauer R, Mattanovich D, Kramer W, Bayer K, Katinger H. Flow cytometric analysis of bacterial physiology during induction of foreign protein synthesis in recombinant *Escherichia coli* cells. *Cytometry* 1998 Feb 1;31(2):125-9.

[63] Lin HY, Neubauer P. Influence of controlled glucose oscillations on a fed-batch process of recombinant *Escherichia coli. J Biotechnol* 2000 Apr 14;79(1):27-37.

[64] Lara AR, Taymaz-Nikerel H, Mashego MR, van Gulik WM, Heijnen JJ, Ramirez OT, et al. Fast dynamic response of the fermentative metabolism of *Escherichia coli* to aerobic and anaerobic glucose pulses. *Biotechnol Bioeng* 2009 Dec 15;104(6):1153-61.

[65] Xu B, Jahic M, Enfors SO. Modeling of overflow metabolism in batch and fed-batch cultures of *Escherichia coli. Biotechnol Progr* 1999 Jan-Feb;15(1):81-90.

[66] Abel C, Hubner U, Schugerl K. Transient behaviour of baker's yeast during enforced periodical variation of dissolved oxygen concentration. *J Biotechnol* 1994 Jan 15;32(1):45-57.

[67] Andersson L, Yang S, Neubauer P, Enfors SO. Impact of plasmid presence and induction on cellular responses in fed batch cultures of *Escherichia coli. J Biotechnol* 1996 May 15;46(3):255-63.

[68] Teich A, Meyer S, Lin HY, Andersson L, Enfors S, Neubauer P. Growth rate related concentration changes of the starvation response regulators sigmaS and ppGpp in glucose-limited fed-batch and continuous cultures of *Escherichia coli. Biotechnol Prog 1999* Jan-Feb;15(1):123-9.

[69] Delvigne F, Goffin P. Microbial heterogeneity affects bioprocess robustness: Dynamic single-cell analysis contributes to understanding of microbial populations. *Biotechnol J* 2014 Jan;9(1):61-72.

[70] Grunberger A, Wiechert W, Kohlheyer D. Single-cell microfluidics: opportunity for bioprocess development. *Curr Opin Biotechnol* 2014 Oct;29C:15-23.

In: New Trends in *Corynebacterium glutamicum*
Editor: Carlos Barreiro
ISBN: 978-1-63483-280-9

Chapter 5

CAROTENOIDS BIOSYNTHESIS IN *CORYNEBACTERIUM GLUTAMICUM*

Miriam Martínez-Castro, Esther Bahíllo, Sonia Martínez-Cámara and José-Luis Barredo*
Gadea Biopharma, Parque Tecnológico de León, León, Spain

ABSTRACT

Carotenoids are a family of yellow to orange-red terpenoid pigments synthesized by photosynthetic organisms and by many bacteria and fungi. They are involved in diverse functions such as photo-protection, light harvesting and as membrane stabilizers. They show beneficial health effects, protecting against oxidative damage, and being responsible of the colours associated with many plants and animals. Animals must obtain the carotenoids through their diet, having some of them important functions as precursors of vitamin A and different hormones. Carotenoids are commercially used as food colorants, animal feed supplements, and cosmetics as well as in the nutraceutical and pharmaceutical industries. Only a few of the hundreds of identified carotenoids are industrially produced, being β-carotene the most prominent. *Corynebacterium glutamicum* is a known producer of decaprenoxanthin, a glycosylated C_{50} yellow pigment. After elucidation of the decaprenoxanthin biosynthetic pathway, metabolic engineering was used to construct a lycopene producing strain of *C. glutamicum* as a platform for the production of different C_{50} and C_{40} carotenoids. The productivities of lycopene, β-carotene, zeaxanthin, C.p.450, sarcinaxanthin and decaprenoxanthin by recombinant strains of *C. glutamicum* are in the range of milligram per gram of dry weight.

INTRODUCTION

Carotenoids are yellow to red-coloured pigments that can be found ubiquitous in nature and are synthesized by plant, fungi, algae and bacteria [1]. They are a subfamily of

* Corresponding author: José-Luis Barredo; joseluis.barredo@gadea.com

terpenoids, also referred to as isoprenoids, and can be differentiated in two groups: (i) the carotenes, a group of hydrocarbon molecules like the lycopene and the β-carotene, and (ii) the xanthophylls, the oxygenated molecules of carotenes including lutein, zeaxanthin, canthaxanthin and astaxanthin [2]. Additionally, the carotenoids can be classified according to the length of their carbon backbone; although most of them have a C_{40} backbone, also C_{30} and C_{50} are known. All carotenoids are formed from a common C_5 molecule of isopentenyl pyrophosphate (IPP) which is synthesized by two different pathways: the mevalonic acid (MVA) pathway and the non-mevalonate pathway or methylerythritol phosphate (MEP) pathway. The MVA pathway is found in eukaryotes, archaea and in a few bacteria, whereas in most bacteria IPP is synthesized from pyruvate and glyceraldehyde-3-phosphate via the MEP pathway [3-6].

More than 95% of the carotenoids are based on a C_{40} backbone and only a rare number of them present the C_{50} or $C_{30/35}$ structures. Some of the products with C_{30} and C_{35} backbone are apocarotenoids derived by cleavage from the oxygenated C_{40} carotenoids [7, 8] and others are C_{30} carotenoids synthesized by condensation of two molecules of C_{15} farnesyl pyrophosphate (FPP). Biosynthesis of C_{30} carotenoids is relatively restricted in nature but it has been described in *Staphylococcus* and in methylotrophic bacteria [9].

The C_{50} carotenoids are formed by a diprenylation step involving the addition of two C_5 units attached to the C(2) and C(2´) of the C_{40} carotenoid. According to their end group, the C_{50} carotenoids can be acyclic or cyclic compounds. This kind of carotenoids is present mainly in extremely halophilic bacteria, a few species of Gram-positive bacteria of the order *Actinomycetales*, and in a Gram-negative *Pseudomonas* strain [10-12]. The first identified C_{50} carotenoid was decaprenoxanthin, isolated from *Flavobacterium dehidrogenans* [13, 14]. To date, only three natural C_{50} carotenoid biosynthetic pathways have been described. The biosynthetic pathways of the ε-cyclic C_{50} carotenoid decaprenoxanthin [2,2-bis-(4-hydroxy-3-methybut-2-enyl)-ε,ε-carotene], the β-cyclic C_{50} carotenoid C.p.450 [2,2-bis-(4-hydroxy-3-methybut-2-enyl)-β,β-carotene], and the γ-cyclic C_{50} carotenoid sarcinaxanthin [2,2-bis-(4-hydroxy-3-methyl-2-butenyl)-γ,γ-carotene)] have been elucidated in *C. glutamicum* [15], *Dietzia* sp. CQ4 [16] and *Micrococcus luteus* [11], respectively.

Decaprenoxanthin and its glucosylated forms are the predominant carotenoids in *C. glutamicum* [15, 17]. Besides *C. glutamicum*, a few other Corynebacteria are able to produce carotenoid pigments, including *Corynebacterium michiganense*, *Corynebacterium erythrogenes*, *Corynebacterium fascians* and *Corynebacterium poinsettiae* [18-21]. Interestingly, among the *C. glutamicum* isolates there are some able to produce decaprenoxanthin derivatives, being others unable to synthesize these carotenoids. Therefore, carotenoids are not essential for the growth of *C. glutamicum* [22].

The carotenoids have diverse biological functions in nature, such as photo-protection and light-harvesting molecules in photosynthetic organisms, and membrane stabilizers. Additionally, the conjugated double bond system of the carotenoids protects against the oxidative damage caused by oxidizing agents and free radicals, conferring important benefits on human and animal health [22, 23]. Animals are not able to synthesize carotenoids so they have to obtain them through the diet because they have important functions, such as precursors in the biosynthesis of vitamin A [24] and some hormones like retinoic acid [25].

The carotenoids are commercially used as food colorants (ice creams, orange juice, butter, candies, etc.), animal feed supplements, cosmetics, and in nutraceutical and pharmaceutical industries due to their health benefits [23]. Some studies proposed that carotenoids can prevent cancer and reduce the cardiovascular risk and Alzheimer, maybe due to their antioxidant properties [26, 27].

The C_{50} carotenoids have longer conjugated double bonds than most carotenoids, including at least one hydroxyl group; both features contribute to their strong antioxidant properties [11]. Thus, the γ-cyclic C_{50} carotenoid sarcinaxanthin, and particularly its glucosides, were described to be effective as photoprotectors, with applications in sunscreens and other light-protecting cosmetics [28]. Therefore, the C_{50} carotenoids have a promising application in cosmetics, nutraceuticals and pharmaceuticals [29].

The commercial value of the carotenoids was reported in 2010 to be $1.2 billion, with an estimated annual growth of 2.3% of the global market [30]. The carotenoid production through chemical synthesis or extraction from plants is limited by low yields, which results in high production costs. This led to the research on microbial production of carotenoids as an alternative, showing higher yields and being more environmentally friendly. In addition, the microbial production of carotenoids can decrease costs using low-cost raw materials such as agro-industrial wastes [2, 23, 31]. Moreover, the use of recombinant DNA technology allows to obtain unnatural carotenoids and to increase their productivity, getting some advantages against the chemical synthesis of carotenoids [32]. However, although more than 600 microbial carotenoids have been identified, only a few are produced at an industrial scale. The global market of carotenoids 2010 was headed by β-carotene with a market value of $261 million [33].

DECAPRENOXANTHIN BIOSYNTHESIS IN *C. GLUTAMICUM*

All the carotenoids are derived from the isoprenoid (or terpenoid) pathway. The basic unit of these molecules is the branched C_5 carbon skeleton IPP, and the number of repetitions of this motif, cyclization reactions, rearrangements and further glucosylation of the carbon skeleton are responsible for the high diversity of structures [6].

Isoprenoids are biosynthesized by condensation of IPP and its isomer DMAPP to form a polyprenyl chain which is further processed [3]. There are two distinct and independent pathways leading to IPP, the MVA pathway and the MEP pathway, also known as deoxyxylulose phosphate (DXP) pathway [5]. The MVA pathway is common in eukaryotes (mammals and fungi), nearly all archaea and a short number of bacteria. Most bacteria synthesize IPP using the MEP pathway, whereas only a few bacteria use both pathways. In higher plants, the MEP pathway is located in plastids, whereas the MVA pathway is found in the cytosol.

The MVA pathway starts from the sequential condensation of three molecules of acetyl-CoA, whereas the first step of the MEP pathway is the formation of DXP by the condensation of pyruvate and glyceraldehyde-3-phosphate, in a reaction catalyzed by the DXP synthase (Figure 1). Then, MEP is synthesized in the presence of NADPH by rearrangement and reduction of DXP in a single step catalyzed by the DXP reductoisomerase. DXP is a precursor not only for IPP and DMAPP but also for thiamine and pyridoxal, and MEP is the first biosynthetic intermediate specific of the MEP pathway. The sequential action of four enzymes (MEP cytidylyltransferase, CDP-ME kinase, MECDP synthase and HMBDP synthase) transforms MEP consecutively into CDP-ME (4-(cytidine 5′-diphospho)-2-C-methyl-D-erythritol), CDP-ME2P (2-phospho-4-(cytidine 5′-diphospho)-2-C-methyl-D-erythritol) and HMBPP (4-hydroxy-3-methylbut-2-enyl diphosphate). Finally, the HMBPP reductase catalyzes the simultaneous formation of IPP and DMAPP in a proportion of about 5:1. DMAPP is also produced by isomerization of IPP catalyzed by the IPP isomerase [34, 35].

Like in most bacteria, the MEP pathway supplies the carotenoid precursors in *C. glutamicum*. This is confirmed by the absence of a gene for hydroxymethylglutaryl-CoA reductase from the MVA pathway and the evidence of the genes encoding deoxyxylulose 5-phosphate synthase and deoxyxylulose 5-phosphate reductoisomerase in the genome of *C. glutamicum* [22].

The starting point for the assembly of the backbone of carotenoids is the isomerization of the five-carbon (C_5) compound IPP to its allylic isomer dimethylallyl pyrophosphate (DMAPP) by the IPP isomerase. DMAPP is condensed with IPP to generate geranyl pyrophosphate (GPP; C_{10}), which undergoes further condensation with IPP leading to the C_{15} carotenoid precursor farnesyl pyrophosphate (FPP). Geranylgeranyl pyrophosphate (GGPP; C_{20}) is formed from FPP by a further addition of IPP. These sequential elongation reactions that lead to GGPP are catalyzed by the GGPP synthase (encoded by *crtE* gene). The head-to-head condensation of two GGPP molecules catalyzed by the phytoene synthase (encoded by *crtB* gene) forms the C_{40} phytoene, a colourless molecule which is the first carotenoid of the pathway. The phytoene desaturase (encoded by *crtI* gene) introduces four double bonds in the phytoene molecule to yield the red carotenoid lycopene [36].

The lycopene is a common precursor of different biosynthetic pathways of carotenoids. Based on transposon mutants and gene direct deletion mutagenesis, the reactions involved in decaprenoxanthin biosynthesis were elucidated. The elongation of lycopene via the C_{45} intermediate nonaflavuxanthin into the acyclic C_{50} carotenoid flavuxanthin by the addition of two molecules of DMAPP at positions C-2 and C-2′ is catalyzed by the lycopene elongase encoded by the *crtEb* gene. The cyclation of flavuxanthin by the formation of the ε-ionine rings leading to decaprenoxanthin is catalyzed by the heterodimeric cyclases encoded by the *crtYe* and the *crtYf* genes. These cyclases must function together for the active cyclation of the linear flavuxanthin to the bicyclic decaprenoxanthin [15]. Similarly, the biosynthesis of sarcinaxanthin in *M. luteus* proceeds via flavuxanthin followed by γ-cyclation catalyzed by the products of the *crtYg* and *crtYh* genes [11]. Based on sequence similarity, the C_{50} carotenoid cyclases from *Dietzia sp.*, *M. luteus* and *C. glutamicum* were proposed to be members of the CrtYcd type cyclases, one of the three different groups of carotenoid cyclases. In contrast to other CrtYcd-type cyclases, the C_{50} carotenoid cyclases cannot use the lycopene as a substrate [37].

The subsequent synthesis of mono- and di-glucosylated decaprenoxanthin in *C. glutamicum* is mediated by the CrtX glucosyltransferase. The physiological function of the decaprenoxanthin glucosylation remains unknown. The deletion of the *crtX* gene in *C. glutamicum* does not cause any phenotypical change. The water solubility of the di-glucosylated form of the zeaxanthin is higher than the non-glucosylated one, being probably that the glucosylation of other carotenoids affects its water solubility similarly [38].

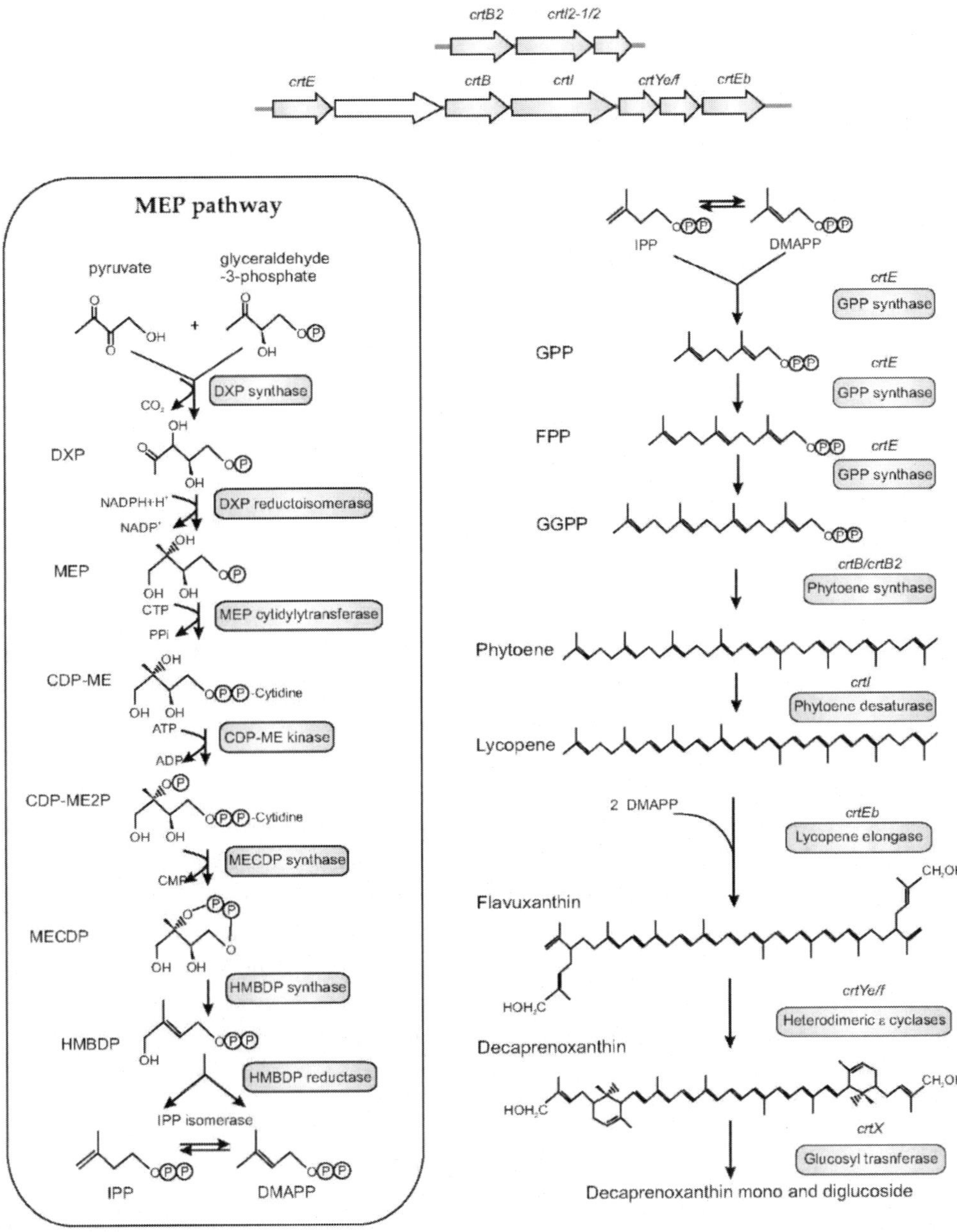

Figure 1. Genomic organization of the two carotenogenic gene clusters and the decaprenoxanthin biosynthesis pathway in *C. glutamicum*. Upper part: the two carotenogenic gene clusters in *C. glutamicum*; the carotenogenic genes are in gray and the open reading frame with unknown function is in white. Decaprenoxanthin biosynthetic pathway is represented in two schemes: in the left is the MEP pathway for IPP and DMAPP, and in the right is the pathway from IPP and DMAPP to decaprenoxanthin via the common C_{40} intermediate lycopene.

CAROTENOGENIC GENES IN *C. GLUTAMICUM*

In most bacteria, the carotenogenic genes are organized in gene clusters. *C. glutamicum* has a main cluster of carotenogenic genes and also a small cluster, separated by almost 2 Mbp, with genes showing similarity to the genes of the main cluster. The main cluster is formed of seven genes: (I) *crtE* (GGPP synthase), *crtB* (phytoene synthase) and *crtI* (phytoene desaturase) are the first three genes for lycopene biosynthesis from GGPP. Heterologous expression of these genes showed that they are involved in lycopene biosynthesis. (II) Within the gene cluster there is also an open reading frame (ORF) that is not involved in the biosynthesis of carotenoids (Figure 1) [15]. The product of this gene shares some similarities with lipid transport proteins from mycobacteria; therefore it may be involved in the transport of carotenoids [22]. (III) The products of the next three genes of the cluster are involved in the subsequent steps from lycopene to decaprenoxanthin. The elongase encoded by *crtEb*, which converts lycopene to the acyclic C_{50} carotenoid flavuxanthin, and the products of the *crtYe* and *crtYf* genes form the heterodimeric ε-cyclase, which catalyzes the cyclation to decaprenoxanthin.

The small cluster consists of the *crtB2* gene (phytoene synthase), and two genes showing similarity to an N-terminal fragment *crtI1-2* and a C-terminal fragment *crtI2-2* of the phytoene desaturase/dehydrogenase. The overexpression of *crtB2* almost complements the *crtB* deletion, showing that *crtB2* encodes a functional phytoene synthase. Therefore, *C. glutamicum* has two phytoene synthases: CrtB and CrtB2. The inducers affecting *crtB2* gene expression and the regulation of these two paralog genes have not yet been elucidated. The other two ORFs of the small cluster do not show phytoene desaturase activity [39].

The glucosyltransferase responsible for the biosynthesis of glucosylated forms of decaprenoxanthin has been recently characterized. Previously, the CrtX glucosyltransferase from the carotenoid gene cluster of *M. luteus* was shown to glucosylate the C_{50} carotenoid sarcinaxanthin, and a similar gene was found in the carotenoid gene cluster of *Dietzia* sp. [11]. Based on the sequence of these genes, a putative *crtX* was found about 4.5 Kb upstream of the decaprenoxanthin gene cluster of *C. glutamicum.* Deletion of this *crtX* homolog results in the absence of carotenoid glucosylation, showing the function of *CrtX* in the glucosylation of decaprenoxanthin. Besides, the *CrtX* from *C. glutamicum* is a glucosyltransferase also active on sarcinaxanthin and C.p.450, but it does not catalyze the glucosylation of the C_{40} zeaxanthin [38].

The genes of the main carotenoid gene cluster (*crtE-ORF-crtB-crtI-crtYe/f-crtEb*) are co-transcribed [39], and, similarly, the second cluster is transcriptionally organized as an operon. However, the transcriptional regulation of both operons has not been yet elucidated.

PATHWAY ENGINEERING OF *C. GLUTAMICUM* FOR CAROTENOIDS BIOSYNTHESIS

The increasing interest for naturally produced carotenoids led to investigations on large-scale production using microbial hosts. Overproduction of carotenoids has been studied in natural producers as well as in non-carotenogenic hosts with biotechnological relevance, such as the Gram-negative bacterium *Escherichia coli* and the yeast *Saccharomyces cerevisiae.*

Both species have been successfully engineered for lycopene and β-carotene production [40, 41]. In addition, the Gram-positive *Bacillus subtilis* is a promising candidate for the production of pharmaceutical isoprenoids due to its GRAS (Generally Recognized As Safe) status and its rapid growth rate. *B. subtilis* engineered with several *Staphylococcus aureus* carotenoid biosynthetic genes may serve as a host for production of C_{30} carotenoids [42].

The main metabolic engineering strategies towards an effective carotenoid production are based on the balanced expression of the suitable target genes, the optimization of isoprenoid precursors supply, as well as the access to cheap carbon sources. Since isoprenyl compounds are needed for cell growth, terpenoids biosynthesis and precursor supply need to be fine-tuned for an effective metabolic engineering [43]. In addition to engineering of isoprenoid pathways, balancing the supply of the starters of the MEP pathway (pyruvate and glyceraldehyde-3-phosphate) is also critical to the production of carotenoids.

C. glutamicum, a natural producer of the C_{50} carotenoid decaprenoxanthin and its glucosides, represents an advantageous production host for terpenoids in general and carotenoids in particular, since it is a GRAS organism and has been used for more than 50 years at the large-scale production in the food and feed biotechnology. Besides, *C. glutamicum* has a well-established tool box for genetic engineering, is able to metabolize many carbon sources for growth and production, and has been additionally engineered for the use of alternative cheaper carbon sources, i.e. wastes resulting from agro-industrial activities, like crude glycerol from biodiesel factories [44], molasses [45], arabinose, which is abundant in hemicellulosic wastes [46], and raw corn starch [47].

Up to now, major efforts in *C. glutamicum* metabolic engineering have been focused on the development of a lycopene producing strain as a platform for the production of other carotenoids under a controlled glucosylation. Moreover, some of the carotenogenic genes of *C. glutamicum* have been heterologously expressed in other species to obtain novel carotenoids.

Overproduction and Controlled Glucosylation of C_{50} and C_{40} Carotenoids

C. glutamicum was engineered for overproduction of its naturally produced carotenoid decaprenoxanthin and its intermediate lycopene, as well as for the production of the carotenoids β-carotene, C.p.450, sarcinaxanthin and zeaxanthin [38, 39]. Since lycopene is a precursor of most natural carotenoids, the lycopene producing strain may serve as a platform for carotenoids production. Additionally, *C. glutamicum* has been engineered for the biosynthesis of glucosylated carotenoids.

Overproduction of different carotenoids in *C. glutamicum* was achieved by engineering a strain that accumulates lycopene with different carotenogenic genes to produce a variety of carotenoids. Although with slight differences, three different lycopene producing strains were engineered: LYC, LYC1 and LYC2. C. *glutamicum* LYC was constructed by deleting the *crtEb* gene, coding for the lycopene elongase, coupled to the overexpression of the *crtE, crtB* and *crtI* genes involved in conversion of GGPP to lycopene. It accumulates about 2.4 ± 0.3 mg/g of dry weight of lycopene [39]. To construct *C. glutamicum* LCY1, the *crtEb* and *crtYe/f* genes involved in the flavuxanthin cyclation to decaprenoxanthin were deleted, and the *crtE*, *crtB* and *crtI* genes were expressed as a synthetic operon. The resulting strain produced up to 0.79 ± 0.09 mg/g of dry weight of lycopene in minimal medium. Afterwards, the potential of *C. glutamicum* as a carotenoid factory was tested modifying the lycopene accumulating strain LYC1 for the production of C_{50} carotenoids (C.p.450 and sarcinaxanthin), C_{40} carotenoids (β-carotene and zeaxanthin), and for the overproduction of its naturally produced carotenoid decaprenoxanthin [38].

The lycopene processing genes for the production of the C_{50} carotenoids were expressed in LYC1. Thus, the heterologous expression of *crtE2*, *crtYg*, and *crtYh* from *M. luteus* and the codon-optimized genes *lbtA* and *lbtBC* from *Dietzia sp.* CQ4 entailed the production of sarcinaxanthin and C.p.450, respectively. Overproduction of decaprenoxanthin was successfully done overexpressing the *crtEb* and *crtYe/f* genes in the lycopene accumulating strain LYC1. LYC2 was constructed by the knockout of the *crtX* gene of LYC1. The expression of the *crtEb* and *crtYe/f* genes of LYC2 entailed the production of non-glucosylated C_{50} carotenoids [38].

After showing the ability of LYC1 and LYC2 to produce C_{50} carotenoids in its glucosylated and non-glucosylated forms, the subsequent attempt was to produce C_{40} carotenoids in LYC2. Heterologous expression of the *crtY* gene from *Pantoea ananatis*, encoding lycopene β-cyclase, generated a β-carotene producing strain, and zeaxanthin was accumulated upon additional expression of the β-carotene hydroxylase encoding gene (*crtZ*), also from *P. ananatis*. The expression of the *crtX* gene of *P. ananatis* in the zeaxanthin producing strain allowed the production of the glucosylated form, whereas the *crtX* genes of *C. glutamicum* and *M. luteus* did not encode active zeaxanthin glucosyltransferases.

The productivities of lycopene, β-carotene, zeaxanthin, C.p.450, sarcinaxanthin and decaprenoxanthin by recombinant strains of *C. glutamicum* are in the range of milligram per gram of dry weight, favorably compared with other production hosts. The controlled glucosylation of carotenoids is remarkable, since the hydrophobic compounds like carotenoids show higher water solubility upon glycosylation. Increased water solubility may allow higher accumulation in microbial cultures, might be beneficial for some water-based industrial applications, and could simplify the downstream processing and the purification of carotenoids [38].

Production of Carotenoids by Heterologous Expression of *C. glutamicum* Carotenogenic Genes

Metabolic engineering of microorganisms by heterologous expression of carotenogenic genes from *C. glutamicum* and other microorganisms allowed the production of carotenoids structurally different from those found in nature [11, 48]. Thus saprenoxanthin is a novel asymmetric ε- and γ-cyclic C_{50} carotenoid produced by a genetically modified strain of *E. coli* expressing some carotenogenic genes from *C. glutamicum* and *M. luteus* [11]. A hybrid gene cluster was constructed using the sarcinaxanthin biosynthetic genes from *M. luteus* and replacing the cyclase (*crtYg* and *crtYf*) genes of *M. luteus* with the cyclase (*crtYe* and *crtYf*) genes of *C. glutamicum*. The expression of this cluster in *E. coli* resulted in the accumulation of the ε-cyclic C_{50} decaprenoxanthin, the γ-cyclic C_{50} sarcinaxanthin, and the asymmetric ε- and γ-cyclic C_{50} saprenoxanthin [11].

The novel lycopene derivative named 3,4,3´,4´-tetradehydrolycopene was produced by the heterologous expression of various combinations of *crtEBI* genes from different bacterial origins, including *C. glutamicum*. The resulting enzymes showed unusual activities when they were expressed in *E. coli*, producing novel lycopene derivatives [48].

The perspective to produce new valuable carotenoid structures not found naturally was created with these approaches. The specificity of the carotenogenic enzymes may appear different in heterologous hosts and combined with other heterologous enzymes, thus opening a great deal of possibilities of generating novel carotenoids.

Production of Carotenoids in *C. glutamicum* Using Cheaper Carbon Sources

A *C. glutamicum* strain able to produce lycopene from N-acetyl-glucosamine (GlcNAc) as the sole substrate was recently constructed. The amino sugar GlcNAc is the monomer of chitin, the second most abundant polysaccharide in nature after cellulose. Chitin is found in the exoskeleton of crustaceans and constitutes the main waste generated from the shrimp processing industries. The assimilation of GlcNAC by *C. glutamicum* was achieved by the overexpression of the endogenous *nagAB* genes encoding the N-acetyl-glucosamine-6-phosphate deacetylase and glucosamine-6-phosphate deaminase, and heterologous overexpression of the *nagE* gene encoding the PTS element specific for GlcNAc uptake in *C. glutamicum* [49].

C. glutamicum Strains for Production of Carotenoids Related Molecules such as Terpenes

The potential of *C. glutamicum* as a microbial factory goes beyond the production of carotenoids. A strain was recently engineered for the production of the terpenoid called sesquiterpene (+)-valencene, an aroma compound of citrus fruits synthesized from farnesyl pyrophosphate, an intermediate of the biosynthesis of carotenoids that is used to flavor foods and drinks. To overproduce sesquiterpene, two endogenous prenyltransferase genes were deleted, and the (+)-valencene synthase gene from the sweet orange *Citrus sinensis* and the farnesyl pyrophosphate synthase gene from *Nootka cypress* were heterologously expressed [50].

Future Approaches in the Metabolic Engineering of *C. glutamicum*

Up to now, most of the genetic modifications in *C. glutamicum* related with carotenoids were focused on the production of different carotenoids. Modulated expression of genes in the TCA cycle and optimization of the terpenoid pathway precursors supply lead to an increase in the production of carotenoids compared to the parental *E. coli* strain [43, 51]. In the same way, the production of carotenoids by the *C. glutamicum* strains may be further increased by optimizing the precursor supply and by modulating the central metabolism.

Norisoprenoids or apocarotenoids, carotenoids-derived molecules, are a commercially important field. One of the most relevant for the human diet is vitamin A, which is a cleavage product of the β-carotene (also called pro-vitamin A). Other norisoprenoids are compounds that contribute to the flavor of food and beverages and to fragrance products, or plant regulatory molecules such as abscisic acid and strigolactones. The formation of norisoprenoids from carotenoids may occur through photo-oxidation, or through enzymatic dioxygenation. When the latter occurs, a carotenoid cleavage dioxygenase (CCD) is often involved. These CCDs have structural features such as hydrophobic patches, which permit them to access carotenoids in a hydrophobic environment such as membranes, though the enzymes themselves are usually water soluble.

There are industrially relevant compounds among the norisoprenoids that are currently commercially produced by chemical synthesis, since their content in plant materials is too low. The industrial production processes using dioxygenases in a microbial production system have not been implemented at a large scale yet. An *E. coli* strain has been constructed by the heterologous expression of the dioxygenase responsible of the breakdown of carotenoids during the ripening of raspberry, being able of convert carotenoids into norisoprenoids [52].

Conclusion

C. glutamicum naturally produces the rare C_{50} carotenoids decaprenoxanthin and its mono- and di-glucosylated forms. The biosynthetic pathway of decaprenoxanthin has been completely elucidated allowing the construction of a lycopene accumulating strain as a platform for C_{40} and C_{50} carotenoids production. The main advantages of *C. glutamicum* are (i) the well described genetics, metabolism and physiology, (ii) the well-stablished genetic tool box, and (iii) the well-known growth requirements; it was engineered for assimilation of alternative carbon sources such as agro-industrial wastes. Additionally, the water solubility of carotenoids can be improved by glucosylation, opening new industrial applications that until now involve the use of starches to get water dispersible preparations of the water insoluble carotenoids. Thus, although *C. glutamicum* has a privileged position as a host for carotenoids, since it produces lycopene, β-carotene, zeaxanthin, C.p.450, sarcinaxanthin and decaprenoxanthin in the range of milligram per gram of dry weight, it has to compete with well stablished strains industrially used for the production of carotenoids, including *Blakeslea trispora* for β-carotene [53] and lycopene [54], *Yarrowia lipolytica* for lycopene [55] and *Xanthophyllomyces dendrorhous* for astaxanthin [56, 57].

References

[1] Britton, G; Liaanen-Jensen, S; Pfander, H; editors. Carotenoids handbook. Basel, Switzerland, Birkhauser, 2004.

[2] Das, A; Yoon, SH; Lee, SH; KimJY, Oh; DK, Kim; SW. An update on microbial carotenoid production: application of recent metabolic engineering tools. *Appl Microbiol Biotechnol*, 2007, 77(3), 505–12.

[3] Daum, M; Herrmann, S; Wilkinson, B; Bechthold, A. Genes and enzymes involved in bacterial isoprenoid biosynthesis. *Curr Opin Chem Biol*, 2009, 13(2), 180-8.

[4] Kirby, J; Keasling, JD. Biosynthesis of plant isoprenoids: perspectives for microbial engineering. *Annu Rev Plant Biol*, 2009, 60, 335–55.

[5] Lange, BM; Rujan, T; Martin, W; Croteau, R. Isoprenoid biosynthesis: the evolution of two ancient and distinct pathways across genomes. *Proc Natl Acad Sci U S A*, 2000, 97(24), 13172-7.

[6] Rohmer M. The discovery of a mevalonate-independent pathway for isoprenoid biosynthesis in bacteria algae and higher plants. *Nat Prod Rep*, 1999, 16(5), 565-74.

[7] Estrada, AF; Maier, D; Scherzinger, D; Avalos, J; Al-Babili, S. Novel apocarotenoid intermediates in *Neurospora crassa* mutants imply a new biosynthetic reaction sequence leading to neurosporaxanthin formation. *Fungal Genet Biol*, 2008, 45(11), 1497-505.

[8] Saelices, L; Youssar, L; Holdermann, I; Al-Babili, S; Avalos, J. Identification of the gene responsible for torulene cleavage in the *Neurospora* carotenoid pathway. *Mol Genet Genomics*, 2007, 278(5), 527-37.

[9] Tao, L; Schenzle, A; Odom, JM; Cheng, Q. Novel carotenoid oxidase involved in biosynthesis of 44'-diapolycopene dialdehyde. *Appl Environ Microbiol*, 2005, 71(6), 3294-301.

[10] Pfander H. C-45-carotenoids and C-50-carotenoids. *Pure Appl Chem* 1994, 66(10-11), 2369-74.

[11] Netzer, R; Stafsnes, MH; Andreassen, T; Goksøyr, A; Bruheim, P; Brautaset, T. Biosynthetic pathway for γ-cyclic sarcinaxanthin in *Micrococcus luteus*: heterologous expression and evidence for diverse and multiple catalytic functions of C(50) carotenoid cyclases. *J Bacteriol*, 2010, 192(21), 5688-99.

[12] Miki, W; Otaki, N; Yokoyama, A; Izumida, H; Shimidzu, N. Okadaxanthin, a novel C50-carotenoid from a bacterium, Pseudomonas sp. KK10206c associated with marine sponge, *Halichondria okadai*. *Experientia*, 1994, 50(7), 684–6.

[13] Weeks, OB; Andrewes, AG; Brown, BO; Weedon, BCL. Occurence of C40 and C45 carotenoids in the C50 carotenoid system of *Flavobacterium dehydrogenans*. *Nature*, 1969, 224, 879-82.

[14] Liaaen-Jensen, S; Hertzberg, S. Bacterial carotenoids: C50-carotenoids 3. Structure determination of dehydrogenans-P439. *Acta Chem Scand*, 1968, 22, 1171–86.

[15] Krubasik, P; Kobayashi, M; Sandmann, G. Expression and functional analysis of a gene cluster involved in the synthesis of decaprenoxanthin reveals the mechanisms for C50 carotenoid formation. *Eur J Biochem*, 2001, 268(13), 3702-8.

[16] Tao, L; Yao, H; Cheng, Q. Genes from a *Dietzia* sp. for synthesis of C40 and C50 beta-cyclic carotenoids. *Gene*, 2007, 386(1-2), 90-7.

[17] Krubasik, P; Takaichi, S; Maoka, T; Kobayashi, M; Masamoto, K; Sandmann, G. Detailed biosynthetic pathway to decaprenoxanthin diglucoside in *Corynebacterium glutamicum* and identification of novel intermediates. *Arch Microbiol*, 2001, 176(3), 217-23.

[18] Saperstein, S; Starr, MP. The ketonic carotenoid canthaxanthin isolated from a colour mutant of *Corynebacterium michiganense*. *Biochem J*, 1954, 57(2), 273-5.

[19] Hodgkiss, W; Liston, J; Goodwin, TW; Jamikorn, M. The isolation and description of 2 marine micro-organisms with special reference to their pigment production. *J Gen Microbiol*, 1954, 11(3), 438–50.

[20] Prebble, JN. Carotenoid production in *Corynebacterium fascians*. *Biochem J*, 1962, 85(2).
[21] Starr, MP; Saperstein, S. Thiamine and the carotenoid pigments of *Corynebacterium poinsettiae*. *Arch Biochem Biophys* ,1953, 43(1), 157-68.
[22] Sandmann, G; Yukawa, H. Vitamin synthesis: carotenoids, biotin and pantothenate. In: Eggeling L, Bott M, editors. Handbook of *Corynebacterium glutamicum*. *Boca Raton, Florida, USA*, CRC Press, 2005
[23] Lee, PC; Schmidt-Dannert, C. Metabolic engineering towards biotechnological production of carotenoids in microorganisms. *Appl Microbiol Biotechnol*, 2002, 60(1–2), 1–11.
[24] Johnson, EA; Schroeder, WA. Microbial carotenoids. *Adv Biochem Eng Biotechnol*, 1996, 53, 117–9.
[25] Vershinin, A. Biological functions of carotenoids: diversity and evolution. *Biofactors*, 1999, 10(2–3), 99–104.
[26] Kirsh, VA; Mayne, ST; Peters, U; Chatterjee, N; Leitzmann, MF; Dixon, LB; Urban, DA; Crawford, ED; Hayes, RB. A prospective study of lycopene and tomato product intake and risk of prostate cancer. *Cancer Epidemiol Biomarkers Prev*, 2006, 15(1), 92-8.
[27] Wang, W; Shinto, L; Connor, WE; Quinn, JF. Nutritional biomarkers in Alzheimer's disease: the association between carotenoids, n-3 fatty acids, and dementia severity. *J Alzheimers Dis*, 2008, 13(1), 31-8.
[28] Goksoyr, A. Carotenoid sunscreen. US 2013/0078203 A1 2013.
[29] Heider, SA; Peters-Wendisch, P; Wendisch, VF; Beekwilder, J; Brautaset, T. Metabolic engineering for the microbial production of carotenoids and related products with a focus on the rare C50 carotenoids. *Appl Microbiol Biotechnol*, 2014, 98(10), 4355-68.
[30] Cutzu, R; Coi, A; Rosso, F; Bardi, L; Ciani, M; Budroni, M; Zara, G; Zara, S; Mannazzu, I. From crude glycerol to carotenoids by using a *Rhodotorula glutinis* mutant. *World J Microbiol Biotechnol*, 2013, 29(6), 1009–17.
[31] Harada, H; Misawa, N. Novel approaches and achievements in biosynthesis of functional isoprenoids in *Escherichia coli*. *Appl Microbiol Biotechnol*, 2009, 84(6), 1021–31.
[32] Rodney, L; Ausich. Commercial opportunities for carotenoid production by biotechnology. *Pure &App. Chem,* 1997, 69(10), 2169-73.
[33] BCC Research. The global market for carotenoids. 2011 Available from: http://www.bccresearch.com/market-research/food-and-beverage/carotenoids-global-market-fod025d.html.
[34] Kuzuyama, T; Seto, H. Two distinct pathways for essential metabolic precursors for isoprenoid biosynthesis. *Proc Jpn Acad Ser B Phys Biol Sci*, 2012, 88(3), 41-52.
[35] Pérez-Gil, J; Rodríguez-Concepción, M. Metabolic plasticity for isoprenoid biosynthesis in bacteria. *Biochem J*, 2013, 452(1), 19-25.
[36] Sandmann, G. Carotenoid biosynthesis and biotechnological application. *Arch Biochem Biophys* , 2001, 385(1), 4-12.
[37] Klassen, JL. Phylogenetic and evolutionary patterns in microbial carotenoid biosynthesis are revealed by comparative genomics. *PLoS One*, 2010, 5(6), e11257.

[38] Heider, SA; Peters-Wendisch, P; Netzer, R; Stafnes, M; Brautaset, T; Wendisch, VF. Production and glucosylation of C50 and C40 carotenoids by metabolically engineered *Corynebacterium glutamicum*. *Appl Microbiol Biotechnol*, 2014, 98(3), 1223-35.

[39] Heider, SA; Peters-Wendisch, P; Wendisch, VF. Carotenoid biosynthesis and overproduction in *Corynebacterium glutamicum*. *BMC Microbiol*, 2012, 12, 198-208.

[40] Li, Q; Sun, Z; Li, J; Zhang, Y. Enhancing beta-carotene production in *Saccharomyces cerevisiae* by metabolic engineering. *FEMS Microbiol Lett*, 2013, 345(2), 94-101.

[41] Lin, Y; Jain, R; Yan, Y. Microbial production of antioxidant food ingredients via metabolic engineering. *Curr Opin Biotechnol*, 2014, 2671-8.

[42] Yoshida, K; Ueda, S; Maeda, I. Carotenoid production in *Bacillus subtilis* achieved by metabolic engineering. *Biotechnol Lett*, 2009, 31(11), 1789-93.

[43] Zhao, J; Li, Q; Sun, T; Zhu, X; Xu, H; Tang, J; Zhang, X; Ma, Y. Engineering central metabolic modules of *Escherichia coli* for improving β-carotene production. *Metab Eng*, 2013, 1742-50.

[44] Meiswinkel, TM; Rittmann, D; Lindner, SN; Wendisch, VF. Crude glycerol-based production of amino acids and putrescine by *Corynebacterium glutamicum*. *Bioresour Technol*, 2013, 145, 254 8.

[45] Xu, J; Zhang, J; Guo, Y; Zai, Y; Zhang, W. Improvement of cell growth and L-lysine production by genetically modified *Corynebacterium glutamicum* during growth on molasses. *J Ind Microbiol Biotechnol*, 2013, 40(12), 1423-32.

[46] Chen, T; Zhu, N; Xia, H. Aerobic production of succinate from arabinose by metabolically engineered *Corynebacterium glutamicum*. *Bioresour Technol*, 2014, 15, 1411-4.

[47] Tateno, T; Fukuda, H; Kondo, A. Direct production of L-lysine from raw corn starch by *Corynebacterium glutamicum* secreting *Streptococcus bovis* alpha-amylase using *cspB* promoter and signal sequence. *Appl Microbiol Biotechnol*, 2007, 77(3), 533-41.

[48] Song, GH; Kim, SH; Choi, BH; Han, SJ; Lee, PC. Heterologous carotenoid-biosynthetic enzymes: functional complementation and effects on carotenoid profiles in *Escherichia coli*. *Appl Environ Microbiol*, 2013, 79(2), 610-8.

[49] Matano, C; Uhde, A; Youn, JW; Maeda, T; Clermont, L; Marin, K; Krämer, R; Wendisch, VF; Seibold, GM. Engineering of *Corynebacterium glutamicum* for growth and L-lysine and lycopene production from N-acetyl-glucosamine. *Appl Microbiol Biotechnol*, 2014, 98(12), 5633-43.

[50] Frohwitter, J; Heider, SA; Peters-Wendisch, P; Beekwilder, J; Wendisch, VF. Production of the sesquiterpene (+)-valencene by metabolically engineered *Corynebacterium glutamicum*. *J Biotechnol*, 2014.

[51] Vadali, RV1; Fu, Y; Bennett, GN; San, KY. Enhanced lycopene productivity by manipulation of carbon flow to isopentenyl diphosphate in *Escherichia coli*. *Biotechnol Prog.*, 2005, 21(5), 1558-61.

[52] Beekwilder, J; van der Meer, IM; Simic, A; Uitdewilligen, J; van Arkel, J; de Vos, RC; et al. Metabolism of carotenoids and apocarotenoids during ripening of raspberry fruit. *Biofactors,* 2008, 34(1), 57-66.

[53] Rodríguez-Sáiz, M; Paz, B; de la Fuente, JL; López-Nieto, MJ; Cabri, W; Barredo, J.L. Genes for carotene biosynthesis from *Blakeslea trispora*. *Appl Environ Microbiol*, 2004, 70, 5589-94.

[54] López-Nieto, MJ; Costa, J; Peiro, E; Méndez, E; Rodríguez-Sáiz, M; de la Fuente, JL; et al. Biotechnological lycopene production by mated fermentation of *Blakeslea trispora*. *Appl Microbiol Biotechnol*, 2004, 66, 153-9.

[55] Matthäus, F; Ketelhot, M; Gatter, M; Barth, G. Production of lycopene in the non-carotenoid-producing yeast *Yarrowia lipolytica*. *Appl Environ Microbiol*, 2014, 80(5), 1660-9.

[56] de la Fuente, JL; Rodríguez-Sáiz, M; Schleissner, C; Díez, B; Peiro, E; Barredo, JL. High-titer production of astaxanthin by the semi-industrial fermentation of *Xanthophyllomyces dendrorhous*. *J Biotechnol*, 2010, 148, 144-6.

[57] Rodríguez-Sáiz, M; de la Fuente, JL; Barredo, JL. *Xanthophyllomyces dendrorhous* for the industrial production of astaxanthin. *Appl Microbiol Biotechnol*, 2010, 88, 645-58.

In: New Trends in *Corynebacterium glutamicum*
Editor: Carlos Barreiro
ISBN: 978-1-63483-280-9

Chapter 6

BIO-BASED ORGANIC AND HYDROXY ACID PRODUCTION WITH *CORYNEBACTERIUM GLUTAMICUM*

***Stefan Wieschalka**[*1] **and Boris Litsanov**[2]*
[1] Technical University of Denmark, NNFCfB, Hørsholm, Denmark
[2] Swiss Federal Institute of Technology Zurich, Institute of Microbiology, Zürich, Switzerland

ABSTRACT

Organic acids are used in several industrial sectors, such as food or pharmaceutical industry and serve as building block chemicals for a vast variety of products. Up to date many of them are obtained from petrochemical resources, implementing energy intensive chemical refining and waste production with a high environmental impact [1]. The enormous annual demand requests the development of efficient and sustainable bio-processes for organic acid production.

Recent studies explored the potential of *Corynebacterium glutamicum* for the production of several organic acids and demonstrated different strategies to improve its performance. In the last decade, also the production of hydroxy and polyhydroxy acids with *C. glutamicum* attracted attention. Especially, because of the high economic value of polyhydroxy acids, this field receives extensive research attention. Being biodegradable and biocompatible, they are an attractive alternative to classical petrochemical-derived plastics. The advantage in producing such substances with *C. glutamicum* is the use of simple media, free of complex components, in contrast to e.g. recombinant yeast. This cheapens the overall fermentation process, and simplifies in addition the product purification, resulting in lower waste production. Furthermore, the production of food-related compounds, or cosmetics and chemicals directly applied to humans, is not as critical with *C. glutamicum*, while with Gram-negative bacteria, the production is accompanied by contamination with pyrogenic compounds (endotoxins).

This chapter summarizes the current knowledge and recent achievements on metabolic and process engineering approaches to tailor *C. glutamicum* with the focus on

[*] Corresponding Author: Stefan Wieschalka; stefan.wieschalka@gmx.de

bio-based production of succinate, the alpha hydroxy acids L-lactate, D-lactate, and glycolate, as well as the biodegradeable polyester polyhydroxybutyrate.

INTRODUCTION

Glycolate ($C_2H_4O_3$) and L-/D-lactate ($C_3H_6O_3$) are prominent representatives of the chemical class of alpha hydroxy acids [AHAs], which are carboxylic acids substituted with a hydroxyl group on the adjacent carbon [2]. The use of AHAs in formulations to treat skin disorders, such as acne, ichthyosis, keratosis, warts, psoriasis or photoaged skin [3], has a long tradition. Treatment with such acids and their salts depends mainly on concentration, pH, formulation, as well as application time and is an important non-surgical therapy of skin aging [2, 4]. Their use is easy since concentrations up to 20% are suitable even for daily self-applications. Professionals may speed up the success of therapy, utilizing concentrations up to 70% [2]. Succinate, a C4-dicarboxylic acid, gains increased attention from the industry (Succinity GmbH, Reverdia, BioAmber, Myriant Technologies LLC) in recent years as a chemical building block, because it has the potential to replace the petroleum-derived maleic anhydrite for sustainable and crude oil independent economy [5–7]. One of the major advantages of succinic acid is the possibility for simple integration as precursor in chemical product chains to produce common bulk chemicals such as gamma-butyrolactone, 1,4-butanediol and its derivative tetrahydrofuran with total annual market of 500 kt, estimated in 2007 [8]. However, additional drive for the increasing industrial interest in succinate is the production of bio-based polymers, such as polybutylene succinate and polybutylene terephalate, as well as the recent advances in the development of new efficient biotechnological producer microorganisms to fulfil the criterion for cheap and fast accumulation of bulk amounts of succinate from diverse carbon sources [7]. Polyhydroxyalkanoates (e.g. poly-3-hydroxybutyrate) may be produced in microorganisms under certain circumstances [9]. Besides being used as energy storage under nutrient-imbalanced conditions in bacteria [10], these polymers are polyesters with very useful features, including biodegradability, biocompatibility and composition versatility with more than 150 known monomers [11, 12]. Although their use is currently limited owing to the high production costs in comparison to petroleum-derived plastics [13], they have received considerable research attention as product from renewable carbon sources [10]. Hence, the named chemicals are of great economic interest and have been, with gradual growing efforts, set as goal for the production with *C. glutamicum* in the last decades. In these studies *C. glutamicum* presented itself as one of the hosts, with best potential for organic acid production from renewable carbon sources often outcompeting natural producers [14].

GLYCOLATE

The simplest AHA is glycolate (GLA, glycolic acid; alpha hydroxy acetic acid). GLA is commonly used in cosmetic industry for multiple applications, such as peelings, creams, moisturizers, toners, masks and cleaners, because of its excellent skin penetration abilities [4, 15, 16]. Furthermore, GLA can be used for pH adjustment, it forms complexes with metal ions and serves as cleaning and rinsing agent in household and industrial applications [17]. Naturally, GLA occurs e.g. in crude grapes, sugar cane juice, sugar beets and *Parthenocissus* species [4]. For the application in dermatologic and industrial products it is chemically synthesized from carbon monoxide and formaldehyde under high pressure and temperature [18, 19]. Besides, GLA was recently produced by enzymatic conversion of glycolonitrile [20–22] with the advantage of producing fewer impurities and saving energy.

C. glutamicum is neither a natural GLA producer, nor is it able to utilize GLA as sole carbon source or in combination with glucose [17]. However, it had been shown to possess the route as well as the regulatory mechanisms to enable the production of GLA by means of glyoxylate conversion. The prerequisite for initial GLA production is either the heterologous expression of a strong glyoxylate reductase, or the deletion of the *aceB*-gene, encoding malate synthase [MS], both in combination with co-feeding of glucose plus acetate. The uptake of acetate activates the glyoxylate shunt [23], providing glyoxylate as direct precursor for GLA (Figure 1). In 2014 Zahoor and co-workers investigated the plasmid-borne expression of the genes *yiaE* and *ycdW*, encoding the two glyoxylate reductases YiaE and YcdW from *Escherichia coli*, in *C. glutamicum* [17]. In accordance with Núñez et al., who demonstrated already in 2002 a tenfold higher affinity and a fourfold higher catalytic efficiency of YcdW with glyoxylate as substrate in comparison to YiaE [24], the group obtained with 0.47 ± 0.03 $U{\cdot}mg^{-1}$ a fivefold higher glyoxylate reductase activity upon *ycdW* expression, than compared to *yiaE* (0.10 ± 0.00 $U{\cdot}mg^{-1}$). Therefore, a *C. glutamicum* mutant expressing *ycdW* was used to produce GLA. This first mutant *C. glutamicum* (pVWEx1-*ycdW*) was cultivated in minimal medium with a mixture of glucose and acetate. Being the immediate precursor for the reaction catalysed by YcdW, the acetate-induced allocation of glyoxylate resulted in production of 0.42 $g{\cdot}l^{-1}$ GLA [17]. Subsequently, the *aceB* gene, encoding MS, was deleted to increase the glyoxylate supply. The MS catalyses the condensation of glyoxylate with acetyl-CoA to malate [25], and competes with YcdW for the substrate glyoxylate (Figure 1). Deactivation of the MS reaction results in glyoxylate accumulation, which in turn probably activates an alternative route for GLA production. Therefore, deletion of *aceB* in *C. glutamicum* wildtype [WT] enabled GLA production (0.3 $g{\cdot}l^{-1}$) and the combination of both *ycdW* expression and *aceB* deletion resulted in a high GLA titre of 4.0 $g{\cdot}l^{-1}$ [17]. The promising strain *C. glutamicum* Δ*aceB* (pVWEx1-*ycdW*) was then further engineered to improve GLA production by enhancing the availability of isocitrate as substrate for isocitrate lyase [ICL], the initial enzyme of the glyoxylate shunt ([26]; Figure 1). The ICL competes with the reactions of the citric acid cycle, more precisely the decarboxylation of isocitrate to 2-oxoglutarate by the isocitrate dehydrogenase [ICD] ([27]; Figure 1). To decrease the activity of the ICD, the start codon of the corresponding gene was mutated from ATG to the rarely used GTG codon. This reduced the ICD activity from 1.37 ± 0.02 $U{\cdot}mg^{-1}$ (WT) by five times to 0.25 ± 0.05 $U{\cdot}mg^{-1}$ (icd^{GTG}) [17]. In subsequent fermentations with *C. glutamicum* icd^{GTG} Δ*aceB* (pVWEx1-*ycdW*), harbouring all above described modifications, 5.3 $g{\cdot}l^{-1}$ GLA were produced in 51 h from a mixture of glucose and acetate with equivalent carbon moles of each substrate, i.e., 60 and 180 mM, respectively ([17]; Table 1).

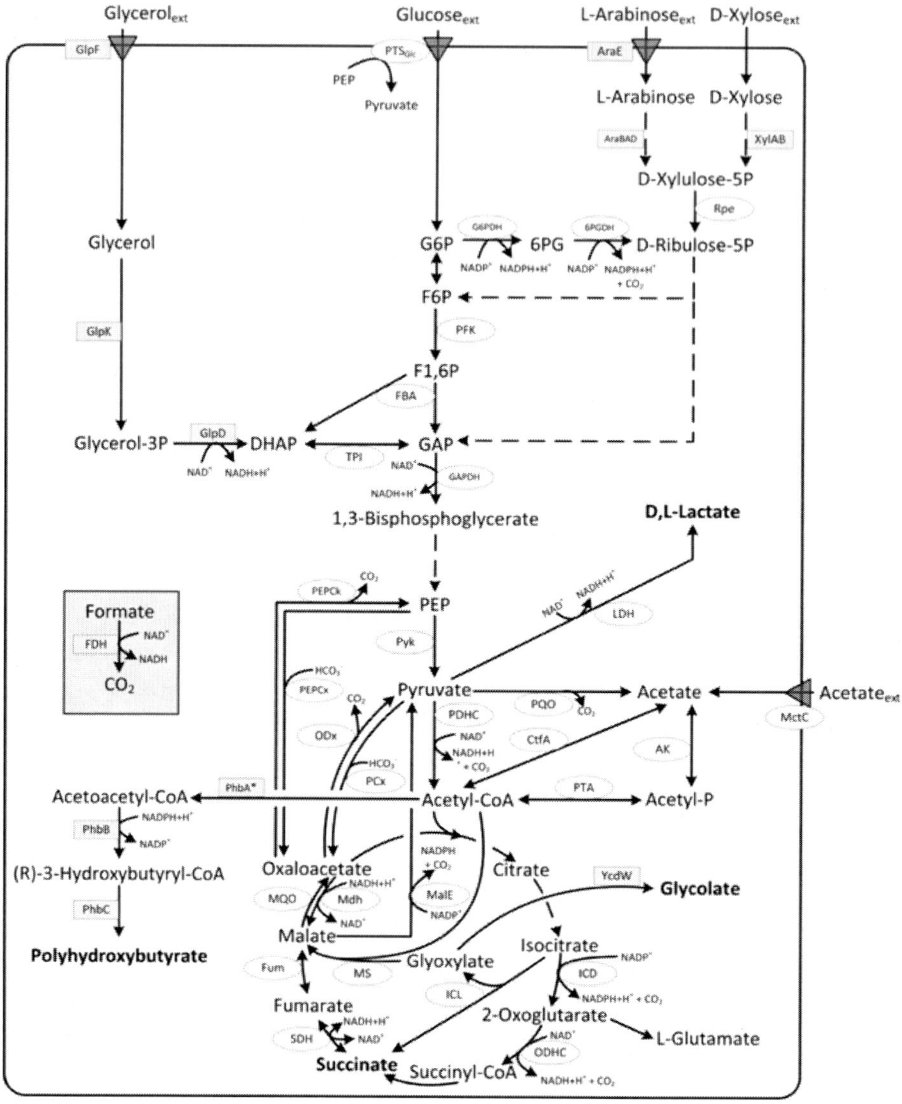

Figure 1. Schematic presentation of the central carbon metabolism of *C. glutamicum* including pathways for the degradation of applied carbon sources (glucose, glycerol, acetate, L-arabinose and D-xylose) for the production glycolate, D,L-lactate, succinate and polyhydroxybutyrate. Ellipses represent enzymes and transport systems present in *C. glutamicum*. Grey rectangles represent heterologous enzymes. Abbreviations: 6PG, 6P-gluconate; 6PGDH, 6PG-dehydrogenase; AK, acetate kinase; AraBAD (from *E. coli*), ribulokinase, arabinose isomerase, L-ribulose-5-phosphate 4-epimerase; AraE (from *E. coli*), L-arabinose transporter; CtfA, CoA transferase A; DHAP, dihydroxyacetone-P; F1,6P, fructose-1,6P; F6P, fructose-6P; FBA, fructosebisphosphate aldolase; FDH (from *M. vaccae*), formate dehydrogenase; FUM, fumarase; GAP, glyceraldehyde-3P; GAPDH, GAP dehydrogenase; GlpD (from *E. coli*), glycerol-3P dehydrogenase; GlpF (from *E. coli*), glycerol facilitator; GlpK (from *E. coli*), glycerol kinase; G6P, glucose-6P; G6PDH, G6P dehydrogenase; ICD, isocitrate dehydrogenase; ICL, isocitrate lyase; LDH (native, or from *L. delbrueckii*), L- and D-lactate dehydrogenase, respectively; MalE, malic enzyme; MctC, monocarboxylic acid transporter; Mdh, malate dehydrogenase; MQO, malate:quinone oxidoreductase; MS, malate synthase; ODHC, 2-oxoglutarate dehydrogenase complex; ODx, oxaloacetate decarboxylase; PCx, pyruvate carboxylase; P, phosphate; PDHC, pyruvate dehydrogenase complex; PEP, phosphoenolpyruvate; PEPCk, PEP carboxykinase; PEPCx, PEP carboxylase; PFK, phosphofructokinase; PhbA (from *R. eutropha*; *AACS

from *Streptomyces* sp.), β-ketothiolase; PhbB (from *R. eutropha*), acetoacetyl-CoA reductase; PhbC (from *R. eutropha*), polyhydroxybutyrate synthase; Pyk, pyruvate kinase; PQO, pyruvate:quinone oxidoreductase; PTA, phosphotransacetylase; PTS, phosphotransferase system; Rpe, ribulose-5-phosphate epimerase; SDH, succinate dehydrogenase; TPI, triosephosphate isomerase; XylAB (from *E. coli*), xylose isomerase, xylulokinase; YcdW (from *E. coli*), glyoxylate reductase.

Although, being the first GLA producing *C. glutamicum* mutant reported so far, the performance of *C. glutamicum icd*^{GTG} Δ*aceB* (pVWEx1-*ycdW*) is already competitive with regard to other microorganisms, such as *E. coli* MG0-GP pHHD/*ycdW-aceA-aceK* which was shown to produce 1.4 $g \cdot l^{-1}$ GLA ([28]; Table 1), but also several recombinant yeasts. In 2013, Koivistoinen et al. succeeded in engineering *Saccharomyces cerevisiae* as well as *Kluyveromyces lactis* for GLA production [29]. By expression of a high affinity glyoxylate reductase from *Arabidopsis thaliana* and deletion of the genes encoding MS and the cytosolic form of ICD the engineered *S. cerevisiae* strain produced just up to about 1.4 $g \cdot l^{-1}$ of GLA in a medium containing D-xylose and ethanol. Similar metabolic engineering in *K. lactis* resulted in a much higher GLA titre of 14.8 $g \cdot l^{-1}$, feeding the same substrates ([29]; Table 1). In summary, with 30% of the maximal theoretical yield [17], GLA production with *C. glutamicum* leaves some room for improvement, but it is on a par with the currently known microbial processes. The high density of recent publications underlines the high demand for sustainable microbial GLA production and therefore, the development of such mutants and processes.

LACTATE

With an additional carbon in the chain in comparison to GLA, is lactate (LA; lactic acid; 2-hydroxypropanoic acid) a chiral AHA with two optical isomers: L-(+)-LA and D-(-)-LA. Both L- and D-isomers are widely used for pharmaceutical, cosmetic, leather, textile, chemical, biomedical and food industries, as well as for green solvent fabrication [30–32]. Naturally, is LA a product of homofermentative, heterofermentative or mixed acid fermentation, and hence a very popular microbial product. Therefore, it is not astonishing that LA was already early used by men to acidify and conserve grocery, e.g. sauerkraut. Though, D-LA, in contrary to L-LA, is at times harmful to human metabolism and can result in acidosis and decalcification [33]. To date, LA is still of great importance, as the annual production of 450000 t [34] and the vast number of publications and patents concerning production and application of this acid show. The major part of industrially produced LA is polymerized to biodegradable fibres and polymers (poly lactic acid; PLA), which exhibit outstanding mechanical properties, such as higher flexibility as polystyrene or easy printability [35, 36]. Besides, PLA is resistant to fatty foods and dairy products, has barrier properties to flavours and aromas, and a good heat sealability, turning it into a substitute for polyethylene terephthalates of great economical interest [37,38]. Biodegradation of PLA is based on its susceptibility to moisture by hydrolysis and subsequent digestion of lower-molecular-weight LA oligomers by microorganisms, a process which can be easily realised by composting [36]. PLA was traditionally produced from L-LA, but the polymer has some limitations due to its low melting temperature of 180°C. Complexes of L-PLA and D-PLA have a higher melting point at 230°C [39]. Therefore, stereocomplexation improves the mechanical performance, thermal resistance, and hydrolysis resistance of PLA based materials [35], elevating the value of D-LA as chemical building block, which is otherwise inappropriate for food applications. Monomeric LA is produced synthetically by hydrolysis of lactonitrile, always yielding an optically inactive racemic mixture of L- and D-forms. The 10% market share of petro-chemically produced LA reported in 2000 [30] did not change over the years, whereas total LA production capacity increased almost 6-fold at the same time [34, 40]. That means, industrial LA production is majorly developed for and carried out by microbial processes. Reasons for this development are environmental concerns, lower production and energy costs, as well as the high product purity of just one isomer that can be obtained in microbial processes, clearly highlighting the high value of LA as product in the production chain from renewable resources [34, 41].

In the past and still today, wildtype and recombinant LA bacteria and recombinant *E. coli* have been mainly employed for the production of both D- and L-LA (reviewed by Lee et al., 2011 [9]). However, lactic acid bacteria usually demand complex nutrients, such as whey, corn steep liquor, or yeast extract, and hence are relatively cost-intensive during cultivation and product purification. The use of less fastidious organisms, such as the above mentioned *E. coli*, but also *S. cerevisiae* and *C. glutamicum* represents a more cost-efficient alternative. Therefore, the named microorganisms have been exploited for efficient LA production in the past [31]. The production of LA with *C. glutamicum* can be achieved by utilizing its reductive metabolism. In general, *C. glutamicum* grows aerobically or anaerobically in the presence of oxygen or nitrate, respectively as final electron acceptor [42, 43]. In absence, or under strong limitation of the acceptor growth is arrested, but the bacteria still have the capability to ferment C6 sugars to L-LA, succinate and acetate as major products [44]. Furthermore, it was observed that the glucose consumption rate per cell was higher than grown under aerobic conditions. This effect is accompanied by an increase in gene expression of several glycolytic and organic acid synthesis related enzymes, including *gapA*, *pgk*, *tpi*, *ppc*, *mdh* and *ldhA* [45]. In return, the flux via the pentose phosphate pathway is reduced from 69% (aerobic) to ~2% [46,47]. Upregulation of some of the glycolytic genes is controlled by SigB, a group-2 sigma factor which is expressed under oxygen deprivation conditions and known to positively regulate glucose metabolism [48].

On basis of this metabolic switch during oxygen limitation, the potential of *C. glutamicum* for industrial production of LA was first studied by Inui et al. in 2004 with wildtype strain R [49]. Aerobically cultivated *C. glutamicum* R cells were transferred to lidded bottles and incubated deprived of aeration. Without any genetic changes, 220 mM (i.e., 19.8 $g \cdot l^{-1}$) L-LA were produced from 130 mM glucose during oxygen-free incubation. Succinate appeared to be the major by-product, with a titre of about 20 mM (i.e., 2.4 $g \cdot l^{-1}$). After deletion of the *ldhA* gene, L-lactate dehydrogenase [LDH] activity, and consequently LA production was completely abolished, showing that *ldhA* encodes the only isoenzyme for the conversion of pyruvate to LA in *C. glutamicum*. Addition of bicarbonate to the medium led to an increased succinate production, but also to increased rates of glucose consumption and LA production. It is presumable that bicarbonate enhances the carbon fixation by PEP carboxylase [PEPCx], which is the main enzyme of anaplerosis under oxygen deprivation conditions [49], and therefore enhances the reductive flux towards succinate (Figure 1). An assumption that was confirmed in 2014 by a study of Radoš et al., revealing a 16-fold enhanced carboxylation of PEP and pyruvate, in presence of CO_2 under oxygen deprivation conditions [47]. The enhanced reductive flow is supposed to result in an increased $NAD^+/NADH$ ratio and, probably as consequence of a derepression of the glyceraldehyde-3-

phosphate dehydrogenase [GAPDH] gene *gapA*, in an increased glucose consumption and eventually, increased LA production rate [49]. However, Okino and co-workers could show by testing different bicarbonate levels in the medium that a maximum yield of LA from glucose (1.79 mol·mol^{-1}; i.e., 0.90 g·g^{-1}) is obtained in absence of bicarbonate [50]. Finally, performing a high cell density (30 g_{CDW}·l^{-1} [CDW = cell dry weight]) fed-batch fermentation with *C. glutamicum* R, in absence of bicarbonate, resulted after 8 h in 574 mM L-LA (i.e., 51.1 g·l^{-1}), with scarce formation (<10 mM) of succinate and acetate as side-products ([50]; Table 1). By addition of 400 mM bicarbonate the L-LA titre could be raised to more than 1 M (i.e., 97.5 g·l^{-1}) within a reduced process time of just 6 h. However, also the concentrations of the by-products succinate and acetate were very high then with 192 mM (i.e., 23 g·l^{-1}) and 50 mM (i.e., 3 g·l^{-1}), respectively ([50]; Table 1).

The results show that even without genetic modification of *C. glutamicum*, the L-LA titre and the $Y_{P/S}$ (1.79 mol L-LA per mol of glucose; i.e., 0.90 g·g^{-1}) from glucose are highly competitive (Table 1). Thus, the best known metabolically engineered L-LA-producing *E. coli* strain SZ85 (*pflB*, *frdBC*, *adhE*, *ackA*, *ldhA*::*ldhL*; with *ldhL* = D-LDH from *Pediococcus acidilactici*) accumulated 505 mM L-LA (i.e., 64 g·l^{-1}) with a $Y_{P/S}$ of 1.9 mol per mol of glucose (i.e., 0.95 g·g^{-1}) ([51]; Table 1). Also recombinant yeast strains from the genera *Saccharomyces* and *Kluyveromyces* have been employed for L-LA production (reviewed by Okano et al., 2009 [31]). To our knowledge, the currently best L-LA producer is *S. cerevisiae* YIBO-*pdc1/5*Δ (devoid of pyruvate decarboxylases 1 and 5; expressing a double copy of the bovine L-LDH). It produced 0.9 M (i.e., 82.3g·l^{-1}) L-LA in 192 h with a $Y_{P/S}$ of up to 1.63 mol L-LA per mol of glucose (i.e., 0.82 g·g^{-1}) and ethanol as major by-product ([52]; Table 1). An example for an effective fermentation process using LA bacteria was reported by González-Vara et al., in 2000 [53]. With the homofermentative *Lactobacillus casei* subsp. *casei* DMS 20011 a continuous fermentation process was established, using cell recycling and near infrared spectroscopy-aided cell culture monitoring. To reduce the growth-inhibiting effects of the product, LA extraction was coupled to the cultivation, using ion-exchange resins. The experimental setup led to a $Y_{P/S}$ of 1.96 mol per mol of glucose (i.e., 0.98 g·g^{-1}) and an excellent productivity of 1.5 M·h^{-1} L-LA (i.e., 138 g·l^{-1}·h^{-1}) ([53]; Table 1). Taking into consideration how big the efforts are to achieve such a good production performance with *L. casei*, being a natural producer, *C glutamicum* is a promising and uncomplicated host alternative for the production of L-LA.

To establish the production of D-LA with *C. glutamicum*, the native L-LDH has to be substituted with an isoenzyme catalysing the stereospecific conversion of pyruvate to the D-isomer (Figure 1). Okino et al. implemented this strategy in 2008 by deletion of the *ldhA* gene in the wildtype of *C. glutamicum* R, and subsequent heterologous expression of the D-LDH from *Lactobacillus delbrueckii* under control of the *tac* promoter (plasmid pCRB204) [54]. Analogue to Inui et al., 2004 [49], the production process was performed under oxygen-deprivation conditions in a high cell density system (60 g_{CDW}·l^{-1}). Within 30 h *C. glutamicum* R produced Δ*ldhA*/pCRB204 about 1.34 M D-LA (i.e., 119.3 g·l^{-1}) with a $Y_{P/S}$ of 1.73 mol per mol of glucose (i.e., 0.86 g·g^{-1}) (Table 1). But also significant amounts of succinate (146 mM) and acetate (52 mM) were detected [54]. Albeit side-products appeared to be substantial, the exchange of the stereospecific LDHs was already sufficient to develop an efficient production process for accumulation of D-LA with optical product purity greater than 99.9% [54]. A similar mutant was developed by Jia and co-workers [55], substituting the L-LDH of *C. glutamicum* Res 167 with the D-LDH of *Lactobacillus bulgaricus*. Also in this case, growth arrested production under oxygen deprivation was the setup of choice, however established as a simplified process: The cells were first grown in an aerated bioreactor and then put under oxygen deprivation by stopping aeration and reducing the stirring speed. With regard to productivity (198.9 M of D-LA in 16 h (i.e., 17.92 $g \cdot l^{-1}$)) [55], the performance was worse compared to other results presented in this chapter. However, simplifying the production process might present a desirable economic advantage and should be further pursued.

In most studies on LA production, LA producer strains are usually further tailored to reduce by-product formation. In contrast to this classical approach, Tsuge et al. intended in 2013 to accelerate LA production by overexpression of glycolytic genes on the chromosome of a strain very similar to the one used by Okino et al. in 2008 [54, 56]. This basic strain, called *C. glutamicum* R CRZ2/pCRB215, is devoid of L-LDH, and PEPCx activity, by deletion of the respective genes, and carries plasmid pCRB215 (2-3fold higher copy number than pCRB204), expressing the D-LDH from *L. delbrueckii* under control of the *gapA* promoter, leading in total to a 5.9-fold increased D-LDH activity when compared to pCRB204 [56]. Direct comparison in a short-term bioprocess (10 h) revealed a 16% higher D-LA concentration using pCRB215 instead of pCRB204 in *C. glutamicum* R CRZ2, probably due to a slight increase in glucose consumption and $Y_{P/S}$. An independent overexpression of the genes encoding the ten glycolytic enzymes glucokinase [GLK], phosphoglucose isomerase, phosphofructokinase [PFK], fructose-bisphosphate aldolase [FBA], GAPDH, triosephosphate isomerase [TPI], phosphoglycerate kinase, phosphoglycerate mutase, enolase and pyruvate kinase in *C. glutamicum* R CRZ2/pCRB215 revealed a beneficial effect on D-LA production by enhancing reactions upstream of glycerate-1,3-bisphosphate during a short-term bioprocess (10 h). Consequently, the effects of overexpressing each one of the genes coding for GLK, PFK, FBA, GAPDH, and TPI (Figure 1, with the exception of GLK) were investigated in a long-term bioprocess (80 h). Strain *C. glutamicum* R LPglc192/pCRB215, harbouring a second copy of the gene for GLK, showed the best performance achieving a titre of almost 1.5 M (i.e., 130 $g \cdot l^{-1}$) D-LA from 870 mM glucose (Table 1), with an optical purity above 99.9% after 80 h [56]. A superior yield (93% compared to 83% of the maximum theoretical possible) was achieved only by strain *C. glutamicum* R LPglc206/pCRB215, harbouring a second copy of the gene coding for TPI [56]. However, the other production parameters were less advantageous. In addition, the results revealed that the LDH reaction is not the rate-limiting step during D-LA production with this *C glutamicum* mutant.

In comparison to other organisms the introduced *C. glutamicum* mutants came up with a good performance even though, e.g. D-LA production with recombinant *E. coli* strains is already further optimized and therefore more efficient. Close to the turn of the millennium, D-LA production reached titres of at 691 mM (i.e., 62.2 $g \cdot l^{-1}$) and 528 mM (i.e., 48 $g \cdot l^{-1}$) with yields of 1.76 and 1.92 mol D-LA per mol of glucose (i.e., 0.89 and 0.95 $g \cdot g^{-1}$), using *E. coli* JP203 (*pta*, *ppc*) and SZ63 (W3110; *pflB*, *frdBC*, *adhE*, *ackA*) [57,58], respectively. These strains formed approximately half of the titres obtained with the best *C. glutamicum* mutants. Meanwhile, to our knowledge the best D-LA producing *E. coli* strain ALS974 (*aceEF*, *pfl*, *poxB*, *pps*, *frdABCD*) achieved yields close to, or even better than many LA bacteria, but required a less demanding medium composition [59]. In detail *E. coli* ALS974 produced 1.5 M D-LA (i.e., 138.1 $g \cdot l^{-1}$) with a productivity of 69.9 $mM \cdot h^{-1}$ (i.e., 6.3 $g \cdot l^{-1} \cdot h^{-1}$) and a $Y_{P/S}$ of 1.98 mol D-LA per mol of glucose (0.99 $g \cdot g^{-1}$) ([59]; Table 1). In contrast, although being resistant to lower pH, yeasts cannot compete with the different bacteria in any relevant process result. For example *S. cerevisiae* strain YILM-2B, which was devoid of pyruvate decarboxylase 1 and expressed a double copy of D-LDH from *Leuconostoc mesenteroides*, accumulated 682.7 mM (i.e., 61.5 $g \cdot l^{-1}$) of 99.9% optical pure D-LA in 72 h with a comparably low $Y_{P/S}$ of up to 1.22 mol D-LA per mol of glucose (i.e., 0.61 g per g of glucose) ([60]; Table 1). Still it is worth to mention these processes, since low pH fermentations facilitate downstream processing and minimize waste products [35]. Amongst LA bacteria, as the classical LA producers, the highest known $Y_{P/S}$ for D-LA so far was reported for the thermotolerant *L. delbrückii* subsp. *lactis* QU 41 with 2 mol D-LA per mol of glucose (i.e., 1.0 g per g of glucose). This represents 100% of the maximum theoretical yield and was achieved by using a continuous culture. First, a high cell density was accumulated by cell recycling and then 223.1 mM (i.e., 20.1 $g \cdot l^{-1}$) D-LA were produced with a productivity of up to 199.8 $mM \cdot h^{-1}$ (i.e., 18 $g \cdot l^{-1} \cdot h^{-1}$) and an optical purity above 99.9% [61].

The economic performance of LA production to obtain a sustainable industrial scale process is in dependence on the use of low-cost carbon sources, in particular from biorenewable resources. The production procedures described above are merely based on glucose as carbon source. Of course is glucose a renewable carbon source - however, the required pre-treatment and purification processes reduce the economic efficiency for glucose and comparable carbon sources as substrate. Therefore, utilization of alternative, crude materials is of great interest [14]. *C. glutamicum* does not naturally consume agro-industrial biomass, such as starch, whey, straw, or hemi- and lignocellulose. But especially lignocellulose, consisting mainly of cellulose (35 – 50% of the plant dry weight), lignin (5 – 30% of the plant dry weight) and hemicellulose (20 – 35% of the plant dry weight), which in turn consists mainly of glucose and to a significant portion of C5 sugars (xylose and arabinose), is the most abundant and potentially attractive source of alternative feedstock for bioconversion processes [62,63]. Some microorganisms, such as *E. coli*, are naturally able to consume the majority of sugars in the mixtures resulting from saccharification from hemicellulose. Nevertheless, *C. glutamicum* is not included in this group and consequently needs to be metabolically engineered to expand its feedstock spectrum. Thus, extension of the substrate acceptance of *C. glutamicum* towards cheap, easily accessible and renewable monomeric as well as polymeric carbon sources is an ongoing field of intensive research [64, 65]. Also in the view of LA production, efforts have been made to extend the substrate spectrum. Subsequently, the consumption of different carbon sources under oxygen deprivation conditions and their influence on the acid production has been determined. Plasmid-borne introduction of the xylose isomerase and xylulokinase genes (*xylA* and *xylB* from *E. coli*), and of L-arabinose isomerase, L-ribulokinase, L-ribulose-5-phosphate-4-epimerase, and L-arabinose transporter genes (*araA*, *araB*, *araD*, and *araE* from *E. coli*), to *C. glutamicum* R (Figure 1), respectively, enabled both aerobic growth and organic acid production with resting cells under oxygen deprivation conditions, using xylose, and accordingly arabinose, as sole carbon source [66,67]. However, the use of these substrates resulted in a shift of the $Y_{P/S}$ in favour of succinate, the major by-product of LA production (reviewed by Wieschalka et al. 2013 [14]). Although production from these C5 sugars is desirable, further engineering of the production strains is recommendable to suppress the negative side-effects. This is especially true also for other feedstock compounds, whose utilization pathways were already successfully introduced by metabolic engineering in *C. glutamicum*, e.g. glycerol, D-cellobiose, mannose and starch [56, 68–70].

SUCCINATE

Succinate is a C4-dicarboxylic acid, which until recently was produced in amounts of 20 - 40 kt per year, mainly from maleic anhydrite, a derivative of the petrochemistry [6]. Only a small portion used in the food industry was derived from biotechnological processes [71]. Nowadays, as the big chemical companies are trying to establish more petroleum-independent production lines based on sustainable feedstock, succinate represents a potent alternative to maleic anhydrite based chemistry, allowing its quick integration into current processes. In 2004, The US Department of Energy officially classified succinic acid as one of the top 12 value added chemicals from biomass [72] and since then a number of companies already established, or are on their way to start biotechnological production of succinate [7, 73]. The range of possible succinate derived products is vast, starting from important solvents and polymers reaching up to surfactants and de-icing salts [6]. The most important application field is the production of key bulk chemicals and organic solvents, such as tetrahydrofuran [THF], 1,4-butanediol [BDO] and γ-butyrolactone [GBL] with a total estimated production volume (in the year of 2007) of 500 kt/a [8]. THF and GBL are commonly used solvents and serve as precursors in numerous chemical syntheses, e.g. in the production of herbicides, rubber additives, and pharmaceuticals. However, BDO is by far the most important succinate-derived product, as it has a constantly growing market with estimated global production of 2 Mt/a in 2012 [73], covering important application field, as the production of high performance resins polybutylene terephthalate (PBT) and polybutylene succinate (PBS) or the synthesis of THF and GBL. The polymer synthesis, in particular of PBS and its modified derivatives (mPBS), is one of the most interesting and promising application fields for succinate-based chemical industry, as their synthesis requires not only succinate-derived BDO, but also pure succinic acid itself [74]. The resulting polymers possess similar characteristics to polypropylene and polyvinylchloride and allow the use of the existing processing equipment. This application would of course increase enormously the volume of the possible succinic acid market by estimated 25 Mt/a [32]. Other possible important derivatives, which can be produced from sustainable succinic acid feedstock, are the groups of pyrrolidones, sulfosuccinates and sulfosuccinamates, as well as those of succinate esters and salts [6]. The pyrrolidones are applied as solvents in chemical synthesis or as precursor/additives in cosmetical, pharmaceutical and food industry, whereas sulfosuccinates and sulfosuccinamates represent interesting succinate-based surfactants. The group of the succinate esters and salts shows a very diverse application field, ranging from pH-modifiers, flavour enhancers, antibacterial agents for food industry, to additives in fuels or de-icers.

Of course, the vast number of different applications for biotechnological derived succinic acid was not the only reason for the industry to focus on this organic acid. Another important issue is the accumulated knowledge and experience with natural or metabolically engineered organisms, which are able to produce succinate fast and cheap from variety of different carbon sources [7]. *C. glutamicum* represents one of these species. The bacterium was extensively studied and engineered to become one of the most promising host for the biotechnological production of succinate [14].

C. glutamicum possesses a complete TCA cycle, which shows high flux levels during aerobic growth on glucose, acetate and the combination of both substrates [75]. Succinate is a major intermediate in this central metabolic pathway and it participates as important metabolite in the glyoxylate shunt during utilization of acetate [23]. In general, for the aerobic production of succinate starting with glucose, the substrate is assimilated via glycolysis and/or the pentose phosphate pathway (Figure 1). At the level of PEP/pyruvate in an optimal case, the resulting C3 backbones are distributed equimolar between the carboxylation and decarboxylation reactions of the PEP/pyruvate node. In particular, one C3 backbone is utilized in the anaplerotic carboxylation reaction either by PEPCx or pyruvate carboxylase [PCx] to form oxaloacetate [OAA]. The second C3 intermediate is decarboxylated by pyruvate dehydrogenase complex [PDHC] to yield acetyl-CoA (Figure 1). In the latter case, citrate synthase [CS] obtains balanced supply of precursor for the condensation of OAA and acetyl-CoA to form citrate. Citrate is than further metabolized in the subsequent reactions of the oxidative TCA cycle going through the intermediates isocitrate, 2-oxoglutarate, and succinyl-CoA to yield succinate. Isocitrate has a special role in this metabolic route as starting substrate for the glyoxylate shunt where ICL can directly cleave isocitrate into succinate and glyoxalate. The resulting glyoxylate is then condensed with acetyl-CoA to form malate by the MS. With the oxidation of malate by malate:quinone oxidoreductase [MQO] the pool of OAA is replenished in anaplerotic fashion (Figure 1). Taken as a standalone metabolic route the glyoxylate shunt represents a second pathway for production of succinate. Independently of the selected pathway, for the aerobic succinate production the further oxidation of succinate to fumarate should be prevented by the obligatory inactivation of the succinate dehydrogenase [SDH] complex via deletion of the corresponding *sdhCAB* genes. This mutation converts succinate from intermediate of the oxidative TCA cycle to its final product [76].

Under anaerobic conditions *C. glutamicum* accumulates succinate as final electron acceptor during fermentation applying the reductive TCA cycle [44, 49]. The production pathway starts from glucose, which is assimilated in glycolysis and delivers per mol substrate 2 mol of PEP/pyruvate together with 2 mol of each NADH and ATP [45]. To proceed to succinate an obligatory carboxylation step is performed, condensing PEP/pyruvate with HCO_3^- to form OAA. The resulting OAA is further reduced to malate by the soluble NADH-dependent malate dehydrogenase [Mdh] regenerating one mol of NAD^+. The fumarase transforms malate into fumarate, which is then reduced by the SDH complex using MQH_2 as electron donor to succinate [77]. In total, this anaerobic succinate production route consumes 2 mol of redox equivalents and HCO_3^- per mol of starting C3 backbone (PEP/pyruvate). To convert the second C3 backbone into succinate additional 2 reducing equivalents are needed. However, the 2 mol of NADH accumulated during glycolysis are already completely used for the production of the first mol of succinate and thus, the anaerobic succinate production is limited, by the availability of reducing cofactors, to 1 mol succinate per mol glucose [7]. An alternative route to utilize the second C3 backbone (in form of pyruvate) is to further oxidize it by the PDHC to acetyl-CoA. In this pathway, the acetyl-CoA can be then used for the generation of ATP, producing stepwise acetate over the phosphotransacetylase [PTA] and acetate kinase [AK] reactions (Figure 1) [78]. The oxidation step from pyruvate to acetyl-CoA delivers an additional mol NADH, which can be reoxidized via production of succinate or by the reduction of 1 mol pyruvate to 1 mol LA [45]. The production of LA is the preferred route for regeneration of NAD^+, as it is a one-step reaction, which can easily balance the reduced NADH output of glycolysis. However, the combination of reductive TCA cycle and functional glyoxylate shunt under anaerobic conditions would also present an optimal metabolic strategy for accumulation of succinate, allowing balanced regeneration of cofactors and increased yields up to theoretically possible 1.66 mol product/mol glucose [79].

The three above-mentioned metabolic pathways present the potential of *C. glutamicum* to offer different strategies for the biotechnological production of succinate, using each of the pathways as stand-alone solutions or combining them, respectively. In the following, the current achievements in the strain development for succinate production with *C. glutamicum* will be presented.

A unique knowledge dataset is available for the aerobic lifestyle and the corresponding central metabolic pathways, fluxes and metabolic regulation in *C. glutamicum* [80–82]. Therefore, it was just a matter of time for the scientific community to take advantage of this knowledge and apply it for the construction of aerobic succinate producer strains of *C. glutamicum*. Litsanov and co-workers were the first in 2012, which presented a *C. glutamicum* strain (BL-1 (pAN6-pycP458Sppc)) capable of accumulating 82 mM succinate in minimal medium from glucose with an yield of 0.36 mol product per mol substrate [76]. The first step to construct this strain was the inactivation of the SDH complex by deletion of the corresponding operon *sdhCAB*. The resulting strain *C. glutamicum* Δ*sdhCAB* showed for the first time aerobic succinate accumulation (40 mM) in the supernatant. However, the major product was acetate (124 mM). Consequent metabolic engineering steps for the reduction of acetate production in *C. glutamicum* Δ*sdhCAB* included the deletion of the genes coding for enzymes involved in acetate production pathways during glucose (overflow-)metabolism, such as PTA in combination with AK as well as CoA transferase [CtfA] and pyruvate:menaquinone oxidoreductase [PQO] (enzymes shown in Figure 1). The resulting strain *C. glutamicum* BL-1 indeed accumulated less acetate (22 mM) and more succinate (66 mM). Although the acetate secretion could be drastically reduced in the resulting strain *C. glutamicum* BL-1 and led also to increased succinate yield (0.28 mol/mol glucose), it could not be eliminated. In addition, three different strategies to increase the anaplerotic flux and to improve the supply of OAA a key metabolite to ensure the optimal flux into the oxidative TCA cycle were tested. At first, a mutant version of the PCx with the known beneficial mutation Pro458Ser, which helped to increase lysine yield [83] in *C. glutamicum* was overexpressed to increase the carboxylation of pyruvate to OAA. However, as this manipulation did not show any positive effect, the PEPCx was additionally overexpressed on the same plasmid,increasing the succinate yield to 0.36 mol per mol of glucose. The activation of the glyoxylate shunt as anaplerotic sequence and additional succinate production pathway by overexpression of the corresponding enzymes ICL and MS was also elucidated as alternative to enhance production, but did not show any positive effects on the accumulation of the desired product. Most probably, the accumulation of succinate and other TCA intermediates, which might be involved in the regulation of the corresponding enzymes, hindered the application of PCx and the glyoxylate shunt for additional improvement of the production. In addition, a process optimization step in which the nitrogen source was limited to restrict cell growth aimed to decouple aerobic succinate production from growth. Indeed this led to additional succinate titre and yield increase with 90 mM and 0.45 mol per mol of glucose, respectively [76]. This study presented for the first time the great potential of *C. glutamicum* for the aerobic production of succinate in minimal medium.

Apparently inspired by this work, Zhu and co-workers put efforts to further develop the concept of aerobic succinate production based on *C. glutamicum*, as they aimed in eliminating carbon loss by acetate secretion and increasing the TCA flux [84]. In this study the group started with the already described metabolic engineering approach, constructing a basic strain ZX1, devoid of the known acetate producing pathways (see BL-1 strains above) and improved anaplerosis by promoter exchange of the PEPCx and PCx coding genes with the strong constitutive promotor of the superoxide dismutase [SOD]. In addition, the *ldhA* gene was deleted to eliminate the possibility of LA formation as by-product after further strain engineering. Of course, an obligatory *sdhCAB* deletion was also accomplished to allow the aerobic succinate accumulation. As shown previously, succinate production with this strain background was high with yield of 0.43 mol per mol of glucose (in CGIII complex medium). However, succinate production was always accompanied by acetate secretion. To counter this loss of carbon, an acetyl-CoA synthetase gene *acsA* from *B. subtilis* was functionally expressed, indeed eliminating acetate accumulation. The succinate production yield of the resulting strain ZX1(pEacsA) was also slightly increased to 0.5 mol per mol of glucose. However, in this case pyruvate accumulated as major by-product in the supernatant. This fact indicated a possible bottleneck at the entrance of the oxidative TCA cycle. To enhance the flux towards succinate the *gltA* gene encoding the CS was overexpressed together with *acsA*. The resulting strain ZX1(pEacsAgltA) produced maximal succinate titre and yield of 0.61 mol per mol of glucose excreting no acetate and only one third of the previously detected pyruvate concentrations. In a fed-batch process this strain reached succinate concentration of 241 mM with yield of 0.63 mol per mol of glucose and average volumetric productivity of 3.55 mM/h performing as the best known aerobic succinate producer based on *C. glutamicum* [84].

The metabolic engineering of suitable aerobic succinate producers based on *C. glutamicum* and the model substrate glucose, proposed many possibilities to test alternative carbon sources for succinate production. Litsanov and co-workers investigated and established the aerobic succinate production from glycerol to convert it from waste product of the biodiesel production to important feedstock for the aerobic succinate production [85]. Therefore, the producer strain BL-1 was equipped with an overexpression plasmid encoding the genes for glycerol facilitator *glpF,* glycerol kinase *glpK* and glycerol-3-P dehydrogenase *glpD* cloned as operon from *E. coli* to enable the utilization of glycerol in *C. glutamicum* (Figure 1). The resulting strain BL-1/pVWEx1-glpFKD produced 79 mM succinate reaching 42% of the theoretical possible yield in minimal medium from glycerol with high volumetric productivity of 3.59 $mM{\cdot}h^{-1}$. In similar fashion the improved aerobic succinate producer ZX1 (pEacsAgltA) was transformed with the plasmid pXaraBAD, harbouring the operon for arabinose utilization from *E. coli* (*araB* - ribulokinase, *araA* - arabinose isomerase, and *araD* - L-ribulose-5-phosphate 4-epimerase) to allow the conversion of arabinose, a sugar highly abundant in hemicellulosic waste into succinate [86]. The resulting succinate producer ZX1 (pXaraBAD, pEacsAgltA) accumulated 74.4 mM succinate with a yield of 0.58 mol per mol of arabinose in minimal medium, presenting again the potential of *C. glutamicum* to convert industrial feedstock into interesting bulk products with high yield. A more exotic concept study recently faced the question how can succinate be produced by *C. glutamicum* from carbon dioxide sequestration and sunlight energy [87]. For that purpose, a consolidated process including phototrophically grown microalgae *Chlamydomonas reinhardtii* as CO_2 fixing element and a starch degrading succinate producer strain based on *C. glutamicum* was developed. The microalgae were grown in photobioreactors with CO_2 supply to accumulated biomass, which after mechanical pre-treatment was fed to *C. glutamicum* based succinate producer strain BL-1 transformed with pBlAmyS overexpression plasmid. The resulting BL-1-pBlAmyS strain was able to utilize starch due to the expression and secretion of amylase AmyS from *B. licheniformis* to the supernatant and accumulated succinate with a yield of

0.28 mol per mol total sugars from microalgae biomass during a batch cultivation [87]. Of course, the described process is far from being economically viable. However, it presents interesting opportunity for integration of the currently developed succinate producer based on *C. glutamicum* for possible future processes.

As the examples above show, *C. glutamicum* has great potential for the aerobic succinate production, achieving high product yield especially on minimal medium and from different biotechnological relevant carbon sources. The aerobic succinate production mode with *C. glutamicum* possesses two obvious advantages, including high production rates due to the general high throughput of the oxidative metabolism in *C. glutamicum*, as well as established process know-how from the amino acid production history, which are often lacking in the described native production hosts [7, 14]. However, the aerobic succinate production in general has a major drawback due to its limited maximal theoretical product yield of 1 mol per mol of glucose related to the oxidative nature of the synthesis pathway [75]. In addition, the above mentioned yield is of course not experimentally possible with growing cells, investing part of the substrate to accumulate biomass and might be only achievable in complex medium as shown for the best *E. coli* based aerobic succinate producer [88]. On the other hand, the anaerobic production pathway for succinate synthesis described as production route for *C. glutamicum* in the beginning of this section allows the accumulation of theoretically 2 mol succinate per mol of glucose, consuming additionally 2 mol of HCO_3^- and 2 mol of NADH [89]. However, as described in the LA section of this chapter *C. glutamicum* shows only limited anaerobic growth using nitrate as final electron acceptor due to the accumulation of toxic nitrite, which cannot be further metabolized [42, 43]. Although nitrate reduction is not sufficient to sustain efficient anaerobic growth to desirable biomass concentrations, aerobically grown *C. glutamicum* cells readily ferment glucose to significant amounts of LA, succinate and acetate under anoxic conditions. The study of Inui and co-workers [49], which determined in more detail the conditions and the parameters of the anaerobic metabolism in *C. glutamicum*, laid the background for the development of the succinic acid production with *C. glutamicum*. This particular study characterized the possible pathways for the accumulation of the different fermentation products using different single deletion mutants. Inactivation of L-LDH and the overexpression of PCx were identified as suitable targets for rational engineering of an anaerobic succinate producer based on *C. glutamicum* as the combination of both factors enhanced two fold the succinate accumulation compared to the WT, simultaneously eliminating LA production. The successor study of Okino et al. defined the use of bicarbonate as additional key parameter to control the ratio of succinate to LA production in WT and developed a reliable production process [50]. The results showed the potential of *C. glutamicum* for accumulation of succinate by demonstrating high volumetric productivity of 99 $mM \cdot h^{-1}$, which however was dependent on the extremely high cell density applied in the experiment. A transcriptome study coupled to crude lysate enzyme measurements from cells incubated under oxygen deprivation proved the previous findings and rounded up the knowledge about the anaerobic central metabolic pathways in *C. glutamicum* [45].

Based on the available data the first anaerobic succinate producer strain *C. glutamicum* R Δ*ldhA*-pCRA717 was constructed by disrupting the gene encoding for the soluble L-LDH and combining the resulting mutant with an overexpression plasmid for the native PCx [90]. Under fed-batch conditions the production strain accumulated 1240 mM succinate with a yield of 1.4 mol per mol of glucose within 48 h, showing high volumetric productivity of 27 mM·h^{-1}. The HCO_3^- supply was provided by 1600 mM of sodium bicarbonate. Besides the main product succinate, 270 mM acetate accumulated as by-product, explaining indirectly the high succinate yield, which based on the described anaerobic succinate production pathways, should be limited by the availability of 2 mol NADH to 1 mol succinate per mol of glucose [89]. It could be assumed that the pyruvate oxidation to acetate in the described strain allowed the additional accumulation of reducing equivalents but obligatory increased the by-product synthesis as supported by the study about the acetate producing pathways in *C. glutamicum* under anaerobic conditions [78]. In addition, the yield was determined considering only the used glucose as substrate for the product accumulation without including the substrate used to grow the applied high-density biomass. Taking the whole process into account, a probable yield of 0.85 mol succinate/mol glucose was calculated [89]. Although, this first study presented already optimal production values for the synthesis of succinate with an anaerobic succinate producer based on *C. glutamicum* R it seemed that the high yield succinate production might always be coupled to increased acetate accumulation, to provide the necessary NADH equivalents for high product yields.

Litsanov and co-workers presented a solution to the above-mentioned problem, which aimed to reduce by-product accumulation level and to increase the supply of NADH equivalents within several metabolic engineering steps introduced in *C. glutamicum* ATCC 13032 WT strain [89]. The strategy involved the independent minimization of acetate production combined with the introduction of formate oxidation reaction [91] as additional electron donor to *C. glutamicum*, delivering one mol NADH for each mol formate oxidized to CO_2. The great advantage of this approach is that supplying 2 mol formate as co-substrate to each mol of glucose utilized by the cells, would enable the maximal succinate yield of 2 mol product per mol of glucose. As comparison, a basic strain based on ATCC 13032 WT was developed, combining the mandatory deactivation of L-LDH, to diminished lactate production, with overexpression plasmid pAN6-*pyc*^{P458S}, coding for the mutant version of the PCx^{P458S} to enhance the pyruvate carboxylation to OAA. This strain accumulated succinate with a yield of 1.16 mol per mol glucose. Consequently, this background was combined with disruptions for all known acetate producing pathway (PTA, AK, PQO, CtfA), as already discussed in the section on aerobic succinate production strains, to minimize the acetate accumulation. The resulting strain BOL-1, equipped with the carboxylation plasmid pAN6-*pyc*^{P458S}, showed 90% reduced acetate formation but also reduced succinate yield of 1 mol per mol of glucose, indicating again the possible connection between acetate formation and succinate yield beyond 1 mol per mol of glucose. The plasmid free strain BOL-2 with improved carboxylation was constructed by chromosomal integration of the *pyc*^{P458S} gene under the control of the strong constitutive promoter P_{tuf} in the *pta-ack* locus and showed no change in the succinate production parameters. To improve further the product yield, an additional supply of reduced cofactor NADH was provided by the oxidation of formate to CO_2 catalysed by FDH from *Mycobacterium vaccae* N10. To enable stable expression of this enzyme, the corresponding gene *fdh* was integrated chromosomally in the *pqo* locus again under the control of the P_{tuf} promoter in the BOL-2 background, resulting in strain BOL-3. The new strain exhibited a significantly increased succinate yield of 1.26 mol per mol of glucose with additional formate supply, when compared to 1.05 mol per mol of glucose, obtained without addition of formate. In fed-batch experiment the BOL-3 pAN6-*gap* strain overexpressing GAPDH to accelerate the glycolytic flux accumulated 1134 mM succinate from 679 mM glucose and 772 mM formate resulting in the highest succinate yield reported with *C. glutamicum* based producer of 1.67 mol per mol of glucose. The acetate by-product accumulation remained with 20 mM very low, proving the possibility to minimize the synthesis of this major by-product but not on costs of succinate yield. The succinate production from glucose with maximal yield was established successfully with a new concept of providing essential NADH supply by formate oxidation instead of pyruvate decarboxylation and subsequent acetate production. An additional advantage of the introduced formate oxidation reaction was the formation of CO_2, which might be used to supply HCO_3^- for the essential PEP/pyruvate carboxylation reactions during succinate synthesis.

In the meantime, as the above described study was prepared two independent reports about the identification of the succinate exporter protein SucE in *C. glutamicum* were released [92,93]. Huhn and co-workers used *in silico* approach to identify SucE comparing the genome of natural succinate producers with that of *C. glutamicum*. In the second study, a differential transcriptome analysis between microaerobic and aerobic conditions helped for the identification of SucE as succinate exporter protein. Both studies came to the conclusion that SucE is responsible for major part of the succinate export under anaerobic conditions in *C. glutamicum*. However, it is not the sole exporter protein as disruption of SucE impaired succinate accumulation in the supernatant but did not eliminate it completely.

Based on the available data, Zhu and co-workers used SucE overexpression to improve the succinate production and combined it with a novel route to optimize the succinate yield by activating the glyoxylate shunt as second product synthesis pathway [94]. This strategy would allow to use the NADH equivalents delivered during pyruvate decarboxylation by the PDHC for succinate production in the reductive TCA cycle whereas the resulting acetyl residues carried by CoA as cofactor would be condensed to produce succinate. The combination of reductive TCA cycle and glyoxylate shunt would reduce the demand of NADH per mol succinate to 1.25 mol per mol product as calculated in a similar study performed for *E. coli* and would allow a maximal succinate yield of 1.66 mol per mol of glucose [79]. In detail, Zhu et al. manipulated a *C. glutamicum* Δ*ldhA* mutant for minimized acetate production and enhanced PEP/pyruvate carboxylation in similar fashion as already described for their corresponding aerobic succinate producer ZX1 (see aerobic succinate production section). The resulting strain SA1 was transformed with plasmid pXaceAB for overexpression of the ICL and MS genes. However the new strain SA3 showed no significant improvement in the succinate synthesis, accumulating nearly the same titre of 115 mM succinate with a yield of 1.23 mol per mol of glucose as the parent strain SA1 although crude lysate activity measurements proved the functional expression of ICL and MS. A plasmid-borne overexpression of the CS gene could improve the flux into the glyoxylate shunt and enhance the succinate yield to 1.34 mol per mol of glucose resulting in strain SA4. The additional overproduction of the succinate exporter SucE in the strain SA4 increased further the succinate yield to 1.43 mol per mol of glucose. In fed-batch studies, the final strain SA5 accumulated 926 mM succinate with a total yield of 1.32 mol per mol of glucose and an average volumetric productivity of 9.4 $mM \cdot h^{-1}$. This was a very good performance for a *C. glutamicum* based succinate producer, but it was not able to outcompete the *E. coli* strain SBS550MG/pHL413, which applied the same strategy of dual pathway utilization for the anaerobic succinate production [79]. Nevertheless, this study proved the feasibility of the glyoxylate shunt activation as alternative succinate production pathway in combination with reductive TCA cycle under anaerobic conditions in *C. glutamicum*.

A different strategy to improve the anaerobic succinate yield was presented to the public during the writing of this chapter. The corresponding study aimed in the utilization of the membrane-bound transhydrogenase PntAB from *E. coli* expressed in L-LDH negative *C. glutamicum* mutant. However, a strain overproducing PntAB could show only 13% improvement of the succinate yield compared to the reference [95]. The same effect was also described previously in a study about organic acid production with *C. glutamicum* under microaerobic conditions, where overproduction of PntAB (membrane-bound) but not of UdhA (soluble) transhydrogenase from *E. coli* could slightly increase the succinate production [96]. A possible reason for the minimal effect of transhydrogenases overproduction on anaerobic succinate accumulation is the low production of NADPH due to the limited flux in the PPP as supported by ^{13}C-NMR analysis of the succinate production under anoxic conditions [77].

A completely different approach to optimize the succinate yield was pursued in the study presented by Jeon and co-workers, who showed that the succinic acid titre could be increased up to 1.5-fold in WT *C. glutamicum* by simple optimization of media components with the help of response surface methodology [97].

The developed set of different approaches for the succinate production with *C. glutamicum* under anaerobic conditions covered almost all possible strategies to optimize product synthesis using the model substrate glucose. However, *C. glutamicum* can grow on a vast number of different carbon sources and it was additionally engineered to increase its substrate spectrum with several industrially relevant substrates such as starch or the hemicellulose sugars arabinose and xylose [69,98–100]. Arabinose was the first alternative carbon sources used to show its general potential for the anaerobic production of organic acid [100]. *C. glutamicum* strains expressing the *araBAD* operon from *E. coli* accumulating 40, 37 and 11% of the respective theoretical yield of succinate, LA and acetate. An alternative study proposed an optimal way to degrade starch by presenting α-amylase (AmyA) from *Streptococcus bovis* on the cell surface of *C. glutamicum* WT cells [56].

The resulting strain was used consequently for the production of organic acid from starch under anaerobic conditions accumulating in total 980 mM LA and 118 mM succinate. A metabolically engineered *C. glutamicum* strain with deletions of the *ldhA*, *pta* and *ack* genes overexpressing the *xylA* (xylose isomerase) and *xylB* (xylulokinase) genes from *E. coli* which enable the utilization of xylose was used in a recent study to elucidate the anaerobic succinate production from corn cob hydrolysate (366 mM xylose and 22 mM glucose) [101]. The respective strain NC-2 was able to accumulate 347 mM succinate within 48 h, supporting the idea that *C. glutamicum* represents an optimal platform for the utilization of hemicellulose derived pentose sugars for the production of succinate.

POLYHYDROXYBUTYRATE

Polyhydroxybutyrate [PHB, poly(3-hydroxybutyrate)] is a representative of polyhydroxyalkanoates [PHA], a class of biologically produced polyesters that may serve as biodegradable plastics, and that can be yielded from renewable resources. According to Jo et al. is PHB the most widely studied and characterized PHA [10].

Table 1. Comparison of the most prominent organic acid production processes with various industrial relevant microorganisms

Strain	Medium, carbon source	Maximal titer in mM ($g \cdot l^{-1}$)	$Y_{P/S}$ in $mol_{product}$ per $mol_{substrate}$ ($g \cdot g^{-1}$)	Productivity[a] in $mM \cdot h^{-1}$ ($g \cdot l^{-1} \cdot h^{-1}$)	Reference
Glycolic acid					
C. glutamicum icd[GTG] Δ*aceB* (pVWEx1-*ycdW*)	minimal medium, glucose, acetate	070 (5.3)	n.s.[b] (0.18)	1.4 (0.10)	[17]
E. coli MG0-GP pHHD/*ycdW-aceA-aceK*	minimal medium, glucose	18.4 (1.4)	0.33 (0.14)	n.s.[b]	[28]
K. lactis H3986	complex medium, D-xylose, ethanol	194.6 (14.8)	[c]	1.02 (0.08)	[29]
L-Lactic acid					
C. glutamicum R	minimal medium, glucose	0574 (51.1)	1.42 (0.70)	71.8 (6.39)	[50]
C. glutamicum R	minimal medium, glucose, bicarbonate	1061 (94.4)	1.79 (0.89)	176.8 (15.74)	[50]
E. coli SZ85	minimal medium, glucose	0505 (44.9)	1.90 (0.94)	7.2 (0.64)	[51]
S. cerevisiae YIBO-*pdc1/5*Δ	complex medium, glucose	914 (82.3)	1.63 (0.82)	4.8 (0.43)	[52]
L. casei	complex medium, glucose	n.s.[b]	1.96 (0.98)	1531 (138)	[53]
D-Lactic acid					
C. glutamicum R Δ*ldhA* pCRB204	minimal medium, glucose	1340 (119.3)	1.73 (0.86)	44.5 (3.96)	[54]
C. glutamicum R LPglc192/pCRB215	minimal medium, glucose	1446 (130.3)	1.66 (0.83)	18.1 (1.63)	[56]
E. coli ALS974	minimal medium, L-isoleucine glucose	1533 (138.1)	1.98 (0.99)	69.9 (6.30)	[59]
S. cerevisiae YILM-2B	complex medium, glucose	682.7 (61.5)	1.22 (0.61)	9.5 (0.85)	[60]
L. delbrückii subsp. *lactis* QU 41	complex medium, glucose	223 (20.1)	2.00 (1.00)	199.8 (18)	[61]
Succinic acid (anaerobic)					
C. glutamicum R Δ*ldhA* pCRA717	minimal medium, glucose, bicarbonate	1240 (146.4)	1.40 (0.92)	27 (3.18)	[90]
C. glutamicum ELB-P	minimal medium, glucose	330 (39)	1.02 (0.67)	5.6 (0.66)	[14]

Strain	**Medium, carbon source**	**Maximal titer in mM ($g \cdot l^{-1}$)**	**YP/S in molproduct per molsubstrate ($g \cdot g^{-1}$)**	**Productivitya in $mM \cdot h^{-1}$ ($g \cdot l^{-1} \cdot h^{-1}$)**	**Reference**
C. glutamicum BOL-3/pAN6-*gap*	saline, glucose, formate, bicarbonate	1134 (133.8)	1.67 (1.09)	21 (2.47)	[89]
C. glutamicum SA5	minimal medium, glucose, bicarbonate	926 (109)	1.32 (0.86)	9.4 (1.11)	[94]
C. glutamicum	minimal medium, corn cob hydrolysate, bicarbonate	345 (40.8)	1.06 (0.69)	7.1 (0.83)	[101]
E. coli SBS550MG/pHL413	complex medium, glucose	330 (39)	1.61 (1.05)	10 (1.18)	[79]
E. coli KJ134	minimal medium, glucose	606 (71.5)	1.53 (1.00)	6.4 (0.76)	[102]
Succinic acid (aerobic)					
C. glutamicum BL-1-pSbAmyA	minimal medium, algal biomass	4.2 (0.5)	0.28 (0.18)	n.s.[b]	[87]
C. glutamicum ZX pX*araBAD* pE*acsAgltA*	minimal medium, arabinose	74 (8.7)	0.58 (0.38)	0.36 (0.04)	[86]
C. glutamicum ZX pE*acsAgltA*	minimal medium, glucose	241 (28.4)	0.63 (0.41)	3.55 (0.42)	[84]
C. glutamicum BL-1/pAN6-*pyc*^{P458S}*ppc*	minimal medium, glucose	90 (10.6)	0.45 (0.29)	0.8 (0.09)	[76]
C. glutamicum BL-1 pVWEx1-*glpFKD*	minimal medium, glycerol	79 (9.3)	0.21 (0.13)	3.6 (0.42)	[103]
E. coli HL27659k(pKK313)	complex medium, glucose,	60 (7.1)	0.95 (0.62)	2.3 (0.27)	[104]

[a] during production phase; [b] n.s. = not specified; c = see reference for detailed information

Although, up to now Gram-negative bacteria were majorly deployed as a workhorse for PHB-production [10], the Gram-positive organisms are emerging recently as preferred hosts. Especially the production of endotoxins by Gram-negatives complicates or even prevents the use of the product for medical or food applications, the major fields intended for PHAs [105]. The alternative production hosts used so far, are e.g. *Bacillus*, *Streptomyces* [106], and *C. glutamicum*.

Incipient with the first report on PHB production with *C. glutamicum* in 2006, the standard production system was for several years based on establishing a biosynthetic pathway by introduction of the *phbCAB* operon derived from *Ralstonia eutropha* [105]. Starting substance for PHB formation are two molecules of acetyl-CoA to generate acetoacetyl-CoA (reversibly catalysed by β-ketothiolase; PhbA). Acetoacetyl-CoA is then reduced by acetoacetyl-CoA reductase (PhbB) to 3-hydroxybutyryl-CoA. Finally, PHB synthase (PhbC) polymerizes 3-hydroxybutyryl-CoA to PHB via ester bond formation (Figure 1). In the very first approach from 2006, plasmid pPS-*phbCAB*, carrying the *phbCAB* operon from *R. eutropha*, encoding the named enzymes, was transformed into *C glutamicum*. Expression of the genes under control of a native *C glutamicum* cell surface protein gene promoter led to intracellular accumulation of PHB inclusion granules of 22.5% (*w/w*) with a number average molecule weight [M_n] of $2.1 \cdot 10^5$ and a polydispersity of 1.63 ([105]; Table 2). Since the PHB content with the recombinant *C glutamicum* strains was still lower than with *R. eutropha* or recombinant *E. coli*, Jo et al. intended in 2007 to further engineer their initial production strain for improved PHB formation [10]. First, the *phbC* gene and its gene product PhbC were modified in parallel approaches by codon optimization and point mutation, respectively. Both the codon optimization for the first five N-terminal amino acids, and the amino acid exchange G4D (a product of evolutionary engineering and previously described to be beneficial for PHB production in *E. coli* [107–109]), improved PHB formation with *C glutamicum* from 22.5% (*w/w*) to 28.6% (*w/w*) and 29.9% (*w/w*), respectively [10]. Subsequently, the production process was further optimized by tuning the dosage of all three genes in the operon. It was discovered that fortification of the monomer-supply by additional *phbA* and *phbB* expression on a second vector was highly effective. Combination of monomer-supply reinforcement with *phbC* codon optimization (yielding strain *C. glutamicum* pPS-*phbC'AB* + pVC7-*AB*) resulted in total in a 2.3-fold increase (in comparison to *C. glutamicum* pPS-*phbCAB*) of PHB production to 52.5% (*w/w*) (Table 2). Additional upturn of the *phbC* expression was assumed to be negative for PHB production. With elevation of the PHB synthase amount in the cell, the molecular weights of the polymers decreased [10], which is equivalent with a decrease of the product quality for practical use. Recently, further studies on PHB have been made on basis of *C. glutamicum* pPS-*phbCAB*. Under the premise that substituting the reversible condensation of acetoacetyl-CoA with an irreversible reaction, the loss of PHB precursor should be prevented, and consequently PHB production improved. For this purpose Matsumoto et al. exchanged *phaA* with *nphT7* (*AACS*) from *Streptomyces* sp. strain CL190, encoding an acetyl-CoA synthase which catalyses the irreversible condensation of acetyl-CoA and malonyl-CoA to acetoacetyl-CoA under formation of CO_2 ([110]; Figure 1). The new mutant produced PHB amounts (19.7% (*w/w*)) slightly lower than the parental strain (Table 2). But especially the application of the same strategy to *E. coli* revealed a strong disadvantage of AACS in comparison to PhaA [110], probably due to an inefficient supply with malonyl-CoA. Anyway, *E. coli* is to date a very efficient production host for PHB. Recombinant strains can produce up to 90% (*w/w*) (reviewed by Leong et al., 2014 [111]). E.g. accumulated *E. coli* JM109 pGEM-*phaR*$_{YB4}$*C*$_{YB4}$*AB* (PhaRC$_{YB4}$) 80% (*w*/*w*) PHB with a M_n of $20 \cdot 10^3$ and a polydispersity of 2.1 ([112]; Table 2). And even PHB production from glycerol, as example for a low-cost carbon source, was very successfully performed under microaerobic conditions with strain *E. coli* CT1061-PHB, giving 51% (*w*/*w*) [113].

In 2007 it was shown that the production of PHB in recombinant strains, similar to *C. glutamicum* pPS-*phbCAB*, glutamate production is promoted due to increased glucose and oxygen consumption, as well as a reduction in the formation of intermediates and by-products [114]. In contrast, the group of Jo et al. reported in 2009 that glutamate and PHB production could be combined indeed, but the process was either favouring glutamate or PHB as product, in dependence on the conditions the cells were exposed to. In detail, a two-step cultivation was established with *C. glutamicum* pPS-*phbCAB*, shifting the production pattern from glutamate production under biotin limitation to PHB production, by increasing biotin levels in the medium [115]. The intension behind this setup is to enhance economic viability of PHB as commercial bioployester, reducing the production costs due to parallel formation of a second high-value compound. An important advantage for this strategy is that after separation of the cells from the broth, PHB remains in the biomass fraction and glutamate in the spent culture broth, facilitating the purification of the desired products. In a dual production process 113.3 mM glutamate (i.e., 18 $g \cdot l^{-1}$) and after biotin induction 36% (*w*/*w*) PHB were obtained. The overall performance of PHB production in this setup was thereby 1.5-fold increased in comparison to a single production system (23% (*w*/*w*)). As reason for the metabolic switch, Jo et al. presumed a strong influence of biotin on the flux of acetyl-CoA, the starting substance of PHB formation, towards glutamate [115].

The field of PHB production with *C glutamicum* needs, and is still under the focus of research. It is for sure a long way to establish industrially rentable PHB production processes with *C glutamicum*. The above described approaches show that the intension is to develop a setup that is commercially and ecologically worthwhile and therefore in general valuable for the global community. Hence, first studies face already the fact that production from inexpensive, untreated substrates is also desirable and will reduce the production costs. E.g. produced Song et al. successfully PHB from starch, and even slightly more effective than from glucose, using an α-amylase cell-surface displaying system [116]. Furthermore, several efforts have been made to produce copolymers consisting of PHB and several other PHAs such as 3-hydroxyvalerate or LA [117,118]. Therefore, an interesting future can be predicted in the area of economic PHB production with *C glutamicum*.

Table 2. PHB biosynthesis by recombinant *C. glutamicum* and *E. Coli*

Strain	Cell mass in $g_{CDW} \cdot l^{-1}$	PHB content in % (*w*/*w*)	M_W	M_n	Polydispersity ($M_W \cdot M_n^{-1}$)	Reference
Polyhydroxybutyrate						
C. glutamicum pPS-*phbCAB*	12.5	22.5	$3.4 \cdot 10^5$	$2.1 \cdot 10^5$	1.6	[105]
C. glutamicum pPS-*phbC'AB* + pVC7-*AB*	13.6	52.5	$14 \cdot 10^5$	$7.1 \cdot 10^5$	2.0	[10]
C. glutamicum pPS-*phbCACSB*	9.0	19.7	n.s.[a]	n.s.[a]	n.s.[a]	[110]
E. coli JM109 pGEM-*phaR*$_{YB4}$*C*$_{YB4}$*AB* (PhaRC$_{YB4}$)	9.3	80.0	$20 \cdot 10^3$	$42 \cdot 10^3$	2.1	[112]

[a] n.s. = not specified

Conclusion

To sum up, the above overview showed the immense potential of *C. glutamicum* to either naturally accumulate or to produce after metabolic optimization each of the industrially important organic and hydroxy acids - glycolate, lactate, succinate and polyhydroxyalkanoate. Although the production yields in the cases of glycolate and polyhydroxyalkanoates are still lagging behind the reference hosts, showing the necessity for further optimisations, those strains established to produce lactate and succinate compete very efficiently with natural producers and currently used biotechnological hosts (Table 1). Moreover, the constant race to improve the organic and hydroxy acid production in *C. glutamicum* gave the scientific community the opportunity to access some fundamental questions about the global metabolism, transcriptome or proteome of *C. glutamicum* in otherwise not considered physiological situations, such as anaerobic or microaerobic conditions, as well as pH or dicarboxylic acid induced stress [45, 77, 119, 120]. Especially the last study presents an important contribution in understanding and enhancing the stress resistance of *C. glutamicum* against organic and hydroxy acid induced stress during biotechnological production.

In conclusion, it seems that the field for the production of organic and hydroxy acids with *C. glutamicum* will soon compete for at least the same attention from the scientific and industrial community as the classical amino acid production research area, considering the constantly increasing industrial attention, the number of recent studies and the high quality of the described results. It is obvious that the presented achievements in the chapter above, proved once more that *C. glutamicum* is indeed a unique platform organism for the tailor-made metabolic engineering and accumulation of desired products.

References

[1] Hatti-Kaul R, Törnvall U, Gustafsson L, Börjesson P. Industrial biotechnology for the production of bio-based chemicals – a cradle-to-grave perspective. *Trends Biotechnol* 2007;25:119–24. doi:10.1016/j.tibtech.2007.01.001.

[2] Babilas P, Knie U, Abels C. Cosmetic and dermatologic use of alpha hydroxy acids. *JDDG J Dtsch Dermatol Ges* 2012;10:488–91. doi:10.1111/j.1610-0387.2012.07939.x.

[3] Kornhauser A, Coelho SG, Hearing VJ. Applications of hydroxy acids: classification, mechanisms, and photoactivity. *Clin Cosmet Investig Dermatol CCID* 2010;3:135–42. doi:10.2147/CCID.S9042.

[4] Fratila A, Uerlich M. Fruchtsäurepeeling Die dermatologische und ästhetisch-korrektive Anwendung von Alpha-Hydroxysäuren (AHA). *Hautarzt* 1999;50:448–60. doi:10.1007/s001050050942.

[5] Hermann BG, Patel M. Today's and tomorrow's bio-based bulk chemicals from white biotechnology: a techno-economic analysis. *Appl Biochem Biotechnol* 2007;136:361–88.

[6] Cukalovic A, Stevens CV. Feasibility of production methods for succinic acid derivatives: a marriage of renewable resources and chemical technology. *Biofuels Bioprod Biorefining* 2008;2:505–29. doi:10.1002/bbb.105.

[7] Litsanov B, Brocker M, Oldiges M, Bott M. Succinic Acid. In: Bisaria V, Kondo A, editors. *Bioprocess. Renew. Resour. Commod. Bioprod. 1st ed., John Wiley & Sons, Inc.;* 2014, p. 435–72.

[8] McKinlay JB, Vieille C, Zeikus JG. Prospects for a bio-based succinate industry. *Appl Microbiol Biotechnol* 2007;76:727–40. doi:10.1007/s00253-007-1057-y.

[9] Lee JW, Kim HU, Choi S, Yi J, Lee SY. Microbial production of building block chemicals and polymers. *Curr Opin Biotechnol* 2011;22:758–67. doi:10.1016/j.copbio.2011.02.011.

[10] Jo S-J, Matsumoto K, Leong CR, Ooi T, Taguchi S. Improvement of poly(3-hydroxybutyrate) [P(3HB)] production in *Corynebacterium glutamicum* by codon optimization, point mutation and gene dosage of P(3HB) biosynthetic genes. *J Biosci Bioeng* 2007;104:457–63. doi:10.1263/jbb.104.457.

[11] Keshavarz T, Roy I. Polyhydroxyalkanoates: bioplastics with a green agenda. *Curr Opin Microbiol* 2010;13:321–6. doi:10.1016/j.mib.2010.02.006.

[12] Lee SY, Chang HN. Production of poly(3-hydroxybutyric acid) by recombinant Escherichia coli strains: genetic and fermentation studies. *Can J Microbiol* 1995;41 Suppl 1:207–15.

[13] Taguchi S, Doi Y. Evolution of polyhydroxyalkanoate (PHA) production system by "Enzyme Evolution": successful case studies of directed evolution. *Macromol Biosci* 2004;4:145–56. doi:10.1002/mabi.200300111.

[14] Wieschalka S, Blombach B, Bott M, Eikmanns BJ. Bio-based production of organic acids with *Corynebacterium glutamicum. Microb Biotechnol* 2013;6:87–102. doi:10.1111/1751-7915.12013.

[15] Rendon M, Cardona LM, Bussear EW, Benitez AL, Colón LE, Johnson LA. Successful treatment of moderate to severe melasma with triple-combination cream and glycolic acid peels: a pilot study. *Cutis* 2008;82:372–8.

[16] Bergfeld W, Tung R, Vidimos A, Vellanki L, Rernzi B, Stanton-Hicks U. Improving the cosmetic appearance of photoaged skin with glycolic acid. *J Am Acad Dermatol* 1997;36:1011–3. doi:10.1016/S0190-9622(97)80290-3.

[17] Zahoor A, Otten A, Wendisch VF. Metabolic engineering of *Corynebacterium glutamicum* for glycolate production. *J Biotechnol n.d.* doi:10.1016/j.jbiotec.2013.12.020.

[18] Loder JD. *Process for manufacture of glycolic acid, n.d.*

[19] Shattuk MT. *Continuous glycolic acid process.* 2443482, 1948.

[20] He Y-C, Xu J-H, Su J-H, Zhou L. Bioproduction of glycolic acid from glycolonitrile with a new bacterial isolate of *Alcaligenes sp.* ECU0401. *Appl Biochem Biotechnol* 2010;160:1428–40. doi:10.1007/s12010-009-8607-y.

[21] Panova A, Mersinger LJ, Liu Q, Foo T, Roe DC, Spillan WL, et al. Chemoenzymatic synthesis of glycolic acid. *Adv Synth Catal* 2007;349:1462–74. doi:10.1002/adsc.200700061.

[22] Wu S, Fogiel AJ, Petrillo KL, Jackson RE, Parker KN, DiCosimo R, et al. Protein engineering of nitrilase for chemoenzymatic production of glycolic acid. *Biotechnol Bioeng* 2008;99:717–20. doi:10.1002/bit.21643.

[23] Gerstmeir R, Wendisch VF, Schnicke S, Ruan H, Farwick M, Reinscheid D, et al. Acetate metabolism and its regulation in *Corynebacterium glutamicum. J Biotechnol* 2003;104:99–122. doi:10.1016/S0168-1656(03)00167-6.

[24] Núñez MF, Kwon O, Wilson TH, Aguilar J, Baldoma L, Lin ECC. Transport of L-lactate, D-lactate, and glycolate by the LldP and GlcA membrane carriers of *Escherichia coli*. *Biochem Biophys Res Commun* 2002;290:824–9. doi:10.1006/bbrc.2001.6255.

[25] Reinscheid DJ, Eikmanns BJ, Sahm H. Malate synthase from *Corynebacterium glutamicum*: sequence analysis of the gene and biochemical characterization of the enzyme. *Microbiol Read Engl* 1994;140 (Pt 11):3099–108.

[26] Reinscheid DJ, Eikmanns BJ, Sahm H. Characterization of the isocitrate lyase gene from *Corynebacterium glutamicum* and biochemical analysis of the enzyme. *J Bacteriol* 1994;176:3474–83.

[27] Eikmanns BJ, Rittmann D, Sahm H. Cloning, sequence analysis, expression, and inactivation of the *Corynebacterium glutamicum icd* gene encoding isocitrate dehydrogenase and biochemical characterization of the enzyme. *J Bacteriol* 1995;177:774–82.

[28] Martin CH, Dhamankar H, Tseng H-C, Sheppard MJ, Reisch CR, Prather KLJ. A platform pathway for production of 3-hydroxyacids provides a biosynthetic route to 3-hydroxy-γ-butyrolactone. *Nat Commun* 2013;4:1414. doi:10.1038/ncomms2418.

[29] Koivistoinen OM, Kuivanen J, Barth D, Turkia H, Pitkanen J-P, Penttila M, et al. Glycolic acid production in the engineered yeasts *Saccharomyces cerevisiae* and *Kluyveromyces lactis*. *Microb Cell Factories* 2013;12:82. doi:10.1186/1475-2859-12-82.

[30] Hofvendahl K, Hahn–Hägerdal B. Factors affecting the fermentative lactic acid production from renewable resources1. *Enzyme Microb Technol* 2000;26:87–107. doi:10.1016/S0141-0229(99)00155-6.

[31] Okano K, Tanaka T, Ogino C, Fukuda H, Kondo A. Biotechnological production of enantiomeric pure lactic acid from renewable resources: recent achievements, perspectives, and limits. *Appl Microbiol Biotechnol* 2009. doi:10.1007/s00253-009-2280-5.

[32] Bozell JJ, Petersen GR. Technology development for the production of biobased products from biorefinery carbohydrates—the US Department of Energy's "Top 10" revisited. *Green Chem* 2010;12:539. doi:10.1039/b922014c.

[33] Datta R, Tsai S-P, Bonsignore P, Moon S-H, Frank JR. Technological and economic potential of poly(lactic acid) and lactic acid derivatives. *FEMS Microbiol Rev* 1995;16:221–31. doi:10.1016/0168-6445(94)00055-4.

[34] Christensen CH, Rass-Hansen J, Marsden CC, Taarning E, Egeblad K. The renewable chemicals industry. *ChemSusChem* 2008;1:283–9. doi:10.1002/cssc.200700168.

[35] Straathof AJJ. Transformation of biomass into commodity chemicals using enzymes or cells. *Chem Rev* 2014;114:1871–908. doi:10.1021/cr400309c.

[36] Lunt J. Large-scale production, properties and commercial applications of polylactic acid polymers. *Polym Degrad Stab* 1998;59:145–52. doi:10.1016/S0141-3910(97)00148-1.

[37] Dodds DR, Gross RA. Chemicals from biomass. *Science* 2007;318:1250–1. doi:10.1126/science.1146356.

[38] Lorenz P, Zinke H. White biotechnology: differences in US and EU approaches? *Trends Biotechnol* 2005;23:570–4. doi:10.1016/j.tibtech. 2005.10.003.

[39] Ikada Y, Jamshidi K, Tsuji H, Hyon SH. Stereocomplex formation between enantiomeric poly(lactides). *Macromolecules* 1987;20:904–6. doi:10.1021/ma00170a 034.

[40] Sauer M, Porro D, Mattanovich D, Branduardi P. Microbial production of organic acids: expanding the markets. *Trends Biotechnol* 2008;26:100–8. doi:10.1016 /j.tibtech. 2007.11.006.

[41] Wee Y, Kim J, Ryu H. Biotechnological production of lactic acid and its recent applications. *Food Technol Biotechnol* 2006;44:163–72.

[42] Nishimura T, Vertès AA, Shinoda Y, Inui M, Yukawa H. Anaerobic growth of *Corynebacterium glutamicum* using nitrate as a terminal electron acceptor. *Appl Microbiol Biotechnol* 2007;75:889–97. doi:10.1007/s00253-007-0879-y.

[43] Takeno S, Ohnishi J, Komatsu T, Masaki T, Sen K, Ikeda M. Anaerobic growth and potential for amino acid production by nitrate respiration in *Corynebacterium glutamicum*. *Appl Microbiol Biotechnol* 2007;75:1173–82. doi:10.1007/s00253-007-0926-8.

[44] Dominguez H, Nezondet C, Lindley ND, Cocaign M. Modified carbon flux during oxygen limited growth of *Corynebacterium glutamicum* and the consequences for amino acid overproduction. *Biotechnol Lett* 1993;15:449–54. doi:10.1007/BF00129316.

[45] Inui M, Suda M, Okino S, Nonaka H, Puskás LG, Vertès AA, et al. Transcriptional profiling of *Corynebacterium glutamicum* metabolism during organic acid production under oxygen deprivation conditions. *Microbiol Read Engl* 2007;153:2491–504. doi:10.1099/mic.0.2006 /005587-0.

[46] Bartek T, Blombach B, Lang S, Eikmanns BJ, Wiechert W, Oldiges M, et al. Comparative ^{13}C-metabolic flux analysis of PDHC-deficient L-valine producing *Corynebacterium glutamicum*. *Appl Environ Microbiol* 2011. doi:10.1128/AEM.00575-11.

[47] Radoš D, Turner DL, Fonseca LL, Carvalho AL, Blombach B, Eikmanns BJ, et al. Carbon flux analysis by ^{13}C nuclear magnetic resonance to determine the effect of CO_2 on anaerobic succinate production by *Corynebacterium glutamicum*. *Appl Environ Microbiol* 2014;80:3015–24. doi:10.1128/AEM.04189-13.

[48] Ehira S, Shirai T, Teramoto H, Inui M, Yukawa H. Group 2 sigma factor SigB of *Corynebacterium glutamicum* positively regulates glucose metabolism under conditions of oxygen deprivation. *Appl Environ Microbiol* 2008;74:5146–52. doi:10.1128 /AEM. 00944-08.

[49] Inui M, Murakami S, Okino S, Kawaguchi H, Vertès AA, Yukawa H. Metabolic analysis of *Corynebacterium glutamicum* during lactate and succinate productions under oxygen deprivation conditions. *J Mol Microbiol Biotechnol* 2004;7:182–96. doi:10.1159/000079827.

[50] Okino S, Inui M, Yukawa H. Production of organic acids by *Corynebacterium glutamicum* under oxygen deprivation. *Appl Microbiol Biotechnol* 2005;68:475–80. doi:10.1007/s00253-005-1900-y.

[51] Zhou S, Shanmugam KT, Ingram LO. Functional replacement of the *Escherichia coli* D-(−)-lactate dehydrogenase gene (*ldhA*) with the L-(+)-lactate dehydrogenase gene (*ldhL*) from *Pediococcus acidilactici*. *Appl Environ Microbiol* 2003;69:2237–44. doi:10.1128/AEM.69.4.2237 -2244.2003.

[52] Ishida N, Saitoh S, Onishi T, Tokuhiro K, Nagamori E, Kitamoto K, et al. The effect of pyruvate decarboxylase gene knockout in *Saccharomyces cerevisiae* on L-lactic acid production. *Biosci Biotechnol Biochem* 2006;70:1148–53. doi:10.1271/bbb.70.1148.

[53] González-Vara y R. A, Vaccari G, Dosi E, Trilli A, Rossi M, Matteuzzi D. Enhanced production of L-(+)-lactic acid in chemostat by *Lactobacillus casei* DSM 20011 using ion-exchange resins and cross-flow filtration in a fully automated pilot plant controlled via NIR. *Biotechnol Bioeng* 2000;67:147–56. doi:10.1002/(SICI)1097-0290(20000120)67:2<147::AID-BIT4>3.0.CO;2-F.

[54] Okino S, Suda M, Fujikura K, Inui M, Yukawa H. Production of D-lactic acid by *Corynebacterium glutamicum* under oxygen deprivation. *Appl Microbiol Biotechnol* 2008;78:449–54. doi:10.1007/s00253-007-1336-7.

[55] Jia X, Liu P, Li S, Li S, Wen J. D-Lactic acid production by a genetically engineered strain *Corynebacterium glutamicum. World J Microbiol Biotechnol* 2011;27:2117–24. doi:10.1007/s11274-011-0675-9.

[56] Tsuge Y, Tateno T, Sasaki K, Hasunuma T, Tanaka T, Kondo A. Direct production of organic acids from starch by cell surface-engineered *Corynebacterium glutamicum* in anaerobic conditions. *AMB Express* 2013;3:72. doi:10.1186/2191-0855-3-72.

[57] Chang D-E, Jung H-C, Rhee J-S, Pan J-G. Homofermentative production of D- or L-lactate in metabolically engineered *Escherichia coli* RR1. *Appl Environ Microbiol* 1999;65:1384–9.

[58] Zhou S, Causey TB, Hasona A, Shanmugam KT, Ingram LO. Production of optically pure D-lactic acid in mineral salts medium by metabolically engineered *Escherichia coli* W3110. *Appl Environ Microbiol* 2003;69:399–407. doi:10.1128/AEM.69.1.399-407.2003.

[59] Zhu Y, Eiteman MA, DeWitt K, Altman E. Homolactate fermentation by metabolically engineered *Escherichia coli* Strains. *Appl Environ Microbiol* 2007;73:456–64. doi:10.1128/AEM.02022-06.

[60] Ishida N, Suzuki T, Tokuhiro K, Nagamori E, Onishi T, Saitoh S, et al. D-Lactic acid production by metabolically engineered *Saccharomyces cerevisiae. J Biosci Bioeng* 2006;101:172–7. doi:10.1263/jbb.101.172.

[61] Tashiro Y, Kaneko W, Sun Y, Shibata K, Inokuma K, Zendo T, et al. Continuous D-lactic acid production by a novel thermotolerant *Lactobacillus delbrueckii* subsp. lactis QU 41. *Appl Microbiol Biotechnol* 2011;89:1741–50. doi:10.1007/s00253-010-3011-7.

[62] Aristidou A, Penttilä M. Metabolic engineering applications to renewable resource utilization. *Curr Opin Biotechnol* 2000;11:187–98. doi:10.1016/S0958-1669(00)00085-9.

[63] Lynd LR, Weimer PJ, Zyl WH van, Pretorius IS. Microbial cellulose utilization: fundamentals and biotechnology. *Microbiol Mol Biol Rev* 2002;66:506–77. doi:10.1128/MMBR.66.3.506-577.2002.

[64] Rumbold K, van Buijsen HJ, Gray VM, van Groenestijn JW, Overkamp KM, Slomp RS, et al. Microbial renewable feedstock utilization. *Bioeng Bugs* 2010;1:359–66. doi:10.4161/bbug.1.5.12389.

[65] Blombach B, Seibold GM. Carbohydrate metabolism in *Corynebacterium glutamicum* and applications for the metabolic engineering of L-lysine production strains. *Appl Microbiol Biotechnol* 2010;86:1313–22. doi:10.1007/s00253-010-2537-z.

[66] Kawaguchi H, Vertès AA, Okino S, Inui M, Yukawa H. Engineering of a xylose metabolic pathway in *Corynebacterium glutamicum. Appl Environ Microbiol* 2006;72:3418–28. doi:10.1128/AEM.72.5.3418-3428.2006.

[67] Kawaguchi H, Sasaki M, Vertès AA, Inui M, Yukawa H. Engineering of an L-arabinose metabolic pathway in *Corynebacterium glutamicum. Appl Microbiol Biotechnol* 2008;77:1053–62. doi:10.1007/s00253-007-1244-x.

[68] Rittmann D, Lindner SN, Wendisch VF. Engineering of a glycerol utilization pathway for amino acid production by *Corynebacterium glutamicum. Appl Environ Microbiol* 2008;74:6216–22. doi:10.1128/AEM.00963-08.

[69] Sasaki M, Jojima T, Inui M, Yukawa H. Simultaneous utilization of D-cellobiose, D-glucose, and D-xylose by recombinant *Corynebacterium glutamicum* under oxygen-deprived conditions. *Appl Microbiol Biotechnol* 2008;81:691–9. doi:10.1007/s00253-008-1703-z.

[70] Sasaki M, Teramoto H, Inui M, Yukawa H. Identification of mannose uptake and catabolism genes in *Corynebacterium glutamicum* and genetic engineering for simultaneous utilization of mannose and glucose. *Appl Microbiol Biotechnol* 2011;89:1905–16. doi:10.1007/s00253-010-3002-8.

[71] Zeikus JG, Jain MK, Elankovan P. Biotechnology of succinic acid production and markets for derived industrial products. *Appl Microbiol Biotechnol* 1999;51:545–52. doi:10.1007/s002530051431.

[72] Werpy T, Petersen GR. Top value added chemicals from biomass volume I— results of screening for potential candidates from sugars and synthesis gas 2004.

[73] De Guzman D. Chemical industry awaits for bio-succinic acid potential. 2012. Available from: http://www.icis.com/Articles/2012/01/30 /9527521/chemical-industry-awaits-for-bio-succinic-acid-potential.html n.d.

[74] Bechthold I, Bretz K, Kabasci S, Kopitzky R, Springer A. Succinic acid: a new platform chemical for biobased polymers from renewable resources. *Chem Eng Technol* 2008;31:647–54. doi:10.1002/ceat. 200800063.

[75] Bott M. Offering surprises: TCA cycle regulation in *Corynebacterium glutamicum. Trends Microbiol* 2007;15:417–25. doi:10.1016/j.tim.2007. 08.004.

[76] Litsanov B, Kabus A, Brocker M, Bott M. Efficient aerobic succinate production from glucose in minimal medium with *Corynebacterium glutamicum. Microb Biotechnol* 2011. doi:10.1111/j.1751-7915.2011. 00310.x.

[77] Rados D, Turner DL, Fonseca LL, Carvalho AL, Blombach B, Eikmanns BJ, et al. Carbon flux analysis by ^{13}C nuclear magnetic resonance to determine the effect of CO_2 on anaerobic succinate production by *Corynebacterium glutamicum. Appl Env Microbiol* 2014;80:3015–24. doi:10.1128/AEM.04189-13.

[78] Yasuda K, Jojima T, Suda M, Okino S, Inui M, Yukawa H. Analyses of the acetate-producing pathways in *Corynebacterium glutamicum* under oxygen-deprived conditions. *Appl Microbiol Biotechnol* 2007;77:853–60.

[79] Sanchez AM, Bennett GN, San KY. Novel pathway engineering design of the anaerobic central metabolic pathway in *Escherichia coli* to increase succinate yield and productivity. *Metab Eng* 2005;7:229–39. doi:10.1016/j.ymben.2005.03.001.

[80] Eggeling L, Bott M. Handbook of *Corynebacterium glutamicum. Boca Raton, USA: CRC Press;* 2005.

[81] Burkovski A, editor. Corynebacteria: genomics and molecular biology. Norfolk, U.K.: *Caister Academic Press;* 2008.

[82] Yukawa H, Inui M. *Corynebacterium glutamicum*: biology and biotechnology. Springer-Verlag Berlin Heidelberg; 2012.

[83] Ikeda M, Ohnishi J, Hayashi M, Mitsuhashi S. A genome-based approach to create a minimally mutated *Corynebacterium glutamicum* strain for efficient L-lysine production. *J Ind Microbiol Biotechnol* 2006;33:610–5.

[84] Zhu N, Xia H, Wang Z, Zhao X, Chen T. Engineering of acetate recycling and citrate synthase to improve aerobic succinate production in *Corynebacterium glutamicum. PLoS ONE* 2013;8:e60659. doi:10.1371/journal.pone.0060659.

[85] Litsanov B, Brocker M, Bott M. Glycerol as a substrate for aerobic succinate production in minimal medium with *Corynebacterium glutamicum. Microb Biotechnol* 2012. doi:10.1111/j.1751-7915.2012.00347.x.

[86] Chen T, Zhu N, Xia H. Aerobic production of succinate from arabinose by metabolically engineered *Corynebacterium glutamicum. Bioresour Technol* 2014;151:411–4. doi:10.1016/j.biortech.2013.10.017.

[87] Lee J, Sim SJ, Bott M, Um Y, Oh M-K, Woo HM. Succinate production from CO_2-grown microalgal biomass as carbon source using engineered *Corynebacterium glutamicum* through consolidated bioprocessing. *Sci Rep* 2014;4. doi:10.1038/srep05819.

[88] Lin H, Bennett GN, San KY. Metabolic engineering of aerobic succinate production systems in *Escherichia coli* to improve process productivity and achieve the maximum theoretical succinate yield. *Metab Eng* 2005;7:116–27.

[89] Litsanov B, Brocker M, Bott M. Toward homosuccinate fermentation: metabolic engineering of *Corynebacterium glutamicum* for anaerobic production of succinate from glucose and formate. *Appl Env Microbiol* 2012;78:3325–37. doi:10.1128/AEM.07790-11.

[90] Okino S, Noburyu R, Suda M, Jojima T, Inui M, Yukawa H. An efficient succinic acid production process in a metabolically engineered *Corynebacterium glutamicum* strain. *Appl Microbiol Biotechnol* 2008;81:459–64.

[91] Galkin A, Kulakova L, Tishkov V, Esaki N, Soda K. Cloning of formate dehydrogenase gene from a methanol-utilizing bacterium *Mycobacterium vaccae N10. Appl Microbiol Biotechnol* 1995;44:479–83.

[92] Huhn S, Jolkver E, Kramer R, Marin K. Identification of the membrane protein SucE and its role in succinate transport in *Corynebacterium glutamicum. Appl Microbiol Biotechnol* 2011;89:327–35. doi:10.1007/s00253-010-2855-1.

[93] Fukui K, Koseki C, Yamamoto Y, Nakamura J, Sasahara A, Yuji R, et al. Identification of succinate exporter in *Corynebacterium glutamicum* and its physiological roles under anaerobic conditions. *J Biotechnol* 2011;154:25–34. doi:10.1016/j.jbiotec.2011.03.010.

[94] Zhu N, Xia H, Yang J, Zhao X, Chen T. Improved succinate production in *Corynebacterium glutamicum* by engineering glyoxylate pathway and succinate export system. *Biotechnol Lett* 2014;36:553–60. doi:10.1007/s10529-013-1376-2.

[95] Zhou Z, Wang C, Chen Z, Chen Y, Zhang K, Xu H, et al. Increasing available NADH supply during succinic acid production by *Corynebacterium glutamicum. Biotechnol Prog* 2014. doi:10.1002/btpr.1998.

[96] Yamauchi Y, Hirasawa T, Nishii M, Furusawa C, Shimizu H. Enhanced acetic acid and succinic acid production under microaerobic conditions by *Corynebacterium glutamicum* harboring *Escherichia coli* transhydrogenase gene *pntAB*. *J Gen Appl Microbiol* 2014;60:112–8.

[97] Jeon JM, Rajesh T, Song E, Lee HW, Lee HW, Yang YH. Media optimization of *Corynebacterium glutamicum* for succinate production under oxygen-deprived condition. *J Microbiol Biotechnol* 2013;23:211–7.

[98] Kawaguchi H, Vertes AA, Okino S, Inui M, Yukawa H. Engineering of a xylose metabolic pathway in *Corynebacterium glutamicum*. *Appl Env Microbiol* 2006;72:3418–28.

[99] Seibold G, Auchter M, Berens S, Kalinowski J, Eikmanns BJ. Utilization of soluble starch by a recombinant *Corynebacterium glutamicum* strain: growth and lysine production. *J Biotechnol* 2006;124:381–91.

[100] Kawaguchi H, Sasaki M, Vertes AA, Inui M, Yukawa H. Engineering of an L-arabinose metabolic pathway in *Corynebacterium glutamicum*. *Appl Microbiol Biotechnol* 2008;77:1053–62.

[101] Wang C, Zhang H, Cai H, Zhou Z, Chen Y, Chen Y, et al. Succinic acid production from corn cob hydrolysates by genetically engineered *Corynebacterium glutamicum*. *Appl Biochem Biotechnol* 2014;172:340–50. doi:10.1007/s12010-013-0539-x.

[102] Jantama K, Zhang X, Moore JC, Shanmugam KT, Svoronos SA, Ingram LO. Eliminating side products and increasing succinate yields in engineered strains of *Escherichia coli* C. *Biotechnol Bioeng* 2008;101:881–93.

[103] Litsanov B, Brocker M, Bott M. Glycerol as a substrate for aerobic succinate production in minimal medium with *Corynebacterium glutamicum*. *Microb Biotechnol* 2012. doi:10.1111/j.1751-7915.2012.00347.x.

[104] Lin H, Bennett GN, San KY. Fed-batch culture of a metabolically engineered *Escherichia coli* strain designed for high-level succinate production and yield under aerobic conditions. *Biotechnol Bioeng* 2005;90:775–9.

[105] Jo S-J, Maeda M, Ooi T, Taguchi S. Production system for biodegradable polyester polyhydroxybutyrate by *Corynebacterium glutamicum*. *J Biosci Bioeng* 2006;102:233–6. doi:10.1263/jbb.102.233.

[106] Valappil SP, Boccaccini AR, Bucke C, Roy I. Polyhydroxyalkanoates in Gram-positive bacteria: insights from the genera *Bacillus* and *Streptomyces*. *Antonie Van Leeuwenhoek* 2007;91:1–17. doi:10.1007/s10482-006-9095-5.

[107] Taguchi S, Maehara A, Takase K, Nakahara M, Nakamura H, Doi Y. Analysis of mutational effects of a polyhydroxybutyrate (PHB) polymerase on bacterial PHB accumulation using an in vivo assay system. *FEMS Microbiol Lett* 2001;198:65–71.

[108] Taguchi S, Nakamura H, Hiraishi T, Yamato I, Doi Y. In vitro evolution of a polyhydroxybutyrate synthase by intragenic suppression-type mutagenesis. *J Biochem* (Tokyo) 2002;131:801–6.

[109] Normi YM, Hiraishi T, Taguchi S, Abe H, Sudesh K, Najimudin N, et al. Characterization and properties of G4X mutants of *Ralstonia eutropha* PHA synthase for poly(3-hydroxybutyrate) biosynthesis in *Escherichia coli*. *Macromol Biosci* 2005;5:197–206. doi:10.1002/mabi.200400181.

[110] Matsumoto K, Yamada M, Leong CR, Jo S-J, Kuzuyama T, Taguchi S. A new pathway for poly(3-hydroxybutyrate) production in *Escherichia coli* and *Corynebacterium*

glutamicum by functional expression of a new acetoacetyl-coenzyme A synthase. *Biosci Biotechnol Biochem* 2011;75:364–6. doi:10.1271/bbb.100682.

[111] Leong YK, Show PL, Ooi CW, Ling TC, Lan JC-W. Current trends in polyhydroxyalkanoates (PHAs) biosynthesis: Insights from the recombinant *Escherichia coli*. *J Biotechnol* 2014;180:52–65. doi:10.1016/j.jbiotec.2014.03.020.

[112] Tomizawa S, Hyakutake M, Saito Y, Agus J, Mizuno K, Abe H, et al. Molecular weight change of polyhydroxyalkanoate (PHA) caused by the PhaC subunit of PHA synthase from *Bacillus cereus* YB-4 in recombinant *Escherichia coli*. *Biomacromolecules* 2011;12:2660–6. doi:10.1021/bm2004687.

[113] Nikel PI, Pettinari MJ, Galvagno MA, Méndez BS. Poly(3-hydroxybutyrate) synthesis from glycerol by a recombinant *Escherichia coli arcA* mutant in fed-batch microaerobic cultures. *Appl Microbiol Biotechnol* 2008;77:1337–43. doi:10.1007/s00253-007-1255-7.

[114] Liu S, Liu S-J, Qi S-W, Chaudhry MT, Zhang Y, Meng B, et al. Comparative proteomes of *Corynebacterium glutamicum* grown on aromatic compounds revealed novel proteins involved in aromatic degradation and a clear link between aromatic catabolism and gluconeogenesis via fructose-1,6-bisphosphatase. *Proteomics* 2007;7:3775–87. doi:10.1002/pmic.200700481.

[115] Jo S-J, Leong CR, Matsumoto K, Taguchi S. Dual production of poly(3-hydroxybutyrate) and glutamate using variable biotin concentrations in *Corynebacterium glutamicum*. *J Biosci Bioeng* 2009;107:409–11. doi:10.1016/j.jbiosc.2008.12.003.

[116] Song Y, Matsumoto K, Tanaka T, Kondo A, Taguchi S. Single-step production of polyhydroxybutyrate from starch by using α-amylase cell-surface displaying system of *Corynebacterium glutamicum*. *J Biosci Bioeng* 2013;115:12–4. doi:10.1016/j.jbiosc.2012.08.004.

[117] Matsumoto K, Kitagawa K, Jo S-J, Song Y, Taguchi S. Production of poly(3-hydroxybutyrate-co-3-hydroxyvalerate) in recombinant *Corynebacterium glutamicum* using propionate as a precursor. *J Biotechnol* 2011;152:144–6. doi:10.1016/j.jbiotec.2010.07.031.

[118] Song Y, Matsumoto K, Yamada M, Gohda A, Brigham CJ, Sinskey AJ, et al. Engineered *Corynebacterium glutamicum* as an endotoxin-free platform strain for lactate-based polyester production. *Appl Microbiol Biotechnol* 2011. doi:10.1007/s00253-011-3718-0.

[119] Vasco-Cardenas MF, Banos S, Ramos A, Martin JF, Barreiro C. Proteome response of *Corynebacterium glutamicum* to high concentration of industrially relevant C(4) and C(5) dicarboxylic acids. *J Proteomics* 2013;85:65–88. doi:10.1016/j.jprot.2013.04.019.

[120] Barriuso-Iglesias M, Schluesener D, Barreiro C, Poetsch A, Martin JF. Response of the cytoplasmic and membrane proteome of *Corynebacterium glutamicum* ATCC 13032 to pH changes. *BMC Microbiol* 2008;8:225. doi:10.1186/1471-2180-8-225.

In: New Trends in *Corynebacterium glutamicum*
Editor: Carlos Barreiro

ISBN: 978-1-63483-280-9

Chapter 7

L-VALINE PRODUCTION IN *CORYNEBACTERIUM GLUTAMICUM*

Kazumi Hiraga, Crispinus A. Omumasaba and Masayuki Inui*

Research Institute of Innovative Technology for the Earth,
Kyoto, Japan

ABSTRACT

Corynebacterium glutamicum is the widely recognized leading microorganism for industrial production of important amino acids such as L-glutamate and L-lysine. Recent advances have expanded the potential utility of this microorganism to include the production of organic acids such as lactate and succinate. Besides expanding the product range, these advances have also helped to reveal new insights into mechanisms and pathways that result in the elevated production of industrially useful chemicals in the microorganism. Only slight genetic modifications of the microorganism elevate the chemical productivity to such an extent that its utility in a variety of current industries is likely to be enhanced in the emerging biorefinery future. L-Valine, an essential amino acid in vertebrates that may become limiting under specific feeding conditions, has become an important supplement in animal feeds following the proliferation of protein-reduced diets in animal husbandry. Biosynthesis of this branched chain amino acid from pyruvate in cells of *C. glutamicum* is a four-step reaction catalyzed by acetohydroxyacid synthase (AHAS), acetohydroxyacid isomeroreductase (AHAIR), dihydroxy acid dehydratase (DHAD), and transaminase B. Maximizing L-valine production in *C. glutamicum* consequently entails overexpressing the genes encoding enzymes of this complex while simultaneously modulating the expression of genes involved in the synthesis of competing branched chain amino acids or genes that catalyze the feedback inhibition of AHAS. Further gains in productivity result from deactivation or suppression of other competing pathways coupled with improved intracellular redox balance of the subject L-valine producing strains. These strain modifications appear to work best in oxygen-deprived *C. glutamicum* largely due to the superior metabolic efficiency that

* Corresponding Author: Masayuki Inui; inui@rite.or.jp

oxygen deprivation engenders in this strain. Differing approaches to strain improvement and bioprocess optimization are manifested in the wide differences in the reported productivities, ranging from 195~1940 g/L titers, 0.27~0.88 moles of L-valine per mole of glucose consumed, and 3.7~61.3 mM/h yields. This chapter presents an overview of these approaches with respect to recent advances in development of recombinant *C. glutamicum* strains for L-valine production under both aerobic and under oxygen-deprived conditions.

INTRODUCTION

Corynebacterium glutamicum is a rapidly growing, gram positive soil bacterium and has been the major workhorse of industrial amino acid production for decades. As global population increases, amino acid production continues to see increased demand. Ajinomoto Co. Ltd. estimated the 2011 global demand for monosodium glutamate, the principal amino acid for which the microorganism attained its industrial importance, at 2.6 million metric tons and growing at 4~5% per annum driven mainly by the increasing population in South East Asia and Africa [1]. The 2012 demand for lysine of 1.95 million metric tons and that for threonine of 330 thousand metric tons are estimated to rise with the rising necessity to soybean meal in feed formulations [2]. The excess of 500 metric tons of L-valine produced annually by fermentation or by extraction from acidic hydrolysates of proteins [18] to use as in animal feed supplements should see the same increasing demand as those of other feed supplements. This increased demand for amino acids necessitates urgent advancements in strains and processes of amino acid production. The high natural ability of *C. glutamicum* to produce amino acids drives interest in genetic manipulation of the microorganism to meet the anticipated increased demand for amino acids, resulting in new strains with better production characteristics.

Outside the increasing demand for L-valine in feed supplements, the amino acid attracts attention because of its application in pharmaceutical and cosmetic products, and as a precursor of antibiotics or herbicides [19]. A recent report by the European Commission's Panel on Additives and Products or Substances used in Animal Feed (FEEDAP) affirmed the safety of L-valine *produced* by genetically modified *C. glutamicum* for all animal species [17]. Major advances in corynebacterial branched-chain amino acid (BCAA) productivity in general result to a large extent from modifications within the gene cluster encoding the enzymes of the pyruvate dehydrogenase complex [9, 14] and are accentuated by the capacity of the strains to efficiently metabolize energy [7]. *C. glutamicum* is capable of considerable amino acid production under aerobic as well as oxygen suppressed conditions while cells are actively growing or when growth is suppressed.

Metabolic Engineering of *C. glutamicum* for L-Valine Production

L-Valine is a branched-chain amino acid essential in the diet of all vertebrates. Currently it finds use mostly as an animal feed supplement and in the pharmaceutical industry. Its global demand is several hundred tons annually and growing [3]. Microbial L-valine production in *C. glutamicum* using glucose as a substrate has been reported under aerobic or anaerobic conditions in shaking flask, batch fermentation and fed-batch fermentation (Table 1). Strains based on disruption of the *aceE* gene that encodes the enzyme that converts pyruvate to acetyl-CoA cannot grow on glucose. The production medium must be therefore be supplemented with acetate. Growth will occur on acetate, the depletion of which triggers metabolism of glucose to L-valine.

In Table 1, some of the strains with *ΔaceE* do not require that the medium is supplemented with acetate, in which case brain heart infusion (BHI) is used to promote L-valine productivity. This requirement for supplementation may be a manifestation of the incomplete suppression of *ΔaceE/Δpqo*-based strains.

All the strains are based on enhancement of the activity of more of the four enzymes of the L-valine biosynthetic pathway(*ilvBNCDE*). Discounting the growth phase of the recombinant strains, tremendous variants in the productivity of final strains have been reached in titer (195-1940 g/L), mol of L-val/mol of consumed glucose (0.27-0.88 mol/mol glc) and yield in mM/h of L-val (3.7-61.3).

A detailed biosynthetic pathway of L-valine and by-products in *C. glutamicum* is shown in Figure1. The figure summarizes the major nodes in the metabolic engineering for L-valine production. The metabolic map reveals the genes encoding the enzymes at each step, with cofactor dependency added to emphasize that genetic engineering targets both enzymes and their cofactor dependencies. The direct metabolic pathway for the L-valine production consists of the four reactions which are catalyzed by acetohydroxyacid synthase (AHAS, E.C. 4.1.3.18, *ilvBN* gene product), acetohydroxyacid isomeroreductase (AHAIR, E.C. 1.1.1.86, *ilvC* gene product), dihydroxyacid dehydratase (DHAD, E.C. 4.2.1.9, *ilvD* gene product), and transaminase B (E.C. 2.6.1.42, *ilvE* gene product). The genes are found clustered in *ilvBNCDE* on the *C. glutamicum* chromosome. The first reaction, catalyzed by AHAS, results in the formation of 2-acetolactate upon condensation of two molecules of pyruvate. This represents the key step of L-valine biosynthesis by virtue of the enzyme being subject to feed-back inhibition by other branched-chain amino acids.

Most L-valine production strains, as a rule, exhibit modifications in one or more of the genes of the *ilvBNCDE* cluster to increase carbon flow toward L-valine. This is usually supplemented by additional genetic modifications aimed at suppressing side reactions that otherwise lead to formation of acetate (*aceEF, pqo*), lactate (*ldhA*), and succinate (*ppc*) from pyruvate. The aim is to channel as much of the available pyruvate as possible toward L-valine production. Other side reactions not directly involved in depletion of the pyruvate pool have also been targeted for suppression, with occasionally notable increases in L-valine productivity. Disruption of the *ilvA* gene to suppress L-threonine formation, *panB* to limit D-pantothenate production, and *leuA* curtail L-leucine synthesis, all target branched-chain amino acids which might otherwise inhibit AHAS activity. Moreover, other approaches target modifications within the glycolytic pathway to direct more carbon towards pyruvate (e.g. by modulating *tpi* expression to limit dihydroxyacetone phosphate production) or altogether channeling carbon flow away from the glycolytic to the pentose phosphate pathway (by modulating *pgi* expression) in order to generate the NADPH required downstream in L-valine production.

Table 1. Summary of L-valine producing strains of C. glutamcum and subspecies

Cultivation	Strain with remarks	Culture condition	Titer (mM(g/L))	$Y_{P/S}$ (mol of L-Val/ mol of consumed Glc)	Q_P *(mM/h)*	*Ref.*
Shaking flask or Batch fermentation	*C. glutamicum*13032*ΔilvAΔpanBC*(pJC*ilvBNCD*)	4% glucose, 30°C, 48h	91(10.7)	–	–	Radmacher et al. 2002 [4]
	C. glutamicum ΔilvAΔpanBC ilvN M13(pECKA*ilvBNC*)	4% glucose, 30°C, 48h	130(15.2)	–	–	Elišáková et al. 2005 [5]
	C. glutamicum ΔaceE (pJC4 *ilvBNCE*)	4% glucose, 1% acetate, 0.5% BHI, 30°C, 64h	105(12.3)	–	–	Blombach et al. 2007 [6]
	C. glutamicum A-1 (pVK*ilvN53C*) (H^+*-ATPase defective mutant*)	10% glucose, soybean-meal hydrolysate, 72h	96(11.2)	–	–	Wada et al. 2008 [7]
	*C. glutamicum ilvN*M13 *ΔpanB P-ilvA*M1CG *P-ilvD*M7 *P-ilvE*M6(pECKA *ilvBNC*) (all mutations on genome)	4% glucose, 30°C, 48h	136(15.9)	–	–	Holátko et al. 2009 [8]
Fed-batch fermentation	*C. glutamicum ΔaceE*(pJC4 *ilvBNCE*)	Glucose, acetate, 0.5% BHI**, 30°C	210(24.6)	0.6*	10*	Blombach et al. 2007 [6]
	C. glutamicum ΔaceE(pJC4 *ilvBNCE*)	Glucose, acetate, 0.5% BHI**, 30°C	195(22.8)	0.39*	7.9*	Blombach et al. 2008 [9]
	C. glutamicum ΔaceEΔpqo(pJC4 *ilvBNCE*)	Glucose, acetate, 0.5% BHI**, 30°C	225(26.4)	0.52*	9.5*	Blombach et al. 2008 [9]
	C. glutamicum ΔaceEΔpqoΔpgi(pJC4 *ilvBNCE*)	Glucose, acetate, 0.5% BHI**, 30°C	412(48.3)	0.75*	5.6*	Blombach et al. 2008 [9]
	C. glutamicum ΔaceEΔpqoΔpgi Δpyc (pJC4 *ilvBNCE*)	Glucose, acetate, 0.5% BHI**, 30°C	240(28.1)	0.86*	5.2*	Blombach et al. 2008 [9]
	C. glutamicum aceE A16 *ΔpqoΔppc* (pJC4 *ilvBNCE*)	Glucose, 30°C	738(86.5)	0.36*	13.6*	Buchholz et al. 2013 [10]
	B. flavum ATCC14067 (pDXW-8-*ilvEBN^{r}C*)	Glucose, 37°C, 48h	325(38.1)	0.37	6.8	Hou et al. 2012 [11]
	C. glutamicum ATCC13032 MP*ilvA ΔavtA*(pDXW-8-*ilvEBN^{r}C*)	Glucose, 31°C, 72h	267(31.5)	0.27	3.7	Hou et al. 2012 [12]
	C. glutamicum R ΔldhA (pCRB-*BN^{GE}C^{TM}*)(pCRB-*DLD*)	Glucose, oxygen-deprived condition, 33°C, 24h	1470(172.2)	0.63	61.3	Hasegawa et al. 2012 [13]
		Glucose, oxygen-deprived condition, 33°C, 48h	1940(227.3)	–	–	Hasegawa et al. 2012 [13]
	C. glutamicum R ΔldhAΔppcΔptaΔackAΔctfAΔavtA ilvN^{GE}C^{TM} + gapA + pyk + pfkA + pgi + tpi (pCRB-BN^{GE}C^{TM})(pCRB-DLD)	Glucose, oxygen-deprived condition, 33°C, 24h	1280(150)	0.88	53.3	Hasegawa et al. 2013 [14]

* These values are calculated from the production phase (growth phase is not included). ** BHI: Brain Heart Infusion broth.

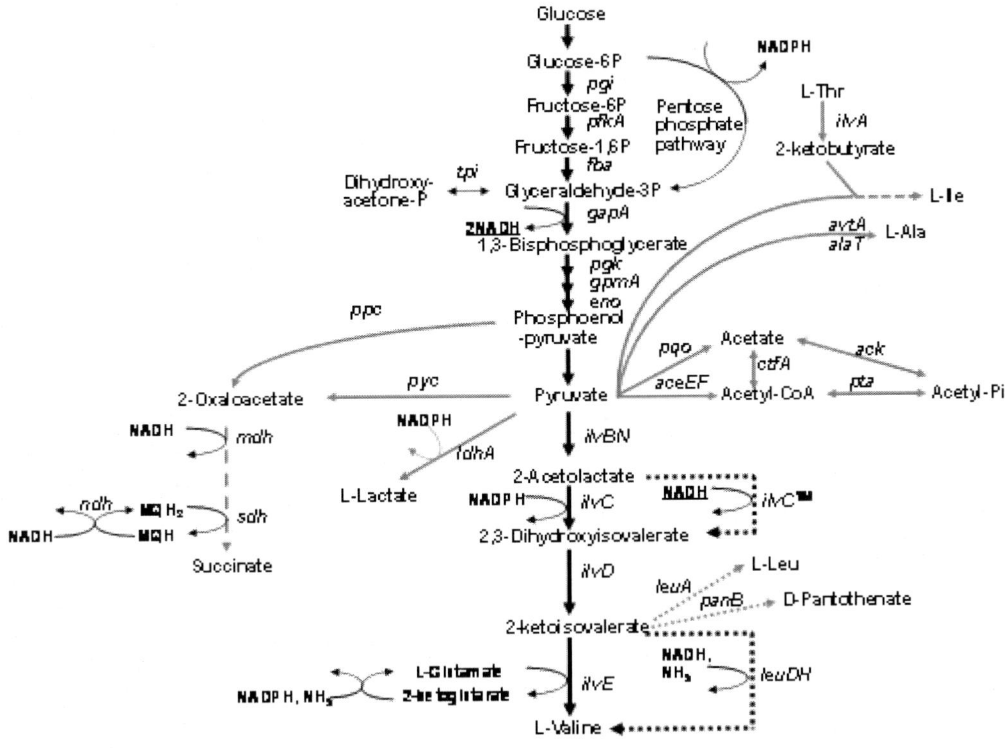

Figure 1. Biosynthetic pathway of L-valine and by-products of *C. glutamicum*. Some important gene names are shown in the biosynthetic pathway of L-valine. *ilvBN*, acetohydroxy acid synthase(AHAS); *ilvC*, acetohydroxy acid isomeroreductase (AHAIR); *ilvD*, dihydroxy acid dehydratase (DHAD); *ilvE*, transaminase B; *leuDH*, leucine dehydrogenase (LeuDH from *Lysinibacillus sphaericus*). See details in the text.

There remains a wide variety in L-valine productivity that cannot be tied to the approach selected or whether the production is under fully aerobic or oxygen-deprived conditions. Final L-valine titers ranging from 195 to 1940 mM, corresponding to 0.27 to 0.88 mole of L-valine per mole of glucose consumed and 3.7 to 61.3 mM L-valine per hour have been reported to date.

Radmacher et al. deleted genomic *ilvA* and *panBC* genes to suppress branched-chain amino acids, and overexpressed the *ilvBNCD* genes in *C. glutamicum* 13032 by use of a plasmid. The engineered strain *C. glutamicum* 13032*ΔilvAΔpanBC*(pJC*ilvBNCD*) produced 91 mM (10.7 g/L) L-valine aerobically [4]. Elisakova et al. eschewed the plasmid and constructed a recombinant strain carrying the *ilvN*M13 mutation within the chromosome. The plasmid-free *C. glutamicumΔilvAΔpanBC ilvNM13* strain production of 90 mM L-valine within 48h of aerobic cultivation was not much changed from the plasmid-carrying strain. However, when recombinant *C. glutamicum ΔilvAΔpanBC ilvN*M13 strain was transformed with a plasmid, pECKA-*ilvBNC*), L-valine productivity increased to 130 mM (15.2 g/L) under the same conditions [5]. That the mutant acetohydroxyacid synthase activity of gained resistance to feedback inhibition by all three of L-valine, L-Isoleucine, and L-leucine accounted for the observed elevated productivity.

Using a different approach, Blombach et al. engineered an aerobic L-valine producer overexpressing the *ilvBNCE* genes in the genetic background of the pyruvate dehydrogenase complex (PDHC)-deficient *C. glutamicum* strain (*ΔaceE*). As the PDHC-deficient strain was unable to grow on glucose, it was grown on a medium with a mixture of glucose and acetate. After depletion of the acetate in the medium during the growth phase, the recombinant strain started producing L-valine from glucose in a production phase. Productivity reached 210 mM (24.6 g/L) L-valine with a yield of 0.6 mole per mole consumed glucose [6] and 195 mM (22.8g/L) L-valine with a yield of 0.39 mole per mole consumed glucose [9] in fed-batch fermentation. The strain's L-valine productivity was further improved with additional deletion of the *pqo* encoding pyruvate:quinone oxidoreductase to suppress acetate production, thus *C. glutamicum* strain(*ΔaceE Δpqo*(pCJ4 *ilvBNCE*)) produced 225 mM (26.4 g/L) L-valine with a yield of 0.52 mole per mole consumed glucose [9]. Furthermore, in order to increase the pool of NADPH required for L-valine pathway, carbon flow was directed away from the glycolytic pathway towards the pentose phosphate pathway by disruption of *pgi* encoding phosphoglucose isomerase. The resultant strain, *C. glutamicum* (*ΔaceE Δpqo Δpgi* (pCJ4 *ilvBNCE*)), double productivity to 412 mM (48.3 g/L) L-valine with a yield of 0.75 mole per mole consumed during fed-batch fermentation [9]. To cap it all, deletion of pyruvate decarboxylase improved the product yield to 0.86 mole per mole glucose but at a decrease titer of L-valine of 240 mM (28.1 g/L).

Yet another approach involves the modulation of the central metabolic capacity of subject strains. Wada et al. used a H^+-ATPase defective mutant to engineer strain *C. glutamicum* A-1 carrying plasmid pVK*ilvN* 53C that produced 96 mM (11.2g/L) L-valine aerobically in 72h [7]. The H^+-ATPase defective mutant was more effective at sugar metabolism than the control strain, with a concomitant higher L-valine productivity. Proteome analyses revealed that the defect triggered upregulation of genes encoding glycolytic enzymes such as 6-phosphofructokinase and pyruvate kinase.

It is likely that a strain harboring all the desired mutations on the chromosome will be a better production strain than one which relies entirely on plasmids. Holátko et al. tested a host strain carrying all the mutations in question on the chromosome. The recombinant *C. glutamicum ilvN*M13 *ΔpanB P-ilvA* M1CG *P-ilvD* M7 *P-ilvE* M6(pECKA *ilvBNC)* produced 136 mM (15.9 g/L) L-valine in 48h [8], well within the range of productivities reported in other studies. The crucial measure of strain stability remains to be determined so long us prolonged and larger-scale production has not been attempted.

The theoretically promising approach of using homologous genes from other organisms but encoding enzymes with superior characteristics has been assayed by various groups. Buchholz et al. exchanged the native corynebacterial *aceE* gene encoding the E1p subunit of the pyruvate dehydrogenase complex (PDHC) with mutated *dapA* promoter variants to construct *C. glutamicum* strains with gradually reduced growth rates and lower PDHC activities. Of these, they selected *C. glutamicumΔaceE* A16 (pJC4 *ilvBNCE*) because it grew to the highest biomass and showed highest L-valine production yields. The strain production characteristics were further improved by deletion of the *pqo* encoding pyruvate:quinone oxidoreductase and *ppc* encoding phosphoenolpyruvate carboxylase, to suppress metabolic flow from pyruvate to acetate. The resultant PDHC-deficient *C. glutamicumΔaceE* A16 *ΔpqoΔppc* (pJC4 *ilvBNCE*) platform strain produced up to 738 mM (86.5 g/L) L-valine at an overall yield of 0.36 mole per mole of glucose consumed [10]. Overexpression of the transaminase B-encoding *ilvE* gene on top of overexpressed *ilvBNCD* instead of *ilvBNCE* transformed the L-valine-producing strain into a 2-ketoisovalerate producer. Productivity reached 303 mM (35 g/L) 2-ketoisovalerate at an overall yield of 0.24 mole per mole of glucose consumed. Since 2-ketoisovalerate is a precursor of isobutanol, this approach is more appropriate for the metabolic engineering of isobutanol producing strains described elsewhere in this book. Regardless, this approach demonstrates how easily the metabolism of *C. glutamicum* can be channeled to another product.

Hou et al. chose *Brevibacterium flavum* ATCC14067 as a host for L-valine production because of its capacity to grow at the high temperature of 40°C compared to that of *C. glutamicum* of 33°C. Thermotolerance is generally desirable to the extent that it renders a process less susceptible to contamination. The *ilvEBN^{r}C* genes from *B. flavum* NV128 provided the best candidate strain for L-valine production. The best recombinant, ATCC14067 (pDXW-8-*ilvEBN^{r}C*) produced 325 mM (38.1 g/L) L-valine at a yield of 0.37 mole per mole of glucose consumed in 48 h at 37°C. The approach of enhancing L-valine production by overexpression of key enzymes in thermotolerant strains represents an attractive alternative approach that could engender desirable process characteristics [11]. A totally different approach by Hou et al. involves replacement of the native promoter of *ilvA* of *C. glutamicum* with the weak MP*ilvA*(P-*ilvA* M1CG) promoter to suppress the biosynthesis of L-isoleucine. To minimize L-alanine accumulation, they disrupted *alaT* and/or *avtA*. The most prolific recombinant strain from this approach was *C. glutamicum* ATCC13032 MP*ilvA* *ΔavtA* (pDXW-8-*ilvEBN^{r}C*). It produced 267 mM (31.5 g/L) L-valine at a yield of 0.27 mole per mole glucose consumed in 72 h at 31°C [12].

For non-aerobic production of L-valine, an efficient L-valine producing strain proved the prime importance of redox balance in overall productivity. The biosynthesis of a mole of L-valine in *C. glutamicum* from glucose generates a net of two moles of NADH in glycolytic pathway but consumes 2 moles of NADPH, one by acetohydroxyacid isomeroreductase (AHAIR) and the other at regeneration of L-glutamate as an amino-group donor for transaminase reaction (Figure1). *C. glutamicum* lacks transhydrogenase activity to convert NADPH to NADH, resulting in overall cofactor imbalance and negatively impacting efficiency under growth arrested oxygen-deprived conditions. But oxygen-deprived production in general engenders important advantages that could mean simpler and more efficient production processes. In order to maintain redox balance, the supply of NADPH through pentose phosphate pathway is one choice. However, it is a choice which leads to a net loss of carbon in the pathway, resulting in depressed yields of L-valine. Hasegawa et al. succeeded in altering the cofactor dependencies within the L-valine biosynthesis pathway from NADPH to NADH. They engineered a *C. glutamicum* R strain with a disrupted *ldhA* encoding lactate dehydrogenase to suppress lactic acid formation into which they overexpressed modified *ilvBN^{GE}C^{TM}D* genes encoding feedback-resistant acetohydroxyacid synthase (encoded by *ilvN^{GE}*(G156E) and a modified acetohydroxyacid isomeroreductase (encoded by *ilvC^{TM}*(*ilvC* (S34G, L48E, and R49F)) with preference for NADH to replace the NADPH dependence of the unmodified gene). In addition use of NH_3 instead of glutamate as an amino donor and the consequent NADH dependence for this reaction were attained by using the gene encoding leucine dehydrogenase from *Lysinibacillus sphaericus* instead of native transaminase B. The best strain from this approach, *C. glutamicum* R *ΔldhA* (pCRB-*BN^{GE}C^{TM}*) (pCRB-*DLD*) produced 1470 mM (0.63 g/L) L-valine at a yield of 0.63 mole per mole glucose consumed in 24h, and 1940 mM (227.3g/L) L-valine in 48h under oxygen-deprived conditions with enhanced glucose uptake and decreased NADH/NAD^+ ratio [13]. To achieve the high productivity, elimination of by-products such as succinate was crucial. They disrupted the *ppc* gene encoding the phosphoenolpyruvate carboxylase to suppress succinate production but it was at the expense of both glucose consumption and L-valine production due to severe elevation of the intracellular NADH/NAD^+ ratio. The perturbed intracellular redox state was compensated for by deletion of the three genes (*pta, ackA,* and *ctfA*) associated with NADH-producing acetate synthesis and overexpression of five glycolytic enzyme genes (*gapA* encoding NADH-inhibited glyceraldehyde 3-phosphate dehydrogenase), *pyk, pfkA, pgi,* and *tpi*) enhanced the rate of glucose consumption. Deletion of the alanine transaminase gene *avtA* suppressed L-alanine production too. The best strain Val9, *C. glutamicum* R *ΔldhAΔppcΔptaΔackAΔctfAΔavtA* $ilvN^{GE}C T^{M}$ + *gapA* + *pyk* + *pfkA*+ *pgi* + *tpi* (pCRB-$BN^{GE}CT^{M}$) (pCRB-*DLD*) produced 1280 mM L-valine (150 g/L) in 24 h at a yield of 0.88 mole per mole of glucose consumed under oxygen-deprived condition [14]. This level of productivity for the relatively younger approach suggests the techniques employed in aerobic L-valine production by *C. glutamicum* are equally applicable to non-aerobic production of the same.

CONCLUSION

The various approaches to enhancing L-valine production in *C. glutamicum* outlined in this chapter confirm the suitability of the microorganism to systematic metabolic engineering to yield strains with excellent production properties (high titer and yield under aerobic and anaerobic conditions) in a well-defined genome context. It is likely that more powerful recombinant strains for industrial production of L-valine and other amino acids will be engineered by integration of regulatory networks to create desired combinations of pathways and production environments. Introduction of heterologous biosynthetic routes or enzymes to create less complex but higher yielding metabolic steps resulting in suppressed by-product formation and reduced carbon loss and optimized redox balance will emerge.

The most available approaches to enhance L-valine production in *C. glutamicum* include deregulation of the key regulatory enzyme acetohydroxyacid synthase ($ilvN^{fbr}$), overexpression of the genes encoding enzymes that convert pyruvate to L-valine, increasing the pool of available pyruvate by suppressing by-product formation from pyruvate in conjunction to enhancing the glycolytic pathway, optimization of redox balance (NADH/NAD^+ or NADPH/$NADP^+$ balance), and growth-independent production. The approach to utilize growth-independent production is particularly promising given its inherent advantages of substrate promiscuity, elevated productivity, and by-product suppression. It is by utilizing cells at a high density under oxygen-deprived conditions that Hasegawa et al. succeeded to produce such a high yield of L-valine. Under oxygen-deprived conditions, cell growth is so suppressed that the vast pool of available glucose is channeled solely toward production of L-valine. Deregulation of acetohydroxyacid synthase ($ilvN^{GE}$), suppression of by-product formation through disruption of *ldhA* (lactate), *ppc* (succinate), *pta, ackA, ctfA* (acetate), and *avtA* (L-Ala), and enhancement of glycolytic pathway (*gapA, pyk, pfkA, pgi,* and *tpi*) all accentuated the productivity. Attaining the right cofactor balance by using NADH-dependent *ilvC* mutant ($ilvC^{TM}$ (S34G, L48E, and R49F)), and by introduction of NADH-dependent leucine dehydrogenase instead of transaminase (*ilvE*) ensured that no unwelcome carbon loss was necessitated.

As industry enters the biorefinery age, the tolerance of biocatalysts of any inhibitory compounds resultant from the necessity to employ mixed and pre-treated substrates such as those derived from in lignocellulosic hydrolysates will start to play a major role in the suitability of any industrial biocatalyst. Prior to enzymatic hydrolysis of lignocellulosic biomass to sugars, the biomass must necessarily be subjected to harsh physical and/or thermochemical treatment to improve substrate availability for enzymatic saccharification. Such pretreatments generate by-products that may inhibit further saccharification and fermentation steps. Sugar metabolism by *C. glutamicum* cells under oxygen-deprived conditions has demonstrated good tolerance to individual inhibitory compounds or mixtures thereof [15]. Yet another advantage that use of *C. glutamicum* presents is the microorganism's capacity for simultaneous utilization of mixed sugars. In mineral media containing a mixture of D-glucose, D-xylose, L-arabinose, and D-cellobiose under oxygen-deprived conditions, *C. glutamicum* strain ACX-*araE* can simultaneously and completely consume all the sugars [16]. These features underline the merits of the growth-arrested bioprocess under oxygen-deprived conditions for the lignocellulosic biorefineries.

REFERENCES

[1] Ajinomoto (2012) Food products business. Available: http://www. ajinomoto. com/jp/ir/pdf/ fact/Food-Oct2012.pdf.

[2] Ajinomoto (2013) Feed-use amino acids business. Available: http://www.ajinomoto. com/jp /ir/pdf/ fact/Feed-useAA-Oct2013.pdf.

[3] Leuchtenberger W, Huthmacher K, Drauz K. Biotechnological production of amino acids and derivatives: current status and prospects. *Appl. Microbiol. Biotechnol.* 2005; 69: 1–8.

[4] Radmacher E, Vaitsikova A, Burger U, Krumbach K, Sahm H, Eggeling L. Linking central metabolism with increased pathway flux: L-valine accumulation by *Corynebacterium glutamicum. Appl. Environ. Microbiol.* 2002; 68: 2246–2250.

[5] Elišáková V, Pátek M, Holátko J, Ňesvera J, Leyval D, Goergen JL, et al. Feedback-resistant acetohydroxy acid synthase increases valine production in *Corynebacterium glutamicum. Appl. Environ. Microbiol.* 2005; 71: 207–213.

[6] Blombach B, Schreiner ME, Holátko J, Bartek T, Oldiges M, Eikmanns BJ. L-valine production with pyruvate dehydrogenase complex-deficient *Corynebacterium glutamicum. Appl. Environ. Microbiol.* 2007; 73: 2079–2084.

[7] Wada M, Hijikata N, Aoki R, Takesue N, Yokota A. Enhanced valine production in *Corynebacterium glutamicum* with defective H^+-ATPase and C-terminal truncated acetohydroxyacid synthase. *Biosci. Biotechnol. Biochem.* 2008; 72: 2959–2965.

[8] Holátko J, Elišáková V, Prouza M, Sobotka M, Nešvera J, Pátek M. Metabolic engineering of the L-valine biosynthesis pathway in *Corynebacterium glutamicum* using promoter activity modulation. *J. Biotechnol.* 2009; 139: 203–210.

[9] Blombach B, Schreiner ME, Bartek T, Oldiges M, Eikmanns BJ. *Corynebacterium glutamicum* tailored for high-yield L-valine production. *Appl. Microbiol. Biotechnol.* 2008; 79: 471–479.

[10] Buchholz J, Schwentner A, Brunnenkan B, Gabris C, Grimm S, Gerstmeir R, et al. Platform engineering of *Corynebacterium glutamicum* with reduced pyruvate dehydrogenase complex activity for improved production of L-lysine, L-valine, and 2-ketoisovalerate. *Appl. Environ. Microbiol.* 2013; 79: 5566–5575.

[11] Hou X, Ge X, Wu D, Qian H, Zhang W. Improvement of L-valine production at high temperature in *Brevibacterium flavum* by overexpressing *ilvEBN^{r}C* genes. *J. Ind. Microbiol. Biotechnol.* 2012; 39: 63–72.

[12] Hou X, Chen X, Zhang Y, Qian H, Zhang W. L-Valine production with minimization of by-products' synthesis in *Corynebacterium glutamicum* and *Brevibacterium flavum. Amino Acids.* 2012; 43: 2301–2311.

[13] Hasegawa S, Uematsu K, Natsuma Y, Suda M, Hiraga K, Jojima T, et al. Improvement of the redox balance increases L-valine production by *Corynebacterium glutamicum* under oxygen deprivation conditions. *Appl. Environ. Microbiol.* 2012; 78: 865–875.

[14] Hasegawa S, Suda M, Uematsu K, Natsuma Y, Hiraga K, Jojima T, et al. Engineering of *Corynebacterium glutamicum* for high-yield L-valine production under oxygen deprivation conditions. *Appl. Environ. Microbiol.* 2013; 79: 1250–1257.

[15] Sakai S, Tsuchida Y, Nakamoto H, Okino S, Ichihashi O, Kawaguchi H, et al. Effect of lignocellulose-derived inhibitors on growth of and ethanol production by growth-arrested *Corynebacterium glutamicum* R. *Appl. Environ. Microbiol.* 2007; 73: 2349–2353.

[16] Sasaki M, Jojima T, Kawaguchi H, Inui M, Yukawa H. Engineering of pentose transport in *Corynebacterium glutamicum* to improve simultaneous utilization of mixed sugars. *Appl. Microbiol. Biotechnol.* 2009; 85:105–15.

[17] EFSA FEEDAP Panel (EFSA Panel on Additives and Products or Substances used in Animal Feed), 2013. Scientific Opinion on the safety and efficacy of L-valine produced by *Corynebacterium glutamicum* (KCCM 80058) for all animal species, based on a dossier submitted by CJ Europe GmbH. EFSA Journal 2013;11(10):3429, 20 pp. doi:10.2903/j.efsa.2013.3429.

[18] Eggeling, L. 2001. Amino acids, p. 281-303. In C. Ratledge and B. Kristiansen (ed.), Basic bio/technology. Cambridge University Press, London, United Kingdom.

[19] Park JH and Lee SY. Fermentative production of branched chain amino acids: a focus on metabolic engineering *Appl. Microbiol. Biotechnol.* 2010; 85:491–506.

[20] Magnus JB. 2007. Metabolic Engineering of the valine pathway in *Corynebacterium glutamicum* – Analysis and Modelling. Schriften des Forschungszentrums Jülich. GmbH, Reihe Lebenswissenschaften/Life Sciences XIX, ISBN: 978-3-89336-499-2.

In: New Trends in *Corynebacterium glutamicum*
Editor: Carlos Barreiro

ISBN: 978-1-63483-280-9

Chapter 8

Isobutanol Production in *Corynebacterium glutamicum*

Kazumi Hiraga, Crispinus A. Omumasaba and Masayuki Inui*
Research Institute of Innovative Technology for the Earth,
Kizugawa-shi Kyoto, Japan

Abstract

In recent times, *Corynebacterium glutamicum* has started to attract attention as an industrial microorganism in the production of organic acids like lactate and succinate. This emerging reputation is based largely on the findings of the last decade or so that have identified the microorganism as a potent producer of organic acids under oxygen-deprived conditions.

The microorganism is naturally an attractive host for production of alcohols from precursors of amino acid biosynthesis. Higher alcohols like isobutanol possess several properties such as a lower hygroscopicity, vapor pressure and a higher energy density, that allow for better safety in handling and more efficient exploitation than ethanol, the first-generation biofuel from renewable resources. Isobutanol production using *C. glutamicum* under aerobic processes at titers ranging from 4.5 g/L to 13 g/L reflect how different approaches to metabolic engineering can lead to significantly different productivities.

These aerobic productivities have been stretched to 25.4 g/L in a growth-arrested bioprocess. Meanwhile, some approaches that are yet to be incorporated in the reported productivities have been reported. In this chapter, we focus on recent advances in the development of recombinant *C. glutamicum* strains for isobutanol and the consequent production of isobutanol under aerobic as well as under oxygen deprived conditions using *C. glutamicum.*

---

* Corresponding Author: Masayuki Inui; inui@rite.or.jp

INTRODUCTION

Corynebacterium glutamicum is a rapidly growing, gram positive soil bacterium with a longstanding reputation as the workhorse for industrial amino acid production. The amino acids find use as dietary and feed supplements, in cosmetics and pharmaceuticals, and as precursors of antibiotics and herbicides [14], industries which are particularly sensitive to global population changes. The recent rapid increase in global population has expectedly fueled a rise in the demand for amino acids, particularly in the developing regions of South East Asia and Africa [1, 2]. An increasing global population not only strains the amino acid supply but also directly drives the increase in global energy demand in an era of declining fossil fuel reserves. Continued excessive reliance on fossil fuel resources is also blamed for the undesirable changes in global climate changes witnessed in recent times. Clearly there exists a need to diversify the global food, feed, and energy resource pool.

Initial efforts to diversify the global energy resource pool focused on ethanol as a renewable fuel. An ensuing jump in global food prices soon revealed that as long as ethanol continued to be derived from food resources including corn and sugar cane, it could not make for a feasible renewable energy resource in times of rising global food demand. Moreover, higher alcohols are simply better fuels than ethanol due to their higher energy density and lower hygroscopicity, vapor pressure, and corrosivity, full compatibility with existing engines and pipelines that allow for safer handling and more efficient use [15]. Furthermore, branched-chain higher alcohols have even higher energy density: isobutanol has an octane number (motor octane value: 90) higher than that of its 1-butanol isomer (motor octane value: 80). As it is less toxic than 1-butanol, isobutanol is more bioprocess friendly given similar metabolic engineering, protein rational design, and process engineering techniques (gas stripping, two-phase partitioning bioreactor etc.) resources.

It happens that essentially the same mechanisms through which *C. glutamicum* synthesizes some amino acids can be employed to produce isobutanol. Development of more sustainable and cost-effective biofuels is feasible if economical bioprocesses can emerge to help alleviate our reliance on petroleum-based fuels. Increased recent attention to production of higher alcohols from renewable resources is exemplified by efforts to develop viable isobutanol production processes with *C. glutamicum* as one of the more attractive strains.

ADVANCES IN METABOLIC ENGINEERING FOR ISOBUTANOL PRODUCTION

An outline of the shared biosynthetic pathway of L-valine and isobutanol in *C. glutamicum* is shown in Figure 1. The finer details of the L-valine pathway are discussed elsewhere in this book. The first three steps of L-valine and isobutanol biosynthesis from pyruvate are catalyzed by acetohydroxyacid synthase (AHAS, E.C. 4.1.3.18, *ilvBN* gene product), acetohydroxyacid isomeroreductase (AHAIR, E.C. 1.1.1.86, *ilvC* gene product), and dihydroxyacid dehydratase (DHAD, E.C. 4.2.1.9, *ilvD* gene product). Thereafter, production of isobutanol is achieved via the action of heterologous of *kivD* encoding aldehyde reductase from *Lactococcus lactis* and *adhA* encoding alcohol dehydrogenase native to *C. glutamicum* or *adhP* encoding alcohol dehydrogenase derived from *Escherichia coli*. Besides these genetic modifications, strategies to enhance the isobutanol productivity of *C. glutamicum* are essentially the same as those to generate strains with high L-valine productivity: channeling as much pyruvate as possible towards isobutanol through suppressing formation of by-products and efficient sugar metabolism to minimize loss of carbon with properly balanced cofactor dependencies.

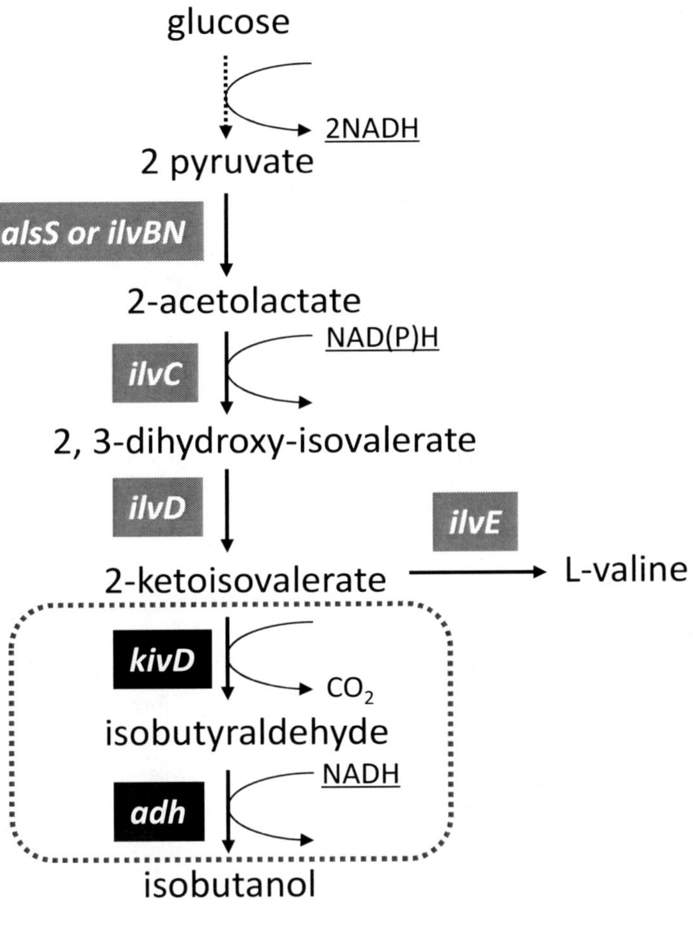

Figure 1. The isobutanol and L-valine biosynthesis pathway from glucose in *C. glutamicum*. Heterologous genes derived from the Ehrlich pathway (dotted line) are introduced into *C. glutamicum* to extend L-valine biosynthetic pathway for production of isobutanol. The need and production of co-factors are also shown.

To maximize the pool of available pyruvate, one approach is to use a platform strain exhibiting a disruption in the *aceE* gene encoding pyruvate dehydrogenase complex (PDHC) enzyme that converts pyruvate to acetyl-CoA. Under aerobic conditions, the acetyl-CoA will consequently enter the TCA cycle to generate energy. As PDHC is essential for growth on glucose, the *aceE*-deficient strain will not grow on glucose as a sole substrate. The medium must therefore be supplemented with acetate to ensure growth. This is the basis that Blombach et al. [6] used in their "two phase cellular process." Aerobic growth phase on acetate is followed by an oxygen-depleted production phase under which growth is arrested.

Pyruvate conversion to lactate by the action of lactate dehydrogenase encoded by *ldhA*, to acetate following catalysis by pyruvate:quinone oxidoreductase encoded by *pqo*, or to oxaloacetate by the *pyc*-encoded pyruvate carboxylase is minimized by disruption the respective genes. Under anaerobic production conditions, anaplerosis can be addressed by disruption of genes including *ppc* encoding PEP carboxylase, *mdh* encoding malate dehydrogenase, and *pyc* encoding pyruvate carboxylase.

Under non-aerobic conditions, a cofactor imbalance similar to the one possible during non-aerobic L-valine production may arise. The isobutanol pathway consumes 1 mole of NADPH and 1 mole of NADH for every mole isobutanol produced. Two moles of NADH are produced upstream in glycolytic (EMP) pathway. Ensuring proper redox cofactor balance under anaerobic conditions is essential to efficient substrate metabolism. Introduction of a gene encoding a transhydrogenase to convert between these cofactors as needed or a change of the cofactor preference of the *ilvC*-encoded AHAIR from NADPH to NADH have been reported.

Atsumi et al. summarized the non-fermentative keto-acid pathways for production of higher alcohols including isobutanol, 1-butanol, 2-methyl-1-butanol, 3-methyl-1-butanol, 2-phenylethanol and 3-methyl1-pentanol as fuel alternatives [3]. Metabolic engineering of *C. glutamicum* for production of isobutanol introduces into the strain the last two steps of the Ehrlich pathway. This imparts the capacity for oxidative decarboxylation of 2-ketoisovalerate, an intermediate of L-valine biosynthesis by keto-acid decarboxylase (KDC). The isobutylaldehyde produced from this reaction is subsequently reduced to isobutanol by alcohol dehydrogenase. KDC activity is critical in this conversion. Common in many organisms, its activity toward 2-keto acids varies markedly among microorganisms. The decarboxylase activity of Kivd from *Lactococcus lactis* was shown to be greater and wider than that of four other KDCs including Pdc6, Aro10, Thi3 from *S. cerevisiae* and Pdc from *Clostridium acetobutylicum* when co-expressed with of Adh2 from *S. cerevisiae* against endogenous 2-keto acids in *E.coli*. The genes encoding the two enzymes turn *C. glutamicum*, a known industrial amino acid producer, into an attractive isobutanol producing strain.

The strategy to utilize the well-developed amino acid production technology in *C. glutamicum* but channel amino acid intermediates toward the 2-keto acid degradation pathway for desired alcohol production does not use CoA-mediated alcohol production pathways as is common in typical alcohol production strains. The strategy may be applicable in eukaryotes like yeasts with better recognized ability for alcohol (ethanol) production by fermentation regardless of the compartmentalization of acetyl-CoA in mitochondria. Based on this strategy, isobutanol has been produced in *C. glutamicum* from glucose by overexpressing genes of the L-valine biosynthetic pathway (*ilvBNCD*), elimination of competing carbon pathways under aerobic or under oxygen-deprived conditions (Figure 1). Table 1 shows the summary of isobutanol production by recombinants of *C. glutamicum*.

Smith et al. (2010) showed that *C. glutamicum* as a host strain is less sensitive to isobutanol toxicity than *Escherichia coli* [4]. After exposure to the same concentration of isobutanol, a greater number of *C. glutamicum* viable cells remained. At concentrations below 12.5 g/L of isobutanol, *C. glutamicum* showed 2-fold better viability than *E. coli*. At 20 g/L of isobutanol, the 18% cell viability of *C. glutamicum* was 4-fold higher. This is evidence for choosing *C. glutamicum* over *E. coli* as host strain for production of isobutanol. Moreover, the growth-arrested bioprocess extensively demonstrated with *C. glutamicum* would be greatly advantageous for the production of such toxic compounds as isobutanol.

Table 1. Isobutanol-producing recombinant strains of *C. glutamicum*

Strain	Genes overexpressed	Host engineering	Titer (g/L)	*Yp/s* (mol/mol glucose) (%)	Production condition	References
C. glutamicum pKS167/ΔpycΔldh	*alsS(B. subtilis), ilvCD, kivD(L. lactis), adhA*	*Δpyc, ΔldhA*	4.9	23	aerobic condition	Smith *et al.* (2010)
C. glutamicum Iso7	*ilvBNCD, kivD(L. lactis), adhA, pntAB(E.coli)*	*ΔaceE, Δpqo, ΔilvE, ΔldhA, Δmdh*	13	77	two phase cellular process (aerobic growth phase, then oxygen-depleted production phase under growth arrested condition)	Blombach *et al.* (2011)
C .glutamicum IBU5	*ilvBNC,ilvD kivD(L. lactis), adhP(E. coli)*	*ΔldhA*	25.4	28	oxygen-deprived, growth arrested condition	Yamamoto *et al.* (2013)
			39.4	58.1	oxygen-deprived, growth arrested condition with two-phase extraction system	
C .glutamicum IBU8	*ilvBNC*^{TM}*,ilvD kivD(L. lactis), adhP(E. coli), gapA(two-copies), pyk, pfk,gpi*	*ΔldhAΔppc*	17.8	56.5	oxygen-deprived, growth arrested condition	Yamamoto *et al.* (2013)
			37.4	78.1	oxygen-deprived, growth arrested condition with two-phase extraction system	

ilvC^{TM}: triple mutant of AHAIR

Since product synthesis is decoupled from the microbial growth, the cells can be utilized as cell factories of sequential enzyme reactions directed from glucose to isobutanol without concern for growth. Minimization of by-product formation pathway by gene knockout enhances product yield from glucose because most of the carbon flux is directed toward the desired final product.

The genetic modifications to *C. glutamicum* described by Smith et al. enabled aerobic production of isobutanol. The recombinant *C. glutamicum* strain was deficient in *pyc* and *ldhA* and overexpressed *Bacillus subtilis alsS* encoding an intrinsically feedback-resistant AHAS, native *ilvCD* and *adhA*, and *Lactococcus lactis kivD*. It produced 4.9 g/L isobutanol following long-term fermentation (120 h) [4]. The choice of *Bacillus subtilis* AHAS was dictated its higher affinity for pyruvate than for 2-ketobutyrate, an intermediate of isoleucine biosynthesis [5]. The same strain was able to produce several other higher chain alcohols as by-products (1-propanol, 1-butanol, 2-methyl-1-butanol, 3-methyl-1-butanol, and 2-phenylethanol).

The approach employed by Blombach et al. (2011) in the production of isobutanol in two-phase cellular process of aerobic growth on acetate followed by oxygen-depleted isobutanol production from glucose utilized the Iso7 strain of *C. glutamicum* overexpressing native *ilvBNCD* and *kivD* derived from *Lactococcus lactis* coupled with native *adhA*. The strain sported additional modification of the same *pntAB* /*ΔaceEΔpqoΔilvEΔldhAΔmdh* genes modified in the construction of the L-valine producing strain [6]. After growth in medium containing a mixture of glucose and acetate, the complete consumption of the acetate triggered a non-aerobic isobutanol production from glucose. The final isobutanol titer of 13 g/L within 39.5h at a 77% yield (mol/mol glucose) [6] is at least 3-fold higher than that obtained in the aerobic production by Smith et al. (2010). The elaborate strain construction techniques that eventually led to the realization of this higher productivity have since been emulated in other approaches to enhance both amino acid and higher alcohol production in *C. glutamicum*. The choice by Blombach et al. of a PDHC-deficient *ΔaceE* platform strain with its abolished carbon flow to TCA cycle has the drawback of growing very poorly on glucose. The addition of acetate to the medium as a carbon source for growth in aerobic growth phase excludes the possibility of employing mineral medium in this approach. This drawback is turned into an advantage when cells shift to oxygen-deprived metabolism in the isobutanol production phase, realizing high productivity. However, resolution of any cofactor imbalances that may accompany oxygen-deprived metabolism is crucial for efficient production of isobutanol. *C. glutamicum* does not possess chromosomally-encoded transhydrogenase to catalyze the interconversion between related cofactors. Even when *E. coli* transhydrogenase is overexpressed in *C. glutamicum*, subsequent inactivation of malic enzyme results in such a severe reduction in the isobutanol production yield as to indicate that NADPH generation in *C. glutamicum* requires malic enzyme. This implies that a transhydrogenase-like metabolic cycle consisting of pyruvate and/or PEP carboxylase, malate dehydrogenase and malic enzyme may contribute to the conversion of NADH + H^+ and $NADP^+$ to NADPH+ H^+ and NAD^+ to maintain the cellular redox balance.

An alternative approach is to engineer a NADH-preferring mutant of AHAIR as described by Hasegawa et al. [12] for use in the high production of isobutanol in *C. glutamicum* described by Yamamoto et al. [7]. Hasegawa et al. used site-directed mutagenesis to reverse the cofactor dependency of AHAIR(IlvC^{TM}) and obviate the requirement of a transhydrogenase in *C. glutamicum* during amino acid production. The triple mutant AHAIR in conjunction with ammonia-dependent NAD-specific leucine dehydrogenase from *Lysinibacillus sphaericus* instead of native Glu-dependent transaminase (*ilvE*) permit the ideal cofactor balance and source of amino-group for oxygen-deprived conditions. The gains in L-valine productivity achieved in *C. glutamicum* were naturally transferable to isobutanol production as described by Yamamoto et al.

BIOPROCESS OPTIMIZATION FOR PRODUCTION OF ISOBUTANOL BY METABOLICALLY ENGINEERED STRAINS

That *C. glutamicum* exhibits high tolerance to isobutanol toxicity implies that it is a strong platform for enhanced isobutanol production. The productivity of the recombinant strain can be stretched much further by optimizing strain construction in concert with processing conditions so that as much product as possible is extracted from the fermentation medium into low toxicity organic solvents during the bioprocess. Yamamoto et al. determined the effects of isobutanol on glucose consumption of *C. glutamicum* under aerobic conditions as well as under oxygen deprived conditions. Glucose consumption under oxygen-deprived conditions continues at isobutanol concentrations beyond which aerobic glucose consumption ceases. The glucose consumption rate under aerobic conditions was 30% suppressed in 1.5% (v/v) isobutanol and exceeded 95% in the presence of 2.5% (v/v) isobutanol. In contrast, the glucose consumption rate under oxygen-deprived conditions was unchanged 2.0% (v/v) isobutanol, beyond which the rate fell to 60% in 2.5% (v/v) isobutanol and to 40% in 3.0% (v/v) isobutanol. The higher tolerance of *C. glutamicum* to isobutanol under oxygen-deprived conditions favors a higher product yield. When combined with use of a more powerful alcohol dehydrogenase derived from *E. coli* MG1655, a production strain highly suited to the bioprocess was the result.

The resultant IBU5 strain deficient in *ldhA* and overexpressing *ilvBNC, ilvD, Lactococcus lactis kivD*, and *E. coli adhP* produced 25.4 g/L (343 mM) of isobutanol from glucose with 28% yield (mol/mol glc) [7]. This figure doubles the productivity reported by Blombach et al. [4] but at a comparatively depressed yield. A different strain IBU8 deficient in *ldhA* and overexpressing *ilvBNC*^{TM}(*ilvC* ^{TM}), *ilvD, Lactococcus lactis kivD, E. coli adhP, gapA, pyk,* and *pfk* produced 17.8 g/L (240 mM) of isobutanol from glucose at a yield of 56.5% mol/mol glucose [7]. It is noteworthy that the titer in the growth-arrested bioprocess is more than 4~5-fold higher than the value reported in the aerobic process by Smith et al. [4].

Yamamoto et al. [7] further optimized the bioprocess by using the metabolically engineered recombinant strain for the growth-arrested bioprocess under oxygen-deprived conditions with two-phase extraction system using oleyl alcohol during the fermentation reaction. The productivity of the IBU5 strain more than doubled to 39.4 g/L (532.5 mM) at a 58.1% yield (mol/mol glucose). The total titer was about 1.6-fold than that without product extraction [7]. The IBU8 strain produced 37.4 g/L (505.3 mM) with 78.1% yield (mol/mol glucose), with a total titer about 2.1-fold higher than that without extraction. Interestingly, the continuous extraction essentially eliminated the differences in total isobutanol production between strains IBU5 and IBU8. Moreover, the obtained extraction-layer isobutanol was almost impurity-free. This is an added advantage of the two-phase extraction system. The

IBU8 strain at higher cell concentration (60 g dry cell weight/L) with two-phase extraction achieved 72.7 g/L (981 mM) of isobutanol.

The Genetic Mechanism of Organic Solvent Resistance and Other Approaches

Elucidation of mechanism of isobutanol resistance potentially should permit the engineering of strains harboring genes that allow for more efficient production of isobutanol in *C. glutamicum*. The body of knowledge of organic solvent resistance in gram-negative bacterium is far larger than that of gram-positive bacteria. Nevertheless, it has been established that isobutanol uptake, efflux, attendant changes membrane composition, and stress-responsive proteins, among others are governed by complex, multigenic phenotypes that have been recalcitrant to rational engineering approaches.

Recent reports of genomic approaches to understand isobutanol resistance have emerged, but only with *E. coli* [8, 9]. Minty et al. (2011) applied experimental evolution followed by genome sequencing and a gene expression study the genetic basis for adaptation to exogenous isobutanol. They reported parallel evolution in the genes *marX, hfq, mdh, acrAB, gatYZABCD* and *rph*. The mechanism of adaptation to isobutanol stress based on remodeling the cell envelope and stress response attenuation through attenuation of the RNA polymerase sigma factor, RpoS, was reported.

Atsumi et al. (2010) isolated an isobutanol-tolerant mutant and whole-genome sequencing followed by gene repair and knockout identified mutations in the five genes *acrA, gatY, tnaA, yhbJ,* and *marCRAB*. They reconstructed the tolerance phenotype by combining deletions of these five loci and identified glucosamine 6-phosphate as an important metabolite for isobutanol tolerance which presumably enhanced membrane synthesis[8]. However, the improved isobutanol tolerance did not mean an increase of the final isobutanol titer when the resistant strain was used in production.

It is instructive that isolated butanol- and isobutanol-tolerant bacteria able to tolerate greater than 2.0% (v/v) solvent are phylogenetically distributed over at least nine genera: *Bacillus, Lysinibacillus, Rummeliibacillus, Brevibacillus, Coprothermobacter, Caloribacterium, Enterococcus, Hydrogenoanaerobacterium, and Cellulosimicrobium* [10]. These are all within the phyla *Firmicutes* and *Actinobacteria*. Strain CM4A able to tolerate 3.5% (v/v) butanol and 4.0% (v/v) isobutanol was identified by Kanno et al. (2013). The authors showed that introduction of the *cfa* gene encoding CFA (cyclopropane fatty acid) synthase from strain CM4A (*Firmicutes*) into *E. coli* conferred butanol and isobutanol resistance to the transformant. While it is possible that elucidation of isobutanol-resistance mechanisms in these microorganisms may give valuable information for construction of recombinant isobutanol–resistant *C. glutamicum* strains, more information that will enable researchers to link the improved solvent tolerance with improved isobutanol production is still needed. It should be noted that high titer production of these compounds at concentrations ordinarily toxic to *C. glutamicum* has already been achieved in the two-phase extraction system earlier describe despite the prevailing lack of genetic information on the mechanism of isobutanol resistance [7]. Gas stripping is alternative method to recover the solvent from the fermentation medium in *E. coli* [11]. But should such bioprocess engineering approaches be combined with multi-omics analyses, *in silico* models, genomic information of isobutanol resistance, high throughput screening (especially highly specific and highly active *kivD* and *adh* combination), and directed evolution, a powerful arsenal for high production of isobutanol should be at hand.

Other approaches to enhance microbial isobutanol production from which important lessons can be learned have been reported. Lin et al. reported isobutanol production at elevated temperatures in thermophilic *Geobacillus thermoglucosidasius* [13]. In general, high temperature production of compounds that are not native to the thermophilic host is limited by enzyme stability and the lack of suitable expression systems. They reported the engineering of *Geobacllus thermoglucosidasius* to produce isobutanol at 50°C. With the best enzyme combination and the expression system, 2.2 g/L of isobutanol was produced from glucose and 0.6 g/L of isobutanol production from cellobiose in *G. thermoglucosidasius* within 48 at 50°C. Cofactor stability at these elevated temperatures could be an issue to examine, but this approach differs markedly from approaches under consideration during organic solvent production by mesophilic microorganisms.

CONCLUSION

Biofuels are expected to play an increasingly important role as to supplement the global energy role of petroleum-based fuels. Higher carbon biofuels such as isobutanol will be the bedrock of next generation molecules of energy, finding broader applications commensurate with their characteristic high energy content, low hygroscopicity, as well as other attractive properties. Metabolic engineering provides the fastest and most rational way of improving the microbial cell's ability to produce these alternative fuels. Expanding the pathway that leads to cellular production of ketoacids has been the most successful way to produce non-natural higher chain alcohols as isobutanol in *C. glutamicum,* representing remarkable progress of metabolic engineering of the biosynthetic pathway.

2-ketoacids are metabolic intermediates of amino acid metabolism that can be used for the production of various branched-chain and linear alcohols. Each 2-ketoacid can be converted into the corresponding 1-aldehyde by decarboxylation catalyzed by 2-ketoacid decarboxylase with broad substrate specificity. The aldehyde can then be converted into the corresponding 1-alcohol by alcohol dehydrogenase. 2-ketoacid independent pathways present a new choice for production of some biofuels, but 2-ketoacid dependent pathways are proven in terms of feasibility. The latter pathways therefore are a more powerful platform for the production of biofuels and some chemicals.

The production of these chemicals has been demonstrated under aerobic as well as under non-aerobic conditions. Non-aerobic production appears to be more efficient at this early stage as has been confirmed in a number or reports. Under oxygen-deprived conditions, the redox cofactor balance is important for efficient production of desired compounds at high yield. The two reported approaches to achieve cofactor balance involve genetic modification to channel carbon away from the glycolytic into the pentose phosphate pathway or to modify the cofactor dependency of the AHAIR enzymes involved in the pathway leading to isobutanol production. The growth-arrested bioprocess is efficient for production of isobutanol mainly because major carbon flow is not needed to support cellular proliferation.

The highest productivity of isobutanol reported in the growth-arrested bioprocess is more than two-fold greater than that obtained in the aerobic process by Smith et al. (2010). Combined with bioprocess optimization techniques, the growth-arrested bioprocess could unlock considerable industrial production of toxic compounds. Through protein rational design, metabolic engineering, and bioprocess engineering techniques, Yamamoto et al. reported efficient production of isobutanol under the growth-arrested condition [7]. Other emerging approaches to improve microbial organic solvent resistance and lessons from thermophilic organic solvent production studies promise a future of intense discovery in the journey to industrial applications.

REFERENCES

[1] Ajinomoto (2012) Food products business. Available: http://www.ajinomoto.com/jp/ir/pdf/fact/Food-Oct2012.pdf

[2] Ajinomoto (2013) Feed-use amino acids business. Available: http://www.ajinomoto.com/jp/ir/pdf/fact/Feed-useAA-Oct2013.pdf

[3] Atsumi S, Hanai T, Liao JC. Non-fermentative pathways for synthesis of branched-chain higher alcohols as biofuels. *Nature*. 2008; 451(7174): 86-89.

[4] Smith KM, Cho KM, Liao JC. Engineering *Corynebacterium glutamicum* for isobutanol production. *Appl. Microbiol. Biotechnol.* 2010; 87:1045–1055.

[5] Gollop N, Damri B, Chipman DM, Barak ZEE V. Physiological implications of the substrate specificities of acetohydroxy acid synthases from varied organisms. *J. Bacteriol.* 1990; 172(6): 3444-3449.

[6] Blombach B, Riester T, Wieschalka S, Ziert C, Youn J-W, Wendisch VF, Eikmanns BJ. *Corynebacterium glutamicum* tailored for efficient isobutanol production. *Appl. Environ. Microbiol.* 2011; 77(10): 3300–3310.

[7] Yamamoto S, Suda M, Niimi S, Inui M, Yukawa H. Strain optimization for efficient isobutanol production using *Corynebacterium glutamicum* under oxygen deprivation. *Biotechnol. Bioeng.* 2013; 110: 2938–2948.

[8] Atsumi S, Wu TY, Machado IM, Huang WC, Chen PY, Pellegrini M, Liao JC. Evolution, genomic analysis, and reconstruction of isobutanol tolerance in *Escherichia coli. Mol. Syst. Biol.* 2010; 6: 449.

[9] Minty JJ, Lesnefsky AA, Lin F, Chen Y, Zaroff TA, Veloso AB, Xie B, McConnell CA, Ward RJ, Schwartz DR, Rouillard JM, Gao Y, Gulari E, Lin XN. Evolution combined with genomic study elucidates genetic bases of isobutanol tolerance in *Escherichia coli. Microb. Cell Fact.* 2011; 10:18.

[10] Kanno M, Katayama T, Tamaki H, Mitani Y, Meng XY, Hori T, Narihiro T, Morita N, Hoshino T, Yumoto I, Kimura N, Hanada S, Kamagata Y. Isolation of butanol- and isobutanol-tolerant bacteria and physiological characterization of their butanol tolerance. *Appl. Environ. Microbiol.* 2013; 79(22): 6998-7005.

[11] Baez A, Cho K-M, Liao JC. High-flux isobutanol production using engineered *Escherichia coli*: a bioreactor study with *in situ* product removal. *Appl. Microbiol. Biotechnol.* 2011; 90:1681–1690.

[12] Hasegawa S, Uematsu K, Natsuma Y, Suda M, Hiraga K, Jojima T, et al. Improvement of the redox balance increases L-valine production by *Corynebacterium glutamicum* under oxygen deprivation conditions. *Appl. Environ. Microbiol.* 2012; 78(3): 865–875.
[13] Lin PP, Rabe KS, Takasumi JL, Kadisch M, Arnold FH, Liao JC. Isobutanol production at elevated temperatures in thermophilic *Geobacillus thermoglucosidasius. Metab. Eng.* 2014; 24:1–8.
[14] Park JH, Lee SY. Fermentative production of branched chain amino acids: a focus on metabolic engineering. *Appl. Microbiol. Biotechnol.* 2010; 85:491–506.
[15] Dürre P. Biobutanol: an attractive biofuel. *Biotechnol. J.* 2007; 2:1525–1534.

In: New Trends in *Corynebacterium glutamicum*
Editor: Carlos Barreiro

ISBN: 978-1-63483-280-9

Chapter 9

ANAEROBIC PRODUCTION OF ORGANIC ACIDS AND ALCOHOLS WITH *CORYNEBACTERIUM GLUTAMICUM*

Stéphane Delaunay*[*1], Abdoul Karim Kaboré[1], Mickaël Cregut[1, 2], Huang Bao Truc Khuat[1] and Emmanuel Guedon[1]

[1] Laboratoire Réactions et Génie des Procédés,
Université de Lorraine, Vandoeuvre-lès-Nancy Cedex, France
[2]UMR CNRS 6144 GEPEA CBAC laboratory,
University of Nantes, La Roche sur Yon, France

ABSTRACT

With an annual production of 2.5 million tons L-glutamate and 1.5 million tons L-lysine, *Corynebacterium glutamicum* is the major amino acid producer. Such processes occur under aerobic condition, at high rates and productivities, enabling *C. glutamicum* to hold the leading position in this field. Recently, the potentiality of anaerobic cultivation of *C. glutamicum* was explored. In this culture condition, its metabolism could be dramatically modified. Using wild type or recombinant strains, organic acids can then be produced in significant amounts. In fact, organic acids, among them mono (*i.e.* pyruvate, acetate, lactate...) and diacids (succinate) are key molecules, generally regarded as building block chemicals. Other organic molecules such as alcohols can also be produced under oxygen-depleted conditions. In this review, actual developments and perspectives of the use of *C. glutamicum* in anaerobic processes will be presented with a focus on the metabolic and scientific challenges.

INTRODUCTION

Exhaustion of oil and energy reserves and the inevitable rise in the price of these resources represent many challenges for chemical industry. As an alternative, chemical

[*] Corresponding author: Stéphane Delaunay; Stephane.Delaunay@univ-lorraine.fr

industry turned to biotechnology in order to transform cost effective and renewable biomass deriving mainly from agro resources. In particular, fermentation processes can provide a large number of biomolecules such as acids (pyruvate, lactate , succinate, fumarate , itaconate…) or alcohols (ethanol, butanol, butanediol…), able to replace the building blocks deriving from petroleum for the synthesis of polymers, lubricants, resins or gasoline [1]. However, most fermentative microorganisms are unable to produce one or more products at industrially attractive concentration and with productivity high enough to make the process competitive compared to their equivalent provided by the petrochemical industry. Therefore, development of new microbial catalysts combining these industrial requirements must be considered.

Since its discovery in 1957 *Corynebacterium glutamicum* became a major industrial microorganism mainly known for its ability to produce amino acids such as glutamate and lysine at very large scale. Besides large knowledge of its central carbon metabolism, physiology and genome [2-4], *C. glutamicum* displays many intrinsic properties, making this biocatalyst a potential cell factory for the production of primary metabolites. This organism has a strong resistance to stressful environments (agitation, salinity, solvent...), an excellent robustness in industrial processes as well as an ability to use, either alone or in co-metabolism, carbon sources such as glucose, fructose, sucrose, acetate, lactate, pyruvate, propionate, glutamate, gluconate and serine. In fact, the multi-purpose capabilities of *C. glutamicum* have already been implemented mainly for the industrial production of glutamate and lysine resulting in high yields and conversion rates. Thus, *C. glutamicum* has emerged for more than ten years, as a new platform for metabolic engineering, for the synthesis of several precursor chemicals [5]. In particular, it has been recently shown that, under conditions of oxygen limitation, metabolism of *C. glutamicum* is mainly oriented towards formation of organic acids whereas growth is almost abolished in this condition. Interestingly, growth arrested condition is also useful to perform microbial processes at high cell densities, leading to high volumetric productivities [6].

This chapter focuses on metabolic engineering approaches of *C. glutamicum* with the objectives to produce valuable metabolites in oxygen limited conditions from renewable substrates.

ANAEROBIC METABOLISM OF C_6-CARBOHYDRATES IN *CORYNEBACTERIUM GLUTAMICUM*

In anaerobiosis, in presence of nitrate ions, *C. glutamicum* is still able to grow thanks to the anaerobic respiration. This growth remains limited because of the accumulation of the nitrites which are not reduced in ammonium ions [7-8]. This metabolic capacity of reducing nitrate ions is due to the presence, in *C. glutamicum* genome, of the operon *narKGHJI* encoding a nitrate/nitrite transporter and a nitrate reductase. However, in deprived oxygen conditions, in absence of nitrate ions, with C_6-carbohydrates as carbon source, *C. glutamicum* retains a metabolic activity and can excrete high amounts of organic acids such as lactic acid, succinic acid and acetic acid (Figure 1). This conversion of the carbon source into organic acids is not associated with a residual growth of the bacterium [6]. It is noteworthy that this lack of growth can be considered as an advantage. Indeed, the accumulation in the culture medium of by-products, which are classically synthesized in aerobiosis, remains highly limited. In this context, high productivities were obtained for the production of lactate, succinate and also of other metabolites whose production implies deprived oxygen conditions.

The transcriptional profiles of *C. glutamicum* R were compared in both aerobic conditions and limiting conditions of oxygen supply. In these last ones, 161 genes were up-regulated (more than 2 fold) and 221 genes were down-regulated (more than 2 fold) [9]. These genes were involved in transcription, translation, DNA replication, metabolite transport or metabolism [10]. The main responsible for the metabolic modifications between aerobic and anaerobic culture conditions could be the non-essential sigma factor, SigB.

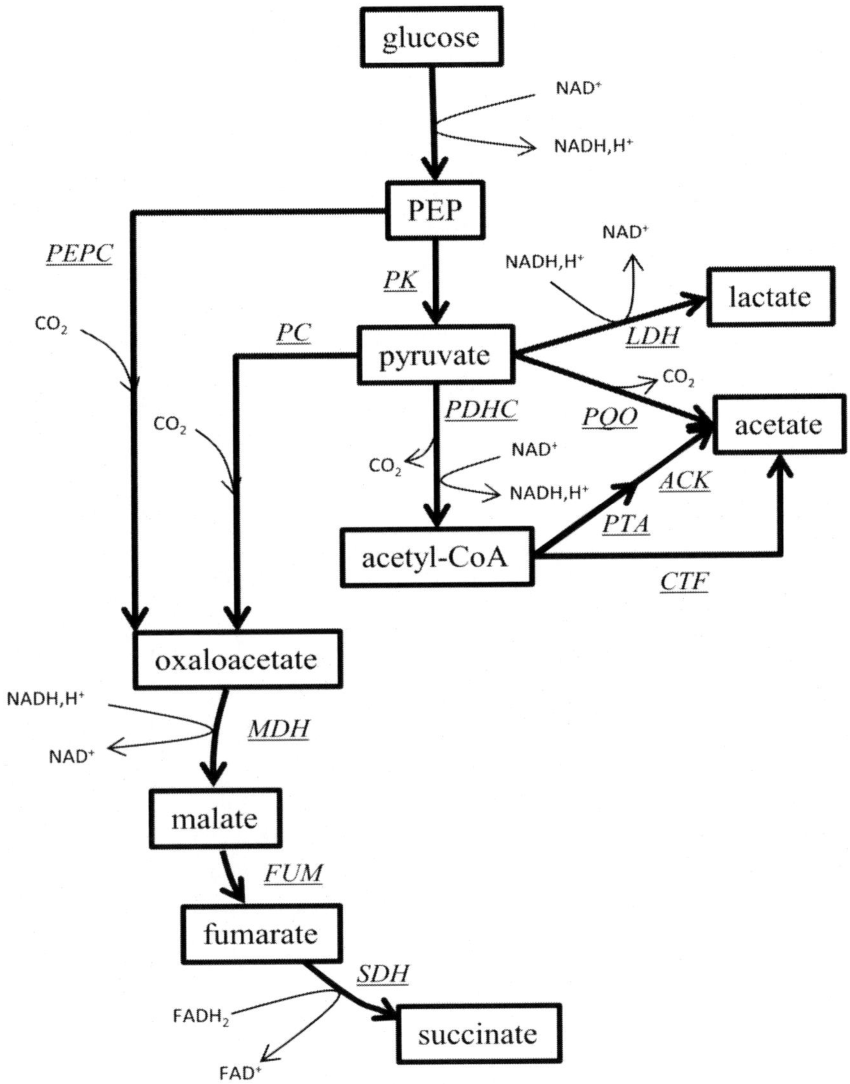

PEP: phosphoenolpyruvate; PEPC: phosphoenolpyruvate carboxylase; PC: pyruvate carboxylase; PK: pyruvate kinase; PDHC: pyruvate dehydrogenase complex; PQO: pyruvate: quinone oxydoreductase; MDH: malate dehydrogenase; FUM: fumarase; SDH: succinate dehydrogenase; LDH: lactate dehydrogenase; CTF: coenzyme A transferase; PTA: phosphotransacetylase; ACK: acetate kinase.

Figure 1. Glucose metabolism in *C. glutamicum*, in deprived oxygen conditions.

The transcription of this factor increases during the transition phase between the exponential growth phase and the stationary growth phase [11]. The expression of SigB is also up-regulated during acid or salt stresses, after temperature variations (cold or heat), after an addition of ethanol [12], in response to a lactate accumulation [13] and in limiting conditions of oxygen supply [9, 14]. In these last conditions, transcriptomic studies show that the deletion of SigB induces a reduction in the transcription of the genes involved in the conversion of glucose into organic acids.

Some common steps of the biosynthetic pathways of metabolites produced under oxygen deprived oxygen are the carbohydrate uptake and the glycolysis. In the first part of this chapter, an overview of these pathways is given. Only glucose, fructose and sucrose, which are the main carbon sources used in industry and in academic works, will be considered as substrates. Then, a focus is made on the impact of oxygen deprived conditions on their regulation.

Glucose, Fructose and Sucrose Uptake

Corynebacterium glutamicum is able to use a high variety of carbohydrates and organic acids as carbon sources. In *C. glutamicum*, glucose, fructose or sucrose can penetrate inside the cell using a phosphotransferase system (PTS). Glucose can also be imported by a permease system (Figure 2). PTS plays a major role in the uptake and the phosphorylation of the glucose and fructose in *C. glutamicum*. It utilizes phosphoenolpyruvate as substrate in the phosphorylation reaction generating glucose-6-P and fructose-1-P, respectively and some pyruvate as cosubstrate [15-16]. However, the consumption of glucose by a *C. glutamicum* strain without PTS suggested the presence of alternative uptake system for this carbohydrate. This system would be a permease with a low affinity [15]. On glucose as the sole carbon source, the permease system may be responsible for up to 15% of the whole uptake flow in *C. glutamicum* ATCC17965 [17] and 25% in *C. glutamicum* 2262 [18]. Some candidates responsible for this permease activity could be the products of *iolT1* and *iolT2* [19]. The uptake of carbohydrates through a permease system necessitates a subsequent phosphorylation. This is achieved by a glucokinase whose activity has been determined in *C. glutamicum* for several decades [16]. Several potential glucokinases have been identified in *C. glutamicum*: Glk [20], PpgK [21] and the product of *cgl2647* [19].

Classically, PTS is composed of 2 cytoplasmic proteins (enzyme I (EI) and HPr protein) and of one protein complex which is specific for the carbohydrate (Enzyme II (EII)). The phosphoryl group from PEP is sequentially transferred to EI, HPr, EII and, at last, to the carbohydrate. EI and HPr which are present in all the PTS from *C. glutamicum* are encoded by *ptsI* and *ptsH*, respectively. The enzymes II, specific for glucose, fructose or sucrose are encoded by ptsG, *ptsF* and *ptsS*, respectively [19, 22-23].

In the case of sucrose consumption, during its uptake, this disaccharide is phosphorylated by PTS^{Suc}. The phosphorylated sucrose is then hydrolyzed into glucose-6-P and fructose. Whereas glucose-6-P is metabolized through the glycolytic pathway, fructose is exported outside. The mechanisms involved in the fructose export remains unknown. Exported fructose can then be up taken by PTS^{Fru} and thus, phosphorylated into fructose-1-P [15].

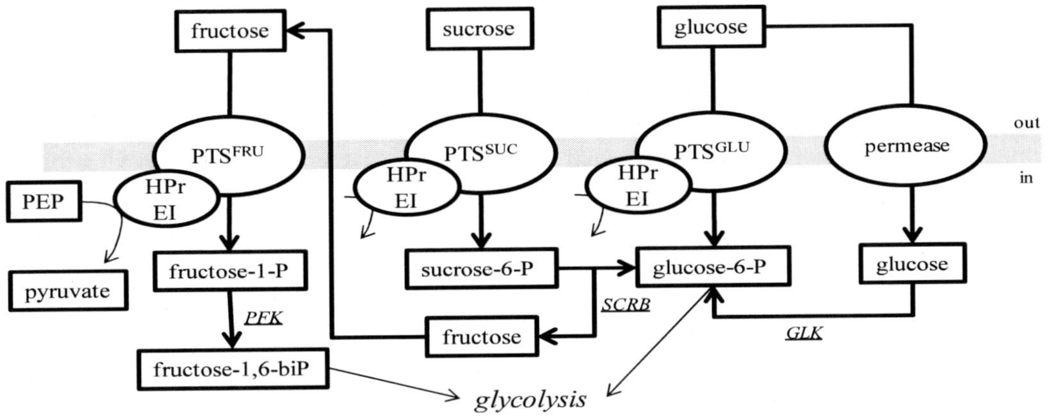

PEP: phosphoenolpyruvate; PTS^{Glc}: glucose PTS; PTS^{Fru}: fructose PTS; PTS^{Suc}: sucrose PTS; Hpr: phosphocarrier protein; EI: PTS phosphotransferase I; PFK: 1-phosphofructokinase; SCRB: sucrose-6-phosphate hydrolase; GLK: glucokinase

Figure 2. Glucose, fructose and sucrose uptake systems in *C. glutamicum.*

C. glutamicum is able to metabolize simultaneously glucose, fructose and sucrose. This can be possible only in the simultaneous presence of several uptake systems responsible for the entry of these three carbohydrates. However, several works show that the respective expression of the PTS is modulated according to the sugars present in the culture medium. Glucose increases the expression of *ptsG*, *ptsH* and *ptsI* whereas fructose and sucrose induce the expression of *ptsF*, *ptsS*, *ptsG*, *ptsH* and *ptsI* [24-25]. Pts G, Pts F and Pts H were also shown to be involved in the uptake of a C_5-carbohydrate, the mannose [26]. These regulations are under the control of transcriptional regulators including SugR and FruR. In *C. glutamicum* genome, *sugR* and *fruR*, encoding DeoR-type transcriptional regulators, are located in the cluster including *ptsF*, *ptsI* and *ptsH*. SugR has been identified as a repressor of *ptsG* and FruR of *ptsF*, *ptsI* and *ptsH* [24, 27-28].

Glycolysis

Upon the sugar uptake and their phosphorylation, the phosphate-sugars metabolism implies 2 pathways: glycolysis and phosphate-pentoses pathway. However, in *C. glutamicum*, in absence of oxygen, glysolysis is mainly responsible for the carbohydrate metabolism leading to the synthesis of pyruvate which is the precursor of most of the metabolites produced under oxygen deprived conditions (Figure 3). Although glycolysis is a pathway of primary importance, all of the glycolytic enzymes have not been precisely characterized in *C. glutamicum*. The main biochemical studies were focused on the irreversible reactions. To date, without considering fructose-1,6-phosphatase and PEP synthase which catalyze the reverse reaction of 6-phosphofructokinase and pyruvate kinase, respectively, and which are involved in the neoglucogenesis, only 4 glycolytic enzymes have been purified or partially purified and 3 enzymes have been biochemically characterized (for review, see [29]). Nevertheless, the sequencing and annotation of *C. glutamicum* genome permitted to conclude that the genes encoding all of the enzymes composing the glycolysis are present [2]. Most of these genes are highly regulated at the transcriptional level (for review, see [30-31]).

To date, pyruvate kinase, which catalyzes the irreversible dephosphorylation of PEP into pyruvate, is recognized as the main control point of glycolysis in *C. glutamicum*. This enzyme has been purified and characterized from *C. glutamicum* ATCC14067 [32] and from *C. glutamicum* ATCC13059 [33]. This tetrameric enzyme is allosterically activated by AMP and strongly inhibited by ATP. It is encoded by *pyk*. The other irreversible enzyme of the glycolytic pathway, the phosphofructokinase, has been characterized from *C. glutamicum* ATCC14067 [34]. In *C. glutamicum*, this enzyme, encoded by *pfkA*, was shown to be inhibited by ADP. No allosteric control mechanism has been identified so far. This explains why this enzyme exerts a weak control on the carbon flux within the glycolysis.

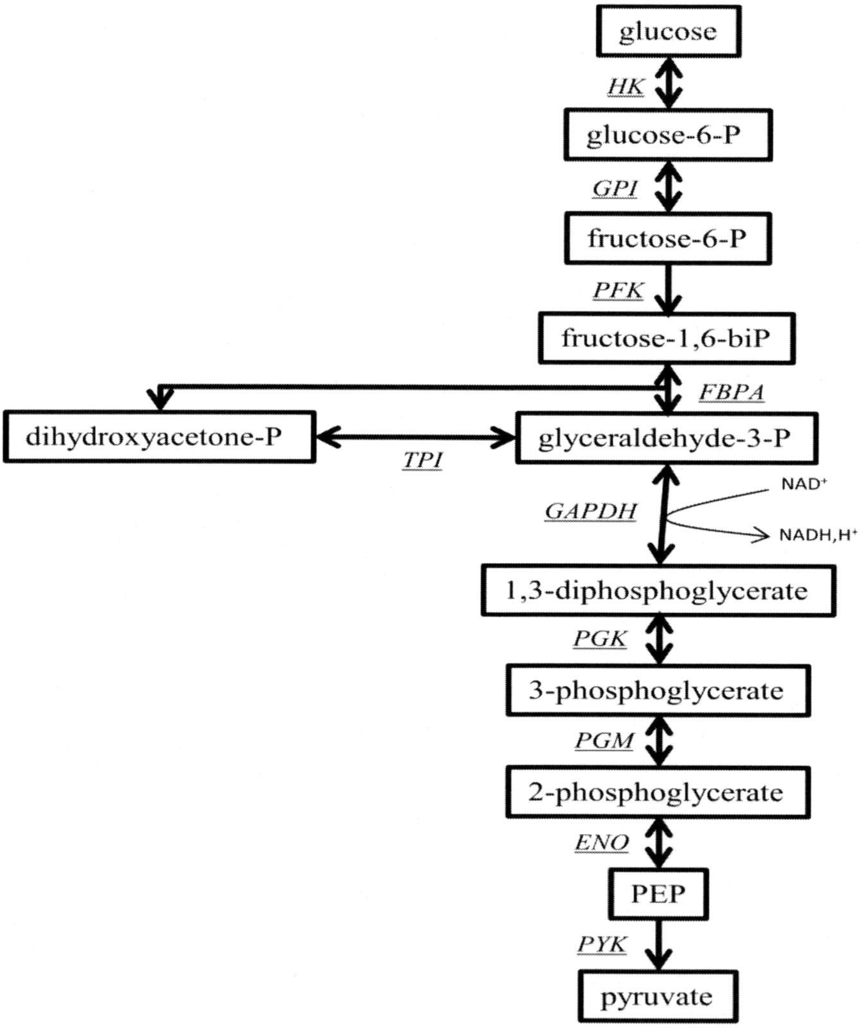

HK: hexokinase; GPI: glucose-6-phosphate isomerase; PFK: phosphofructokinase; FBPA: fructose biphosphate aldolase; TPI: triose phosphate isomerase; GAPDH: glyceraldehyde-3-phosphate dehydrogenase; PGK: phosphoglycerate kinase; PGM: phosphoglycerate mutase; ENO: enolase; PYK: pyruvate kinase.

Figure 3. Glycolysis pathway in *C. glutamicum.*

The other glycolytic enzyme that has been characterized to date, in *C. glutamicum*, is the glucose-6-phosphate isomerase, encoded by *pgi*. Considering the affinity constants of this enzyme for its substrates and their respective intracellular concentrations, it has been estimated that this enzyme would be, *in vivo*, close to its maximal activity in *C. glutamicum*, and, would exert consequently a weak control on the glycolytic flux [35-36].

Two genes encoding a glyceraldehyde-3-phosphate dehydrogenase have been identified in *C. glutamicum*, *gapA* and *gapB* [37]. Only the first one is expressed during the growth on glycolytic carbon sources while the second one is expressed during culture on neoglucogenic carbon sources [38].

As previously mentioned for some *pts* genes, some genes encoding glycolytic enzymes are controlled by SugR. *gapA*, *pfkA*, *fba* (fructose biphosphate aldolase gene), *eno* (enolase gene) and *pyk* are repressed by this transcriptional regulator [39-40].

Regulation of the Genes Encoding Glycolytic Enzymes in *C. glutamicum* Grown in Limiting Conditions of Oxygen Supply

In *C. glutamicum* R, during a culture in limiting conditions of oxygen supply, the glucose consumption is 1.4 fold increased compared to a culture without oxygen limitation [9]. In these oxygen deprived conditions, the genes encoding the main enzymes of the glycolysis, comprising *gapA*, *pgk* (phosphoglycerate kinase gene) and *tpi* (triose phosphate isomerase gene) are 2 fold up-regulated [9]. These genes belong to a same cluster in the genome of *C. glutamicum* R [37]. Moreover, the enzyme activity of the products of these three genes is increased in these culture conditions [9]. One consequence of this change in these enzyme activities is the increase in the carbon flow through the glycolysis and then, a subsequent increase in the specific rate of glucose consumption. However, the respective influence of the three enzymes in this modification of the carbon flow in the glycolysis remains to clarify.

In the oxygen deprived conditions, *sigB* which encodes a group 2 sigma factor of RNA polymerase acting as a transcriptional regulator, is overexpressed [9].

This determination suggested that SigB would be of importance in the regulation of glycolytic flux in limiting conditions of oxygen supply. The *sigB* deletion induces the reduction in the transcription of *tpi*, *gapA*, *pgk* and also *fba* during the culture of the mutant strain on glucose. As a consequence, a decrease in the glucose consumption by the *C. glutamicum* mutant strain was determined in limiting oxygen supply. On the contrary, the growth and the glucose consumption were not affected in the same mutant but in strictly aerobic conditions. This behavioral difference, according to the oxygen supply, suggests that SigB would be a positive regulator of the genes encoding glycolytic enzymes [14].

Lactate Biosynthesis Pathway

Currently, the production of lactate has been observed in *C. glutamicum* in limiting oxygen supply but also in aerobic conditions, at the beginning of the stationary growth phase or during the exponential growth phase during the glutamate production process with a biotin limitation or with a variation of the culture temperature [6, 41-45]. However, the productivity of the lactate production is the highest in anaerobic conditions. In these conditions, the lactate production is related to the production of succinate and acetate [6, 44].

The lactate biosynthetic pathway (Figure 1) is thus considered as the main metabolic pathway in anaerobiosis in *C. glutamicum* wild type strains.

In the lactate biosynthesis pathway, pyruvate is reduced into lactate by the NAD-dependent lactate dehydrogenase. During the lactate biosynthesis, lactate dehydrogenase catalyses the concomitant oxidation of NADH,H^+ in NAD^+. This contributes to the cellular redox balance since NAD^+ is reduced into NADH,H^+ by the glyceraldehyde-3-phosphate dehydrogenase in the glycolysis pathway.

In *C. glutamicum*, lactate dehydrogenase, encoded by *ldhA*, is a 314 amino-acids protein. It catalyzes the reduction of pyruvic acid into lactic acid in presence of NADH, H^+ as the exclusive cofactor.

The optimal pH for this reaction is 6.6. No activity for the L-lactate oxidation was measured with this enzyme whatever the reaction pH and the cofactor nature [42]. Moreover, a *ldhA*-deleted mutant of *C. glutamicum* is unable to produce L-lactate [6]. Both results suggest that no L-lactate dehydrogenase isoenzyme is present in *C. glutamicum*. Km and V_{max} values were determined for pyruvic acid. They were 7.3 mM and 120 mmol/min, respectively, in *C. glutamicum* ATCC13032 [42]. In numerous Gram-positive bacteria, L-lactate dehydrogenases are activated by fructose-1,6-biphosphate [46].

The expression of *ldhA* and the activity of lactate dehydrogenase are increased at the beginning of the stationary growth phase [9], during a fructose-1,6-biphosphate intracellular accumulation, in presence of high ratio of NADH,H^+/NAD^+ [42] or during a limitation in the oxygen supply. In this last condition, the level of *ldhA* mRNA is 9 fold increased [6-9]. The overexpression of *ldhA* at the beginning of the stationary growth phase is determined only in the presence of carbohydrates. Without this type of carbon sources, *ldhA* expression is entirely repressed. This suggests that the presence of carbohydrates is essential for the induction of *ldhA* expression [9].

Whereas a *ldhA*-deleted mutant strain of *C. glutamicum* R is able to produce concentrations of succinate and acetate similar than the wild type strain, its specific rate of glucose consumption decreases and the intracellular ratio of NADH,H^+/NAD^+ is significantly increased. This last measurement has to be correlated to the decrease in the NADH,H^+ oxidation due to the lack of the enzyme. As a consequence to the intracellular accumulation of NADH,H^+, the glyceraldehyde-3-phosphate dehydrogenase activity is inhibited and thus, the glycolytic flux decreases [6].

It has been shown that SugR, which, as indicated previously, is considered as a regulator of PTS and glycolysis genes, is also a transcriptional repressor of *ldhA*. The presence of carbohydrates (glucose, fructose, sucrose or ribose) decreases this repression. This regulation allows the cells to use efficiently the carbon source according to the environmental conditions [24, 27-28, 40, 42]. The repression of *ldhA* by SugR is important for the production of lactate in anaerobiosis but also in aerobiosis. Indeed, it has been shown that the deletion of SugR induced an increase in the L-lactate dehydrogenase activity in aerobiosis and in absence of oxygen. In this last condition, an 8-fold increase was measured. Moreover, the sugR over-expression induces a 25% reduction of the lactate production in anaerobiosis whereas the deletion of this gene was responsible for a 3-time increase in the production of lactate in absence of oxygen [39]. However, fructose-1-phosphate and fructose-1,6-biphosphate exert a negative control on the interaction of SugR with the promoter-targets, including the promoter of *ldhA* [24, 39]. The maximal production of lactate with *C. glutamicum* is then measured only in both limited conditions of oxygen supply and non-limiting supply of carbohydrates within the glycolysis pathway [39].

Succinate Biosynthesis Pathway

In absence of oxygen, and without exogenous electron acceptors, *C. glutamicum* produces some succinate from carbohydrates, using anaplerotic pathways and the reducing arm of TCA cycle [6, 43-44].

In the succinate biosynthesis pathway, phosphoenolpyruvate, a glycolytic intermediate, is converted into oxaloacetate by phosphoenolpyruvate carboxylase. Oxaloacetate can also result from the reaction catalyzed by the pyruvate carboxylase. Oxaloacetate is then converted into succinate by the enzymes of the reductive arm of the TCA cycle (Figure 1).

Pyruvate carboxylase (PC), a 1,140 amino acids enzyme, is encoded by *pyc* [47, 48]. This enzyme catalyzes the irreversible carboxylation of pyruvate to oxaloacetate using ATP. It appears that PC is strongly inhibited by AMP, ADP, acetyl-CoA and aspartate [49, 50]. PC requires biotin for its activity. The second anaplerotic enzyme involved in the succinate biosynthesis pathway, PEP carboxylase (PEPC), encoded by *ppc*, catalyzes the condensation of one bicarbonate on PEP to form oxaloacetate and inorganic phosphate. The enzyme biosynthesized by *C. glutamicum ssp. flavum* was purified and was shown to be activated by acetyl-CoA and fructose-1, 6-bisphosphate and inhibited by aspartate and α-ketoglutarate [51-52]. The same regulation pattern was observed for the enzyme produced by *C. glutamicum* ATCC 13032 [53]. In *C. glutamicum*, the *ppc* gene is associated to the 3-phosphoglycerate kinase and triosephosphate isomerase genes in an operon [54]. In oxygen limited supply, oxaloacetate (OAA) is synthesized rather via the conversion of PEP catalyzed by PEPC than via the conversion of pyruvate by the PC [6, 9]. Therefore, PEPC is the predominant anaplerotic enzyme for the production of succinate. As a result of overexpression of PEPC in oxygen deprived conditions, an increase in the production of succinate by *C. glutamicum* occurs concomitantly to the increase in the MDH activity [9]. In contrast, in aerobic conditions, PC was demonstrated to be the predominant anaplerotic enzyme involved in the reconstitution of the OAA pool for the production of amino acids such as glutamate or lysine [55].

Two types of L-malate dehydrogenase are present in *C. glutamicum*, one is located in the cytoplasm, and is depending on the NAD for its activity (MDH, encoded by *mdh*), whereas the second, the malate:quinone Oxidoreductase (MQO, encoded by *mqo*), is associated to the cellular membrane. The two enzymes have been purified and characterized [56-57]. It was shown that a mutant deficient in MQO is unable to grow on minimal medium while a mutant deficient in MDH has no apparent phenotype [57]. The *mdh* gene encodes a polypeptide of 328 residues and transcriptomic analyses revealed no indication concerning the influence of the nature of the carbon source on the expression of this gene [58]. However, it has been measured that the specific activity of MDH is three times higher when the cells are cultured on acetate rather than on glucose as the sole source of carbon [57].

Fumarase (FUM), also known as fumarate hydratase, catalyzes the reversible conversion of fumarate into malate. It has been purified from *C. glutamicum* ATCC14067 [59]. This enzyme is a homotetramer of 200 kDa. Its N-terminal sequence corresponds to the amino acid sequence deduced from the putative encoding gene (*fum*) identified in the genome of *C. glutamicum* ATCC 13032 [2]. ATP has been found to be an inhibitor of fumarase activity [59].

Succinate dehydrogenase is an enzyme complex also known as succinate:menaquinone oxidoreductase. It belongs to the dehydrogenases that, in *C. glutamicum*, transfer electrons into the respiratory chain. In *C. glutamicum*, this complex is composed of three subunits, SdhA, SdhB, and SdhC, which are encoded by the operon *sdhCAB* [2, 60]. SdhC subunit is anchored in the membrane of the bacterium. This enzyme complex has been purified from *C. glutamicum* ATCC13869 [61]. The respective molecular masses of SdhA, SdhB and SdhC are 67, 29 and 23 kDa whereas the apparent molecular mass of the non-denaturated complex is 410 kDa. This suggests that the succinate dehydrogenase could be a homotrimeric complex of the heterotrimeric protomer. Sdh A subunit was also shown to be essential for the growth of *C. glutamicum* AS019E12 on acetate as sole carbon source since a Δ*sdhA* mutant strain is unable to grow on this carbon source [62].

Mutant strain of *C. glutamicum* R disrupted in *fum* gene was shown to produce minute amounts of succinate [6]. This indicates that presence of FUM activities is necessary for the synthesis of succinate in *C. glutamicum* in these culture conditions. However, Inui and co-workers [9] showed that under oxygen deprived conditions, *fum*, *sdhA*, *sdhB*, and *sdhC* genes are all down-regulated whereas *mdh* gene is 2-3 times overexpressed as well as *ppc* gene. The disruption of transcriptional regulator, *sigB*, leads to the decrease of the transcription level of *ppc*, *fum* and *pqo*, resulting in a decrease in the production rate of succinate in limited oxygen conditions [14]. This result indicates that sigB could be a positive regulator for the succinate production in *C. glutamicum*.

More recently, a succinate carrier, *sucE1*, has been identified in *C. glutamicum* MJ233 and ATCC13032 and was shown to be functional during the production of succinate. The expression level of *sucE1* is higher under microaerobic and anaerobic conditions than under aerobic conditions. Overexpression or disruption of *sucE1* leads, respectively, to an increase or a decrease in the succinate production. These results suggest that *sucE1* encodes an exporter of succinate in low oxygen conditions which is crucial for the export of succinate in these culture conditions [63-64].

Acetate Biosynthesis Pathway

Acetate is not a major and typical metabolite of the catabolism of glucose in *C. glutamicum*. However, under oxygen limitation conditions, *C. glutamicum* can produce significant amounts of acetate [6, 43]. Nevertheless, most of the studies dedicated to acetate biosynthesis by *C. glutamicum* were aimed to the reduction of the production of this metabolite. Indeed, acetate is considered as a by-product which may induce a decrease in the productivity of more valuable products.

Based on genome sequence data from *C. glutamicum* R [37], putative genes involved in the formation of acetate have been identified. These genes encode the pyruvate dehydrogenase (PDHC), phosphoacyltransferase (PTA) and acetate kinase (ACK) activities (Figure 1). The pyruvate: quinone oxidoreductase (PQO) encoded by the *pqo* gene has also been identified in *C. glutamicum*, and may also be involved in the biosynthesis of acetate [65-

66]. Finally, two homologous gene of the CoA-transferase (CTF) that catalyzes the hydrolysis of the acetyl-CoA to form acetate, were found in the genome of *C. glutamicum* R [67].

The pyruvate dehydrogenase (PDHC) complex contains three subunits known as E1p (pyruvate dehydrogenase) E2p (dihydrolipoamide acetyltransferase) and LPD (lipoamide dehydrogenase) [68]. This activity was detected in various *C. glutamicum* strains [17, 69-70]. The PDHC activity is depending on the nature of substrates, thiamine pyrophosphate and Mg^{2+} cofactors in *C. glutamicum* [69]. E1p is encoded by *aceE*. The product of this gene is essential for *C. glutamicum* growth on glucose as the sole carbon source [65]. The gene encoding E2p in *C. glutamicum* has not been identified so far. In *C. glutamicum* ATCC13032 genome, only the product of NCgl2126 presents a homology with some E2 subunits from other microorganisms [2, 65]. *lpd* gene encoding LPD in *C. glutamicum* ATCC 13032 has been characterized and its product purified [70]. This purified enzyme was able to reversibly catalyze the reoxidation of dihydrolipoic acid and NADH,H^+: NAD^+ transhydrogenation.

Figure 1 shows the two acetate biosynthesis pathways from pyruvate: the direct conversion pathway from pyruvate to acetate, which is catalyzed by the PQO, and the second pathway with acetyl-CoA as an intermediate. In fact, in limited oxygen conditions, most of acetate is produced via acetyl-CoA pathway, since in an *aceE* disrupted *C. glutamicum* R strain, only few amount of acetate is produced [67]. The main route for the production of acetate in *C. glutamicum* is therefore channelized through the PDHC with formation of acetyl-CoA as an intermediate. The simultaneous inactivation of PTA-ACK and CTF pathways leads to a significant reduction of acetate production. However, inactivation of each pathway, separately, has no measurable impact on the production of acetate. This is due to the fact that, the CTF pathway seems to compensate the deficiency of the PTA - ACK pathway and conversely. Strains deleted in *aceE*, *pqo*, *ldhA* or in both *aceE* and *ldhA* remain able to produce acetate, but at a very low level. This suggests that at least another pathway involved in the formation of acetate from acetyl-CoA is probably present in *C. glutamicum* and remains to be identified.

In oxygen deprived conditions, *pqo* gene is down-regulated in a *sigB* disrupted *C. glutamicum* mutant strain, resulting in a decrease of 47% of the specific rate of acetate production [14]. This indicates that, as for the biosynthesis of succinate, SigB could be a positive regulator of acetate biosynthesis in *C. glutamicum* cultivated under oxygen deprived conditions.

PRODUCTION OF ORGANIC ACIDS BY *C. GLUTAMICUM*

Succinic Acid Production

Succinic acid and its derivative salts have a wide variety of applications, such as surfactant, foaming agent, chelating ion additive... They can also be used as a basis for the synthesis of a variety of other chemicals such as tetrahydrofuran, 1,4-butanediol, γ-butyrolactone, N-methyl pyrrolidone, pyrrolidone-2. Moreover, the demand for polymers deriving from lactate and succinate, and their use as raw materials for the manufacture of biodegradable plastics, is continuously increasing [63, 71-72]. Microbial fermentations are widely used for a long time for the production of organic acids. Currently, most of studies focus on bacterial strains able to synthesize succinate as a final product of catabolism, with the objective to establish competitive biotechnological processes for the production of succinate.

The addition of bicarbonate was demonstrated to promote an increase in the specific rate of glucose consumption and lactate production (2.5 times and 2.3 times, respectively) and an increase in the production of succinate by 3.6 times. In this condition, the consumption of the bicarbonate and the production of succinate are almost equimolar. The stimulating effect of bicarbonate was explained by the increase of NAD^+ / $NADH,H^+$ ratio (2.9 vs 1.7 in the absence of bicarbonate) which could reduce the inhibition of GAPDH by $NADH,H^+$. On the other hand, the addition of 10 mM pyruvate was shown to have a similar effect to that of bicarbonate. In addition, although pyruvate was quickly consumed, its stimulating effect is remaining during the process, but this pyruvate effect is still not clearly understood [6]. Despite the predominant role of PEPC for the succinate production in wild-type strains of *C. glutamicum*, the deletion of *ldhA*, encoding the L-lactate dehydrogenase, or the overexpression of the endogenous *pyc* gene, has also led to a significant increase in succinate production [6, 73]. In these conditions, the succinate production yield reached 1.4 mole per mole of glucose. These volumetric productivities, specific to this strain, were much higher than those obtained during the aerobic production of succinate by *E. coli* and succinate production processes involving *C. glutamicum* does not require complex media [74]. Thus, *C. glutamicum* appears as a solid alternative to *E. coli* strains for the biotechnological production of succinate. More recently, the importance of *sucE1*, the succinate carrier encoding gene, was demonstrated during the production of succinate by *C. glutamicum*. However, the suppression of *sucE1* is not sufficient to completely inhibit the production of succinate under microaerobic conditions, suggesting the presence of other functional succinate exporters in *C. glutamicum* [63-64]. As a result, SucE1, and probably any other unidentified succinate exporters could be interesting targets for metabolic engineering of *C. glutamicum* for the production of succinate.

Up to now, studies dedicated to succinate production by *C. glutamicum* had to deal with two challenges: (i) the suppression or reduction of the acetate formation as a competing by-product and (ii) the improvement of succinate production yield from glucose which is limited by the availability of reducing equivalents that are needed for the conversion of oxaloacetate in succinate in the reductive branch of the TCA cycle (Figure 1). These issues have been recently resolved by the disruption of several genes, *i.e.* genes involved in acetic acid (*cat*, *pqo*, *pta-ack*) and lactic acid productions (*ldhA),* in combination with the chromosomal insertion of the deregulated gene encoding the pyruvate carboxylase (pyc^{P458s}) of *C. glutamicum* DM1727. Additional integration of formate dehydrogenase gene (*fdh*) from *Mycobacterium vaccae* as well as the overexpression of *gapA* in *C glutamicum* resulted in an increase in glycolytic flux and CO_2 and $NADH,H^+$ availability [75]. Under anaerobic conditions and in fed batch mode, this modified strain was able to use formate, and carbon dioxide as carbon sources and allowed to accumulate up to 1.13 M of succinate after 53 h of culture, with a yield of 1.67 mole of succinate produced per mole of glucose. In addition, other by-products such as 2-ketoglutarate, malate, acetate, fumarate and pyruvate, accumulated in the fermentation broth but at very low concentrations.

Additional studies also aimed to increase reductive equivalent in *C. glutamicum* since two moles of $NADH,H^+$ are required to synthetize 1 mole of succinate. As mentioned above, this was achieved by expressing formate dehydrogenase [75] but the efficiency of such a method is conditioned by a formate supplementation of the culture medium. Similarly, an heterologous expression of *pntAB*, an *E. coli* transhydrogenase that catalyses the inter-conversion between NADPH,H^+ and NADH,H^+ has been carried out in *C. glutamicum* to supply additional NADH,H^+ to the succinate biosynthesis. Whereas no significant metabolic changes were observed under aerobic conditions, an increase in acetate and succinate productions was observed under microaerobic conditions as confirmed by the increase of NADH,H^+/NAD^+ ratio in this condition [76]. Higher NADH,H^+ availability in *C. glutamicum* can also be obtained without genetic manipulation by decreasing the redox potential of culture. When the redox potential of cultures was set at -300 mV, an increase in succinic acid production yield of 72% was observed compared to production carried out with a redox potential of – 40 mV. Interestingly, the lactate pathway, another metabolic path driven by NADH,H^+ availability, was not influenced equally since succinate /lactate ration increased from 0.27 to 0.48 in the same reductive conditions [77]. Using C-13 NMR, a carbon flux analysis of succinate production by *C. glutamicum* under anaerobic condition indicated that succinate production was more influenced by CO_2 availability than NADH,H^+ and that addition of CO_2 caused up to 2 and 3-fold increases in acetate and succinate production respectively, both at the expense of lactate production. In addition, flux analyses indicated that 97% succinate was produced through the reductive TCA branch and a low but significant activity of the oxidative TCA cycle was observed and provided all the reducing equivalents needed for the redox balance [78]. Such study demonstrated clearly that natural cellular regulation could be overcome to obtain higher succinate yields by *C. glutamicum* without genetic manipulation.

Production of Lactic Acid

Lactic acid is widely used in food industry and also as a precursor for the production of biodegradable plastics [1, 79-80]. Historically, *lactobacillus* were mainly used for the selective production of D-(-), or L-(+) lactic acid [81-82]. However, the lactic acid bacteria display undesirable characteristics, such as their need in amino acids or complex nutrients for their growth that result in a more complex purification of lactic acid. As a consequence, other biocatalysts, such as *E. coli, S. cerevisiae* or *C. glutamicum* have been genetically transformed to make them efficient D - and L-lactic acid producers [83].

In a synthetic medium and without aeration or under oxygen limitation, *C. glutamicum* is able to produce naturally up to 20 g/l lactic acid with a high productivity, generally in association with succinic acid and acetic acid [6]. In this condition, although cell growth arrested immediately, cells kept their metabolic capabilities. In growth arrested condition and under oxygen deprivation, lactate production was largely increased up to 52 g/l but such a production was only obtained with high cell density of *C. glutamicum* R (30 g_{DCW}/l) and in fed batch mode. Concentrations of other products such as acetate or succinate were not significant [44]. In similar condition but with 10 times less biomass, *i.e.* 3 g DCW/l, *C. glutamicum* 2262 was able to produce up 55,8 g/l acid lactic with the highest yield of lactic acid biomass reported (14.3 $g_{lactate}/g_{Biomass}$), demonstrating the importance of the origin of strains to produce various fermentation products from pyruvate [84]. Influence of bicarbonate addition in lactate production by *C. glutamicum* has also been examined since it was shown that it resulted in an increase in glucose consumption and NADH,H^+/NAD^+ ratio [85]. On the basis of these physiological effects, addition of 400 mM bicarbonate in the culture medium led to a higher lactate production up to 97 g/l with a production yield of 0.9 g/g close to the maximal theoretical value [44]. In another attempt to increase lactic acid production, suppression of *pyc* gene allowed an increase in the production of L-lactate in *C. glutamicum* cultivated under biotin limitation [86].

Taking advantage of this high lactate productivity, *C. glutamicum* R was also genetically modified to produce D-lactic acid [87]. In this study, the genetic modifications consisted in the introduction and the overexpression of the gene encoding the D-lactate dehydrogenase from *Lactobacillus delbrueckii* in combination with the suppression of the *ldhA* gene, encoding the L-lactate dehydrogenase. In anaerobiosis and under non growing high cell density (60 g_{dcw}/l), this strain was able to produce up to 120 g/l of D-lactic acid after 30 hours of fermentation with an optical purity greater than 99.9%.

In order to increase D-lactate production by this *C. glutamicum R* derivative strain, ten glycolytic genes involved in the carbon flux channeling toward D-lactacte dehydrogenase were overexpressed. Whereas overexpression of genes located upstream glycerate 1,3-biphosphate in glycolytic pathway (*i.e.* glucokinase, glyceraldehyde 3-phosphate dehydrogenase, phosphofructokinase, triose phosphate isomerase and biphosphate aldolase) had a strong effect on D-lactate production with improved productivity of 98, 39, 15, 13 and 10%, respectively, overexpression of pyruvate kinase gene resulted in a decrease of lactate productity by 70%.

When overexpressed individually, phosphofructokinase and glyceraldehyde 3-phosphate dehydrogenase genes, increased or reduced succinate yield, one of the main by-product, by 52 and 26%, respectively [88].

Production of Pyruvic Acid

Pyruvate is a key metabolite that supplies energy to living cells through the TCA cycle when oxygen is present and alternatively, that is converted into fermentation products under anaerobiosis.

Pyruvate, as a final fermentation product, is not generally considered because this molecule does not accumulate enough to be excreted. Therefore most competing metabolic branches that drain pyruvate have to be shut down. This can be done first by inactivating pyruvate dehydrogenase complex through the deletion of *aceE* gene encoding the E1p subunit of PDHC.

The resulting strain is unable to grow with glucose as the sole carbon source unless acetate or ethanol is added in the medium for complementation [65, 89-90]. Interestingly, *C. glutamicum ΔaceE* is able to accumulate 30 mM pyruvate extracellularly as well as L-alanine and L-valine (30 mM). In a recent study, Wieschalka et al. [5] used this *C. glutamicum ΔaceE* strain as a starting point to improve pyruvate production by deleting gene encoding the pyruvate:quinone oxidoreductase. This additional gene deletion resulted in an increase in pyruvate production up to 55 mM from 150 mM glucose. However, L-valine and L-alanine were still produced significantly with concentrations rising up to 19 mM and 28 mM respectively. Deletion of a third gene, *ldhA* encoding the NAD^+ dependent L-lactate dehydrogenase greatly improved pyruvate production since 193 mM accumulated during a batch culture, as well as L-valine (1 mM) and L-alanine (9 mM). Finally, deletion of genes encoding AlaT and AvtA transaminases resulted in a decrease in 50% of alanine production. When cultivated in fed-batch mode at high cell density and in limited oxygen supply condition, more than 500 mM pyruvate could be produced using this *C. glutamicum* mutant strain, with a yield of 0.97 mol pyruvate per mol of glucose [5].

ALCOHOL PRODUCTION BY *C. GLUTAMICUM*

Ethanol Production

S. cerevisiae and in a lesser extent *Z. mobilis* strains have long been used for the production of ethanol, either for alcoholic beverages or as substitute of gazoline by the petroleum industry. However, in the last decade, a large variety of microorganisms, including *S. cerevisiae* and *Z. mobilis,* was studied and modified in an attempt to improve the performance of ethanol production processes. In addition to these last microbial species, the use of gram-negative bacteria such as *E. coli* and *Klebsiella oxytoca*, and the yeast *Pichia stipidis*, were successful [91-92]. Interestingly, *C. glutamicum* was also examined for its ability to produce ethanol because of its ability to produce many metabolites including amino acids and organic acids at the industrial scale [85]. In addition, ethanol production by *C. glutamicum* at very high yield was assumed to be largely feasible because growth arrested conditions are easy to achieve with this bacterium when a limitation of oxygen is applied. In this condition, carbon fluxes are therefore primarily channelized for the production of metabolites. However, to take advantage of these properties for ethanol production, the central metabolism of *C. glutamicum* must be modified because enzyme activities involved in ethanol biosynthesis are not naturally present in this species.

In *S. cerevisiae* and *Z. mobilis*, the metabolic pathway of ethanol biosynthesis from pyruvate is composed of two enzymes, pyruvate decarboxylase (PDC) and alcohol dehydrogenase (ADH). These two enzymes are sufficient to convert the intracellular pyruvate and NADH,H^+ to ethanol [93]. The PDC activity catalyzes the non-oxidative decarboxylation of pyruvate into acetaldehyde by releasing carbon dioxide. Acetaldehyde generated by this reaction is reduced to ethanol by ADH. While the PDC is widely distributed among the plants, yeasts and fungi, it is absent in animals and rare in bacteria.

PDC of *Z. mobilis* was the first bacterial PDC [94] to be identified, characterized [95-99], and cloned [98-99]. It has therefore become one of the best-studied PDC. This enzyme displays a low Km for pyruvate (0.4 mM) that allows PDC to compete for the use of pyruvate with other central metabolism enzymes and leads to the formation of large amount of acetaldehyde, which is then converted into ethanol by ADH [100]. In addition, PDC of *Z. mobilis* is also the most stable PDC known so far [101]. For these reasons, PDC of *Z. mobilis* was transferred and used for the production of ethanol in a wide variety of microorganisms [102-108].

Contrary to PDC, ADH activities are almost present in all organisms. In alcohol producing microorganisms, ADH catalyzes the reversible synthesis of ethanol from acetaldehyde. Although most anaerobic bacteria and some aerobic bacteria are able to synthesize ethanol as a final product, the amount of ADH present in bacteria, with the exception of *Z. mobilis*, is usually 1 to 2 orders of magnitude lower than ADH activities found in *S. cerevisiae*. Indeed, ADH activity levels in *Z. mobilis* approach those measured in yeast and result from the presence of two ADH isozymes, one containing zinc (ADH I) and the second containing iron (ADH II). Generally, both enzymes require NAD as the cofactor [109-110]. While ADH I is essential in the early phase of ethanol production, ADH II is responsible of the major part of the production and high ethanol concentration [111].

Therefore, PDC and ADH II encoding genes from *Z. mobilis* were selected to introduce the ability to produce ethanol from pyruvate in many microorganisms including *C. glutamicum*. In parallel to the suppression of the competing LDH for pyruvate, alcohologenic *C. glutamicum* was achieved by cloning both *pdc* and *adhII* genes under the control of *ldhA* promoter which is induced under oxygen limitation conditions [6].

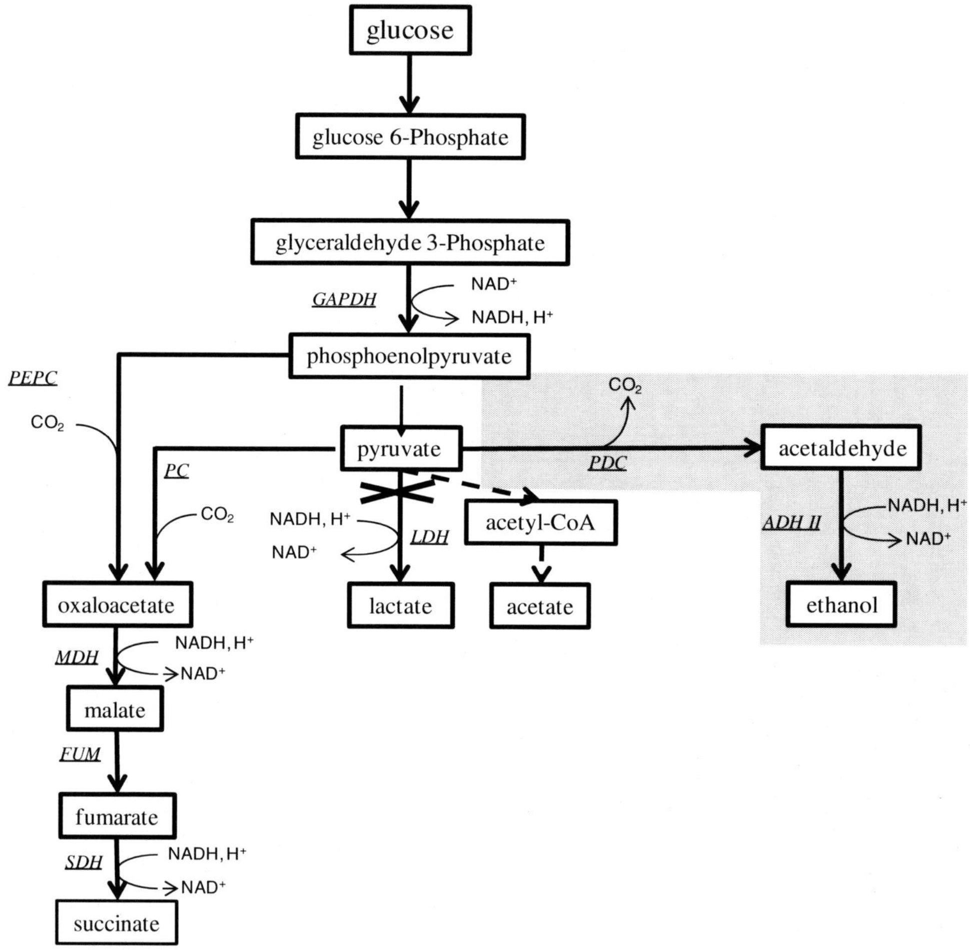

Figure 4. Glucose metabolism in recombinant *C. glutamicum* strain, producing ethanol in deprived oxygen conditions. GAPDH: glyceraldehyde 3-phosphate dehydrogenase; PEPC: phosphoenolpyruvate carboxylase; PC: pyruvate carboxylase; PDC: pyruvate decarboxylase; ADH II: alcohol dehydrogenase; MDH: malate dehydrogenase; FUM: fumarase; SDH: succinate dehydrogenase; LDH: lactate dehydrogenase. The alcohol biosynthesis branch deriving from that of *Zymomonas mobilis* is indicated in grey box. The bold cross indicates the suppression of the lactate synthesis pathway.

Under conditions of oxygen limitation, recombined strains of *C. glutamicum* are unable to grow, but produce a significant quantity of ethanol, 2 g/l, as well as acetate, lactate and succinate as co-products.

The deletion of the endogenous *ldhA* gene led to an increase of 1.4 times of the ethanol production rate and a complete abolition of the formation of L-lactate. With this recombinant strain, under anaerobic and at high cell density conditions, an ethanol productivity of 29.5 g/l/h was obtained [85], but limited information on performance, maximum ethanol productions and by-product concentrations (acetate and succinate) for these experiments were given, making impossible the direct comparison with bacterial cells such as *E. coli*, *K. oxytoca* or *Z. mobilis* [74].

Isobutanol Production

The higher alcohols, including isobutanol, have applications in biofuels, flavors or fragrances synthesis or as chemical feedstocks. As a consequence, their production by genetically modified microorganisms has generated a high interest [112].

Isobutanol production has recently been achieved by taking profit of keto acid production in engineered *C. glutamicum* strains. This strategy was done by overexpressing genes encoding acetolactate synthase (*alsS*) from *B. subtilis*, acetohydroxy acid isomeroreductase (*ilvC*) and dihydroxyacid dehydratase (*ilvD*) from *C. glutamicum*, α-ketoisovalerate decarboxylase (*kivd*) from *L. lactis* and a native alcohol dehydrogenase (*adhA*) [113]. As a result, up to 4 g/L butanol were produced in long-term batch culture. Alternatively, higher butanol production was obtained using a strain of *C. glutamicum* inactivated in pyruvate dehydrogenase complex, pyruvate:quinone oxidoreductase, transaminase B, and overexpressing lactate dehydrogenase, malate dehydrogenase, *ilvBNCD* as well as α-ketoisovalerate decarboxylase, alcohol dehydrogenase and transhydrogenase.

The resulting *C. glutamicum* strain was shown to produce about 175 mM butanol in fed-batch process under oxygen-depleted condition [114]. Improvement of butanol production by *C. glutamicum*, up to 313 mM, was obtained by modulating the activities of α-ketoacid decarboxylase (under the control of *ldhA* promoter) and isobutanol dehydrogenase from *E. coli* (under the control of *gapA* promoter), and by suppressing by-product formation [115].

Conclusion

To date, *C. glutamicum* is industrially used in strictly aerobic processes, for the production of amino acids or nucleotides. Nevertheless, the utilization of the ability of this facultative anaerobic bacterium to retain a metabolic activity in oxygen deprived conditions opens some great opportunities. Coupled with the development of metabolic engineering approaches, this particular metabolism permits an important increase in the range of metabolites produced by *C. glutamicum*. Besides, lactate, acetate and succinate, which are naturally produced by *C. glutamicum* but whose excreted concentration is highly improved using genetic engineering tools, pyruvate, ethanol, isobutanol and other metabolites (alanine, valine…) which are not mentioned in this chapter can be produced. Considering this variety of metabolites produced in anaerobiosis or limiting conditions of oxygen supply, with no doubt, *C. glutamicum* is a promising metabolic platform. In the competition with the petrochemical industry or even with the other microorganisms currently used in industry for the production of the metabolites cited above, some progresses have to be done. The final concentrations, the productivities remain to improve. To date, the most promising process using *C. glutamicum* is probably the one for the production of succinate. Up to now, the improvements in the production of organic acids or alcohols using *C. glutamicum* have been obtained mainly thanks to metabolic engineered strains. To reach the next step towards the industrialization, this methodology will have to be coupled with process engineering approaches. The historical use of *C. glutamicum* in industrial processes, its robustness, and its ability to support high intracellular metabolic fluxes can, nevertheless, make optimistic concerning the potential applications of this bacterium in deprived oxygen conditions.

REFERENCES

[1] Bozell, JJ; Petersen, GR. Technology development for the production of biobased products from biorefinery carbohydrates – US Department of Energy's Top 10' revisited. *Green chemistry*, 2010, 12, 539-554.

[2] Kalinowski, J; Bathe, B; Bartels, D; Bischoff, N; Bott, M; Burkovski, A; Dusch, N; Eggeling, L; Eikmanns, B.J; Gaigalat, L; Goesmann, A; Hartmann, M; Huthmacher, K; Kramer, R; Linke, B; McHardy, A.C; Meyer, F; Mockel, B; Pfefferle, W; Puhler, A; Rey, D.A; Ruckert, C; Rupp, O; Sahm, H; Wendisch, VF; Wiegrabe, I; Tauch, A. The complete *Corynebacterium glutamicum* ATCC 13032 genome sequence and its impact on the production of L-aspartate-derived amino acids and vitamins. *Journal of biotechnology*, 2003, 104, 5-25.

[3] Sauer, U; Eilmanns, BJ. The PEP-pyruvate-oxaloacetate node as the switch point for carbon flux distribution in bacteria. FEMS microbiology reviews, 2005, 29, 764-794.

[4] Eggeling, L; Bott, M. *Handbook of Corynebacterium glutamicum*. Boca Raton: CRC Press, Taylor and Francis group; 2005.

[5] Wieschalka, S; Blombach, B; Eikmanns, BJ. Engineering *Corynebacterium glutamicum* for the production of pyruvate. *Applied microbiology and biotechnology*, 2012, 94,449-59.

[6] Inui, M; Murakami, S; Okino, S; Kawaguchi, H; Vertes, AA; Yukawa, H. Metabolic analysis of *Corynebacterium glutamicum* during lactate and succinate productions under oxygen deprivation conditions. *Journal of molecular microbiology and biotechnology*, 2004, **7**, 182-196.

[7] Nishimura, T; Vertès, AA; Shinoda, Y; Inui, M; Yukawa, H. Anaerobic growth of *Corynebacterium glutamicum* using nitrate as a terminal electron acceptor. *Applied microbiology and biotechnology*, 2007, 75, 889-897.

[8] Takeno, S; Ohnishi, J; Komatsu, T; Masaki, T; Sen, K; Ikeda, M. Anaerobic growth and potential for amino acid production by nitrate respiration in *Corynebacterium glutamicum*. *Applied microbiology and biotechnology*, 2007, 75, 1173–1182.

[9] Inui, M; Suda, M; Okino, S; Nonaka, H; Puskas, LG; Vertes, A; Yukawa, H. Transcriptional profiling of *Corynebacterium glutamicum* metabolism during organic

acid production under oxygen deprivation conditions. *Microbiology*, 2007, 153, 2491-2504.

[10] Suda, M; Teramoto, H; Imamiya, T; Inui, M; Yukawa, H. Transcriptional regulation of *Corynebacterium glutamicum* methionine biosynthesis genes in response to methionine supplementation under oxygen deprivation. *Applied microbiology and biotechnology*, 2008, 81, 505-513.

[11] Larisch, C; Nakunst, D; Huser, AT; Tauch, A; Kalinowski, J. The alternative sigma factor SigB of *Corynebacterium glutamicum* modulates global gene expression during transition from exponential growth to stationary phase. *BMC genomics*, 2007, 8, 4.

[12] Barreiro, C; Nakunst, D; Huser, AT; de Paz, HD; Kalinowski, J; Martin, JF. Microarray studies reveal a 'differential response' to moderate or severe heat shock of the HrcA- and HspRdependent systems in *Corynebacterium glutamicum*. *Microbiology*, 2009 155, 359–372.

[13] Jakob, K; Satorhelyi, P; Lange, C; Wendisch, VF; Silakowski, B; Scherer, S; Neuhaus K. Gene expression analysis of *Corynebacterium glutamicum* subjected to long-term lactic acid adaptation. *Journal of bacteriology*, 2007, 189, 5582–5590.

[14] Ehira, S; Shirai, T; Teramoto, H; Inui, M; Yukawa, H. Group 2 sigma factor SigB of *Corynebacterium glutamicum* positively regulates glucose metabolism under conditions of oxygen deprivation. *Applied and environmental microbiology*, 2008, 74, 5146-5152.

[15] Dominguez, H; Lindley, ND. Complete sucrose metabolism requires fructose phosphotransferase activity in *Corynebacterium glutamicum* to ensure phosphorylation of liberated fructose. *Applied and environmental microbiology*, 1996, 62, 3878-3880.

[16] Mori, M; Shiio, I. Phosphoenolpyruvate: sugar phosphotransferase systems and sugar metabolism in *Brevibacterium flavum*. *Agricultural and biological chemistry*, 1987, 51, 2671-2678.

[17] Cocaign-Bousquet, M; Guyonvarch, A; Lindley, ND. Growth rate dependent modulation of carbon flux through central metabolism and the kinetic consequences for glucose-limited chemostat cultures of *Corynebacterium glutamicum*. *Applied and environmental microbiology*, 1996, 62, 429-436.

[18] Gourdon, P; Raherimandimby, M; Dominguez, H; Cocaign-Bousquet, M; Lindley, ND. Osmotic stress, glucose transport capacity and consequences for glutamate overproduction in *Corynebacterium glutamicum*. *Journal of biotechnology*, 2003, 104, 77-85.

[19] Ikeda, M. Sugar transport systems in *Corynebacterium glutamicum*: features and applications to strain development. *Applied microbiology and biotechnology*, 2012, 96, 1191-1200.

[20] Park, SY; Kim, HK; Yoo, SK; Oh, TK; Lee, J.K. Characterization of *glk*, a gene coding for glucose kinase of *Corynebacterium glutamicum*. *FEMS microbiology letters*, 2000, 188, 209-215.

[21] Lindner, S; Knebel, S; Pallerla, SR; Schobert, SM; Wendisch VF. Cg2091 encodes a polyphosphate/ATP-dependent glucokinase in *Corynebacterium glutamicum*. *Applied microbiology and biotechnology*, 2010, 87, 703-713.

[22] Parche, S; Burkovski, A; Sprenger, GA; Weil, B; Kramer, R; Titgemeyer, F. *Corynebacterium glutamicum*: a dissection of the PTS. *Journal of molecular microbiology and biotechnology*, 2001, 3, 423-428.

[23] Moon, MW; Kim, HJ; Oh, TK; Shin, CS; Lee, JS; Kim, SJ; Lee, JK. Analyses of enzyme II gene mutants for sugar transport and heterologous expression of fructokinase gene in *Corynebacterium glutamicum* ATCC 13032. *FEMS microbiology letters*, 2005, 244, 259-266.

[24] Engels, V; Wendisch, VF. The DeoR-type regulator SugR represses expression of *ptsG* in *Corynebacterium glutamicum*. *Journal of bacteriology*, 2007, 189, 2955-2966.

[25] Tanaka, Y; Okai, N; Teramoto, H; Inui, M; Yukawa, H. Regulation of the expression of phosphoenolpyruvate: carbohydrate phosphotransferase system (PTS) genes in *Corynebacterium glutamicum*. *Microbiology*, 2008, 154, 264-274.

[26] Sasaki, M; Teramoto, H; Inui, M; Yukawa, H. Identification of mannose uptake and catabolism genes in *Corynebacterium glutamicum* and genetic engineering for simultaneous utilization of mannose and glucose. *Applied microbiology and biotechnology*, 2010, 89, 1905-1916.

[27] Gaigalat, L; Schluter, JP; Hartmann, M; Mormann, S; Tauch, A; Puhler, A; Kalinowski, J. The DeoR-type transcriptional regulator SugR acts as a repressor for genes encoding the phosphoenolpyruvate:sugar phosphotransferase system (PTS) in *Corynebacterium glutamicum*. *BMC molecular biology*, 2007, 8: 104.

[28] Tanaka, Y; Teramoto, H; Inui, M; Yukawa, H. Regulation of expression of general components of the phosphoenolpyruvate: carbohydrate phosphotransferase system (PTS) by the global regulator SugR in *Corynebacterium glutamicum*. *Applied microbiology and biotechnology*, 2008, 78, 309-318.

[29] Yokota, A; Lindley ND. Central metabolism: sugar uptake and conversion. In Handbook of *Corynebacterium glutamicum*. In: Eggeling L and Bott M editors. *Handbook of Corynebacterium glutamicum*. Boca Raton: CRC Press, Taylor and Francis Group; 2005; 215-240.

[30] Schröder, J; Tauch, A. Transcriptional regulation of gene expression in *Corynebacterium glutamicum*: the role of global, master and local regulators in the modular and hierarchical gene regulatory network. *FEMS microbiology reviews*, 2010, 34, 685-737.

[31] Teramoto, H; Inui, M; Yukawa, H. Transcriptional regulators of multiple genes involved in carbon metabolism in *Corynebacterium glutamicum*. *Journal of biotechnology*, 2011, 154, 114-125.

[32] Ozaki, H; Shiio, I. Regulation of the TCA and glyoxylate cycles in *Brevibacterium flavum*. II. Regulation of phosphoenolpyruvate carboxylase and pyruvate kinase. *The journal of biochemistry*, 1969, 66, 297-311.

[33] Jetten, MS; Gubler, ME; Lee, SH; Sinskey, AJ. Structural and functional analysis of pyruvate kinase from *Corynebacterium glutamicum*. *Applied and environmental microbiology*, 1994, 60, 2501-2507.

[34] Sugimoto, SI; Shiio, I. Fructose metabolism and regulation of 1-phosphofructokinase and 6-phosphofructokinase in *Brevibacterium flavum*. *Agricultural and biological chemistry*, 1989, 53, 1261-1268.

[35] Dominguez, H; Rollin, C; Guyonvarch, A; Guerquin-Kern, JL; Cocaign-Bousquet, M; Lindley, ND. Carbon-flux distribution in the central metabolic pathways of *Corynebacterium glutamicum* during growth on fructose. *European journal of biochemistry*, 1998, 254, 96-102.

[36] Gourdon, P; Lindley, ND. Metabolic analysis of glutamate production by *Corynebacterium glutamicum*. *Metabolic engineering*, 1999, 1, 224-231.

[37] Yukawa, H; Omumasaba, CA; Nonaka, H; Kós, P; Okai, N; Suzuki, N; Suda, M; Tsuge, Y; Watanabe, J; Ikeda, Y; Vertès, AA; Inui, M. Comparative analysis of the *Corynebacterium glutamicum* group and complete genome sequence of strain R. *Microbiology*, 2007, 153, 1042–1058.

[38] Omumasaba, CA; Okai, N; Inui, M; Yukawa, H. *Corynebacterium glutamicum* glyceraldehyde-3-phosphate dehydrogenase isoforms with opposite, ATP-dependent regulation. *Journal of molecular microbiology and biotechnology*, 2004, 8, 91–103.

[39] Engels, V; Lindner, SN; Wendisch, VF. The global repressor SugR controls expression of genes of glycolysis and of the L-lactate dehydrogenase LdhA in *Corynebacterium glutamicum*. *Journal of bacteriology*, 2008, 190, 8033-8044.

[40] Toyoda, K; Teramoto, H; Inui, M; Yukawa, H. Expression of the *gapA* gene encoding glyceraldehyde-3-phosphate dehydrogenase of *Corynebacterium glutamicum* is regulated by the global regulator SugR. *Applied microbiology and biotechnology*, 2008, 81, 291-301.

[41] Delaunay, S; Gourdon, P; Lapujade, P; Mailly, E; Oriol, E; Engasser, JM; Lindley, N D; Goergen, JL. An improved temperature-triggered process for glutamate production with *Corynebacterium glutamicum*. *Enzyme and microbial technology*, 1999, 25, 762-768.

[42] Dietrich, C; Nato, A; Bost, B; Le Marechal, P; Guyonvarch, A. Regulation of l*dh* expression during biotin-limited growth of *Corynebacterium glutamicum*. *Microbiology*, 2009, 155, 1360-1375.

[43] Dominguez, H; Nezondet, C; Lindley, ND; Cocaign, M. Modified carbon flux during oxygen limited growth of *Corynebacterium glutamicum* and the consequences for amino acid overproduction. *Biotechnology letters*, 1993, 15, 449-454.

[44] Okino, S; Inui, M; Yukawa, H. Production of organic acids by *Corynebacterium glutamicum* under oxygen deprivation. *Applied microbiology and biotechnology*, 2005, 68, 475-480.

[45] Uy, D; Delaunay, S; Germain, P; Engasser, JM; Goergen, JL. Instability of glutamate production by *Corynebacterium glutamicum* 2262 in continuous culture using the temperature-triggered process. *Journal of biotechnology*, 2003, 104, 173-184.

[46] Garvie, EI. Bacterial lactate dehydrogenases. *Microbiological reviews*, 1980, 44, 106-139.

[47] Koffas, MAG; Ramamoorthi, R; Pine, WA; Sinskey, AJ; Stephanopoulos, G. Sequence of the *Corynebacterium glutamicum* pyruvate carboxylase gene. *Applied and environmental microbiology,* 1998, 50, 346-352.

[48] Peters-Wendisch, PG; Kreutzer, C; Kalinowski, J; Patek, M; Sahm, H; Eikmanns, BJ. Pyruvate carboxylase from *Corynebacterium glutamicum*: characterization, expression and inactivation of the *pyc* gene. *Microbiology*, 1998, 144, 915-927.

[49] Koffas, MAG; Jung, GY; Aon, JC; Stephanopoulos, G. Effect of pyruvate carboxylase overexpression on the physiology of *Corynebacterium glutamicum*. *Applied and environmental microbiology*, 2002, 68, 5422-5428.

[50] Peters-Wendisch, PG; Wendisch, VF; Paul, S; Eikmanns, BJ; Sahm, H. Pyruvate carboxylase as anaplerotic enzyme in *Corynebacterium glutamicum*. *Microbiology*, 1997,143, 1095-1103.

[51] Mori, M; Shiio, I. Purification and some properties of phosphoenolpyruvate carboxylase from *Brevibacterium flavum* and its aspartate-overproducing mutant. *The journal of biochemistry*, 1985, 97, 1119-1128.

[52] Mori, M; Shiio, I. Synergistic inhibition of phosphoenolpyruvate carboxylase by aspartate and 2-oxoglutarate in *Brevibacterium flavum. The journal of biochemistry*, 1985, 98, 1621-1630.

[53] Eikmanns, BJ; Follettie, MT; Griot, MU; Sinskey, AJ. The phosphoenolpyruvate carboxylase gene of *Corynebacterium glutamicum*: molecular cloning, nucleotide sequence, and expression. *Molecular and general genetics*, 1989, 218, 330-339.

[54] Schwinde, JW; Thum-Schmitz, N; Eikmanns, BJ; Sahm, H. Transcriptional analysis of the gap-pgk-tpi-ppc gene cluster of *Corynebacterium glutamicum, Journal of bacteriology*, 1993, 175, 3905-3908.

[55] Peters-Wendisch, PG; Schiel, B; Wendisch, VF; Katsoulidis, E; Mockel, B; Sahm, H; Eikmanns, BJ. Pyruvate carboxylase is a major bottleneck for glutamate and lysine production by *Corynebacterium glutamicum. Journal of molecular microbiology and biotechnology*, 2001, 3, 295-300.

[56] Molenaar, D; van der Rest, ME; Petrovic, S. Biochemical and genetic characterization of the membrane-associated malate dehydrogenase (acceptor) from *Corynebacterium glutamicum. European journal of biochemistry*, 1998, 254, 395-403.

[57] Molenaar, D; van der Rest, ME; Drysch, A; Yucel, R. Functions of the membrane-associated and cytoplasmic malate dehydrogenases in the citric acid cycle of *Corynebacterium glutamicum. Journal of bacteriology*, 2000, 182, 6884-6891.

[58] Gerstmeier, R; Wendisch, VF; Schnicke, S; Ruan, H; Farwic, M; Reinscheid, D; Eikmanns, BJ. Acetate metabolism and its regulation in *Corynebacterium glutamicum. Journal of biotechnology*, 2003, 104, 99-122.

[59] Genda ,T; Watabe, S; Ozaki, H. Purification and characterization of fumarase from *Corynebacterium glutamicum. Bioscience, biotechnology and biochemistry*, 2006, 70, 1102-1109.

[60] Wendisch, VF. Genome-wide expression analysis in *Corynebacterium glutamicum* using DNA microarrays. *Journal of biotechnology*, 2003, 104, 273-285.

[61] Kurokawa, T; Sakamoto, J. Purification and characterization of succinate: menaquinone oxidoreductase from *Corynebacterium glutamicum. Archives of microbiology*, 2005, 183, 317-324.

[62] Lee, DS; Park, JS; Kim, Y; Lee, HS. *Corynebacterium glutamicum sdhA* encoding succinate dehydrogenase subunit A plays a role in *cysR*-mediated sulfur metabolism. *Applied microbiology and biotechnology*, 2014, 98, 6751-6759.

[63] Fukui, K; Koseki, C; Yamamoto, Y; Nakamura, J; Sasahara, A; Yuji, R; Hashiguchi, K; Usuda, Y; Matsui, K; Kojima, H; Abe, K. Identification of succinate exporter in *Corynebacterium glutamicum* and its physiological roles under anaerobic conditions. *Journal of biotechnology*, 2011, 154: 25-34.

[64] Huhn, S; Jolkver, E; Kramer, R; Marin, K. Identification of the membrane protein SucE and its role in succinate transport in *Corynebacterium glutamicum. Applied microbiology and biotechnology*, 2011, 89, 327-335.

[65] Schreiner, ME; Eikmanns, BJ. Pyruvate:quinone oxidoreductase from *Corynebacterium glutamicum*: purification and biochemical characterization. *Journal of bacteriology*, 2005, 187, 862-871.

[66] Schreiner, ME; Riedel, C; Holatko, J; Patek, M; Eikmanns, BJ. Pyruvate:quinone oxidoreductase in *Corynebacterium glutamicum*: molecular analysis of the pqo gene, significance of the enzyme, and phylogenetic aspects. *Journal of bacteriology*, 2006, 188, 1341-1350.

[67] Yasuda, K; Jojima, T; Suda, M; Okino, S; Inui, M; Yukawa, H. Analyses of the acetate-producing pathways in *Corynebacterium glutamicum* under oxygen-deprived conditions. *Applied microbiology and biotechnology*, 2007, 77, 853-860.

[68] Eikmanns, B. Central metabolism: tricarboxylic acid cycle and anaplerotic reactions. In: Eggeling L and Bott M editors. *Handbook of Corynebacterium glutamicum.* Boca Raton: CRC Press, Taylor and Francis Group; 2005; 241-276.

[69] Shiio, I; Ujigawa, K. Enzymes of the glutamate and aspartate synthetic pathways in a glutamate-producing bacterium, *Brevibacterium flavum. The journal of biochemistry*, 1978, 84, 647-657.

[70] Schwinde, JW; Hertz, PF; Sahm, H; Eikmanns, BJ; Guyonvarch, A. Lipoamide dehydrogenase from *Corynebacterium glutamicum*: molecular and physiological analysis of the *lpd* gene and characterization of the enzyme. *Microbiology*, 2001, 147, 2223-2231.

[71] Flieger, M; Kantorova, M; Prell, A; Rezanka, T; Votruba, J. Biodegradable plastics from renewable sources. *Folia microbiology (Praha)*, 2003, 48, 27-44.

[72] Zeikus, JG; Jain, MK; Elankovan, P. Biotechnology of succinic acid production and markets for derived industrial products. *Applied microbiology and biotechnology*, 1999, 51, 542–552.

[73] Okino, S; Noburyu, R; Suda, M; Jojima, T; Inui, M; Yukawa, H. An efficient succinic acid production process in a metabolically engineered *Corynebacterium glutamicum* strain. *Applied microbiology and biotechnology*, 2008, 81, 459-464.

[74] Wendisch, VF; Bott, M; Eikmanns, BJ. Metabolic engineering of *Escherichia coli* and *Corynebacterium glutamicum* for biotechnological production of organic acids and amino acids. *Current opinion in microbiology*, 2006, 9, 268-274.

[75] Litsanov, B; Brocker, M; Bott, M. Toward homosuccinate fermentation: metabolic engineering of *Corynebacterium glutamicum* for anaerobic production of succinate from glucose and formate. *Applied and environmental microbiology*, 2012, 78, 3325-3337.

[76] Yamauchi, Y; Hirasawa, T; Nishii, M; Furusawa, C; Shimizu, H. Enhanced acetic acid and succinic acid production under microaerobic conditions by *Corynebacterium glutamicum* harboring *Escherichia coli* transhydrogenase gene *pntAB*. Journal of general and applied microbiology, 2014, 60,112-118.

[77] Chen, XJ; Jiang, ST; Li, XJ; Pan, LJ; Zheng, Z; Luo, SZ. Production of succinic acid and lactic acid by *Corynebacterium crenatum* under anaerobic conditions. Annals of microbiology, 2013, 63, 39-44.

[78] Radoš, D; Turner, DL; Fonseca, LL; Carvalho, AL; Blombach, B; Eikmanns, BJ; Neves, AR; Santos, H. Carbon flux analysis by ^{13}C nuclear magnetic resonance to determine the effect of CO_2 on anaerobic succinate production by *Corynebacterium glutamicum*. *Applied and environmental microbiology*, 2014, 80, 3015-24.

[79] Datta, R; Tsai, SP; Bonsignore, P; Moon, SH; Frank, JR. Technological and economic potential of poly (lactic acid) and lactic acid derivatives. *FEMS microbiology reviews*, 1995, 16, 221-231.

[80] Hofvendahl, K; Hahn-Hagerdal, B. Factors affecting the fermentative lactic acid production from renewable resources (1). *Enzyme and microbial technology*, 2000, 26, 87-107.
[81] Benthin, S; Villadsen, J. Production of optically pure D-lactate by *Lactobacillus bulgaricus* and purification by crystallization and liquid-liquid extraction. *Applied microbiology and biotechnology*, 1995, 426, 826-829.
[82] Kyla-Nikkila, K; Hujanen, M; Leisola, M; Palva, A. Metabolic engineering of *Lactobacillus helveticus* CNRZ32 for production of pure L-(+)-lactic acid. *Applied and environmental microbiology*, 2000, 66, 3835-3841.
[83] Okano, K; Tanaka, T; Ogino, C; Fukuda, H; Kondo, A. Biotechnological production of enantiomeric pure lactic acid from renewable resources: recent achievements, perspectives, and limits. *Applied microbiology and biotechnology*, 2010, 85, 413-23.
[84] Khuat, HB; Kaboré, AK; Olmos, E; Fick, M; Boudrant, J; Goergen, JL; Delaunay, S; Guedon, E. Lactate production as representative of the fermentation potential of *Corynebacterium glutamicum* 2262 in a one-step process. *Bioscience biotechnology and biochemistry*, 2014, 78, 343-349.
[85] Inui, M; Kawaguchi, H; Murakami, S; Vertes, AA; Yukawa, H. Metabolic engineering of *Corynebacterium glutamicum* for fuel ethanol production under oxygen-deprivation conditions. *Journal of molecular microbiology and biotechnology*, 2004, 8, 243-254.
[86] Sato, H; Orishimo, K; Shirai, T; Hirasawa, T; Nagahisa, K; Shimizu, H; Wachi, M. Distinct roles of two anaplerotic pathways in glutamate production induced by biotin limitation in *Corynebacterium glutamicum*. *Journal of bioscience and bioengineering*, 2008, 106, 51-58.
[87] Okino, S; Suda, M; Fujikura, K; Inui, M; Yukawa, H. Production of D-lactic acid by *Corynebacterium glutamicum* under oxygen deprivation. *Applied microbiology and biotechnology*, 2008, 78, 449-454.
[88] Tsuge, Y; Yamamoto, S; Suda, M; Inui, M; Yukawa, H. Reactions upstream of glycerate-1,3-bisphosphate drive *Corynebacterium glutamicum* (D)-lactate productivity under oxygen deprivation. *Applied microbiology and biotechnology*, 2013, 97, 6693-6703.
[89] Blombach, B; Arndt, A; Auchter, M; Eikmanns BJ. L-valine production during growth of pyruvate dehydrogenase complex-deficient *Corynebacterium glutamicum* in the presence of ethanol or by inactivation of the transcriptional regulator SugR. *Applied and environmental microbiology*, 2009, 75, 1197-1200.
[90] Blombach, B; Schreiner, ME; Moch, M; Oldiges, M; Eikmanns, BJ. Effect of pyruvate dehydrogenase complex deficiency on L-lysine production with *Corynebacterium glutamicum*. *Applied microbiology and biotechnology*, 2007, 76, 615-623.
[91] Dien, BS; Cotta, MA; Jeffries, TW. Bacteria engineered for fuel ethanol production: current status. *Applied microbiology and biotechnology*, 2003, 63, 258-266.
[92] Jeffries, TW; Jin, YS. Metabolic engineering for improved fermentation of pentoses by yeasts. *Applied microbiology and biotechnology*, 2004, 63, 495-509.
[93] Talarico, LA; Gil, MA; Yomano, LP; Ingram, LO; Maupin-Furlow, JA. Construction and expression of an ethanol production operon in Gram-positive bacteria. *Microbiology* 2005, 151, 4023-4031.
[94] Dawes, EA; Ribbons, DW; Large, PJ. The route of ethanol formation in *Zymomonas mobilis*. *Biochemical journal*, 1966, 98, 795-803.

[95] Bringer-Meyer, S; Schimz, KL; Sahm, H. Pyruvate decarboxylase from *Zymomonas mobilis*. Isolation and partial characterization. *Archives of microbiology*, 1986, 146, 105-110.

[96] Hoppner, TC; Doelle, HW. Purification and kinetic characteristics of pyruvate decarboxylase and ethanol dehydrogenase from *Zymomonas mobilis* in relation to ethanol production. *European journal of applied microbiology and biotechnology*, 1983, 17, 152-157.

[97] Neale, AD; Scopes, RK; Wettenhall, RE; Hoogenraad, NJ. Pyruvate decarboxylase of *Zymomonas mobilis*: isolation, properties, and genetic expression in *Escherichia coli*. *Journal of bacteriology*, 1987, 169, 1024-1028.

[98] Conway, T; Osman, YA; Konnan, JI ; Hoffmann, EM; Ingram, LO. Promoter and nucleotide sequences of the *Zymomonas mobilis* pyruvate decarboxylase. *Journal of bacteriology*, 1987, 169, 949-954.

[99] Neale, AD; Scopes, RK; Wettenhall, RE; Hoogenraad, NJ. Nucleotide sequence of the pyruvate decarboxylase gene from *Zymomonas mobilis*. *Nucleic acids research*, 1987, 15, 1753-1761.

[100] Ingram, LO; Conway, T; Clark, DP; Sewell, GW; Preston, J.F. Genetic engineering of ethanol production in *Escherichia coli*. *Applied and environmental microbiology*, 1987, 53, 2420-2425.

[101] Dobritzsch, D; Konig, S; Schneider, G; Lu, G. High resolution crystal structure of pyruvate decarboxylase from *Zymomonas mobilis*. Implications for substrate activation in pyruvate decarboxylases. *Journal of biological chemistry*, 1998, 273, 20196-20204.

[102] Arfman, N; Worrell, V; Ingram, LO. Use of the *tac* promoter and *lacIq* for the controlled expression of *Zymomonas mobilis* fermentative genes in *Escherichia coli* and *Zymomonas mobilis*. *Journal of bacteriology*, 1992, 174, 7370-7378.

[103] Bucher, M; Brandle, R; Kuhlemeier, C. Ethanolic fermentation in transgenic tobacco expressing *Zymomonas mobilis* pyruvate decarboxylase. *The EMBO journal*, 1994, 13, 2755-2763.

[104] Deng, MD; Coleman, JR. Ethanol synthesis by genetic engineering in cyanobacteria. *Applied and environmental microbiology*, 1999, 65, 523-528.

[105] Gold, RS; Meagher, MM; Tong, S; Hutkins, RW; Conway, T. Cloning and expression of the *Zymomonas mobilis* "production of ethanol" genes in *Lactobacillus casei*. *Current microbiology*, 1996, 33, 256-260.

[106] Ingram, LO; Aldrich, HC; Borges, AC; Causey, TB; Martinez, A; Morales, F; Saleh, A; Underwood, SA; Yomano, LP; York, SW; Zaldivar, J; Zhou, S. Enteric bacterial catalysts for fuel ethanol production. *Biotechnology progress*, 1999, 15, 855-866.

[107] Ingram, LO; Gomez, PF; Lai, X; Moniruzzaman, M; Wood, BE; Yomano, LP York, SW. Metabolic engineering of bacteria for ethanol production. *Biotechnology and Bioengineering*, 1998, 58, 204-214.

[108] Wood, BE; Ingram, LO. Ethanol production from cellobiose, amorphous cellulose, and crystalline cellulose by recombinant *Klebsiella oxytoca* containing chromosomally integrated *Zymomonas mobilis* genes for ethanol production and plasmids expressing thermostable cellulase genes from *Clostridium thermocellum*. *Applied and environmental microbiology*, 1992, 58, 2103-2110.

[109] Conway, T; Ingram, LO. Similarity of *Escherichia coli* propanediol oxidoreductase (*fucO* product) and an unusual alcohol dehydrogenase from *Zymomonas mobilis* and *Saccharomyces cerevisiae*. *Journal of bacteriology*, 1989, 171, 3754-3759.
[110] Reid, MF; Fewson, CA. Molecular characterization of microbial alcohol dehydrogenases. *Critical reviews in microbiology*, 1994, 20, 13-56.
[111] O'Mullan, PJ; Buchholz, SE; Chase, T; Eveleigh, DE. Roles of alcohol dehydrogenases of *Zymomonas mobilis* (ZADH): characterization of a ZADH-2-negative mutant. *Applied microbiology and biotechnology*, 1995, 43, 675-678.
[112] Nozzi, NE; Desai, SH; Case, AE; Atsumi, S. Metabolic engineering for higher alcohol production. *Metabolic engineering*, 2014, 25, 174-182.
[113] Smith, KM; Cho, KM; Liao, JC. Engineering *Corynebacterium glutamicum* for isobutanol production. *Applied microbiology and biotechnology*, 2010, 87, 1045-1055.
[114] Blombach, B; Riester, T; Wieschalka, S; Ziert, C; Youn, JW; Wendisch, VF; Eikmanns, BJ. *Corynebacterium glutamicum* tailored for efficient isobutanol production. *Applied and environmental microbiology*, 2011, 77, 3300-3310.
[115] Yamamoto, S; Suda, M; Niimi, S; Inui, M; Yukawa, H. Strain optimization for efficient isobutanol production using *Corynebacterium glutamicum* under oxygen deprivation. *Biotechnology and bioengineering*, 2013, 110, 2938-2948.

In: New Trends in *Corynebacterium glutamicum*
Editor: Carlos Barreiro
ISBN: 978-1-63483-280-9

Chapter 10

USE OF *CORYNEBACTERIUM GLUTAMICUM* FOR HEAVY METALS BIOREMEDIATION: ARSENIC AS PROTOTYPE

Almudena F. Villadangos**, *José A. Gil* and *Luis M. Mateos
Area of Microbiology, Department of Molecular Biology,
Faculty of Biology
University of León, León-Spain

ABSTRACT

The use of members from corynebacteria, a Gram-positive group of actinobacteria, for biochemical transformations and/or production of metabolites is well known since long time ago; in fact the microbial production of metabolites such as several essential amino acids and nucleotides are based on members of the coryneform group. However, very little efforts have been devoted to the possible use of non-pathogenic *Corynebacterium* species for bioremediation purposes such as conversion of toxic pollutant compounds, its transformation, and/or the cellular accumulation of the noxious heavy metals (mercury, lead, cadmium or arsenic as an example).

Arsenic is presently considered the most toxic element for life in the Earth crust, as indicated by the agency for toxic substances and disease registry (ATSDR, U.S.A). Unlike organic contaminants, which are degraded into harmless chemical species, metals and metalloids such as arsenic cannot be destroyed, and they have to be immobilized or transformed into less toxic forms in order to sequester the contaminants and avoid health problems associated to heavy metals exposure. Arsenic enters the biosphere by leaching from geological formations or by human actions, which include burning of fossil fuels, wastes generated by mining industries (gold, copper, etc.) and improper agricultural and medical practices.

The non-pathogenic *Corynebacterium glutamicum* was found to be very resistant to both inorganic arsenic species present in the environment [the oxianions arsenate (AsO_4^{3-}) and arsenite (AsO_3^{3-})] due to the presence of two *ars* operons (*ars1* and *ars2*).

* Corresponding Author: Almudena F. Villadangos; aferv@unileon.es

Several engineered *C. glutamicum* mutant strains were constructed and used as efficient strains for arsenic remediation (biocontention) from contaminated environments.

INTRODUCTION

Microorganisms are ubiquitous in nature and are involved in many biological processes of life. Chemical elements have been found in different proportions in all kind of habitats, co-existing with microorganisms since early history of Earth, nearly 4,000 million years ago. In the case of metals, this relation is revealed by a major role either directly or indirectly in almost all metabolic processes, such as the presence of divalent (or transition) metals at the active centers of enzymes [1].

Frequently metal ions have a role in the physiology of cells as (i) catalyzers of key reactions, (ii) preservers of protein structure, or (iii) its vital role in electron transport. Metals such as calcium (Ca), cobalt (Co), chromium (Cr), copper (Cu), iron (Fe), potassium (K), magnesium (Mg), manganese (Mn), sodium (Na), nickel (Ni), zinc (Zn), and vanadium (V)/molybdenum (Mo), in certain cases, are essential and serve as micronutrients in trace amounts for normal cell metabolism, but at high concentration they have noxious effects on various organisms and on human health [2].

Many other metals, mainly heavy metals (name due to its atomic density, which is greater than 5 g cm^{-3}), seem to have no biologically relevant functions. This is the case of mercury (Hg), lead (Pb), cadmium (Cd), silver (Ag), aluminum (Al), arsenic (As) and gold (Au), which are toxic for living beings even at very low concentrations [3]. The heavy metals toxicity is exerted either by displacement of essential metals or ions from its native binding sites when the heavy metal is present or through ligand interactions, mostly due to the heavy metal avidity for the sulfhydryl group of cysteines (mostly in proteins), which resulted blocked and inactivated protein [1].

The ability of living systems to both acquire and reject metals has led to the selection of a whole repertoire of mechanisms of interaction that ensure the adaptation of microorganisms to a changing and frequently hostile environment [4].

The emergence of metal tolerance mechanisms was a very early event in the microbial evolution millions of years ago. Thus different microbial groups exposed to high concentrations of toxic metals have developed a broad diversity of strategies and abilities to cope with these toxic elements (metallic stress) in a variety of environments such as anaerobic, aerobic, thermophilic, halophilic, and so on [1]. These resistance mechanisms have been reviewed frequently over the last 40 years and include: i) prevention of the metal uptake by cells (membrane composition, loss/absence of certain permeases, etc.); ii) intracellular accumulation of the toxic metals inside the cell as inert structures after interaction with cellular compounds; iii) change of the redox state of the incorporated metal such as reduction of the metal to a less/more toxic state; and iv) the cellular release of the metal trough certain efflux channels [5].

Research on microbial resistance mechanisms towards heavy metals is essential for the potential applications of microorganisms in bioremediation [4]. Heavy metal resistant microorganisms might represent the starting material to obtain microbes for bioremediation purposes.

Source of Heavy Metals: Natural and Anthropogenic Origins

Only 24 metals had been discovered early of the nineteenth century, from which 12 were discovered in the eighteenth century. Presently we know 86 different metals and its mankind's association comes from long time ago. Gold and copper were the first discovered metals, and together with mercury and lead form part of the seven metals of antiquity. Arsenic was initially described in the thirteenth century by Albertus Magnus in Germany [6] and it is the 20th element most abundant in the crust and core of the Earth (2 and 4 ppm, respectively). Cadmium was firstly reported in 1817. In general the discovery of a metal was related to its relative presence in the earth's crust; the application of metallurgical processes, two centuries ago, modified the metal environmental equilibrium in many aquatic and terrestrial ecosystems.

Environmental contamination by heavy metals aroses by natural process from the Earth (geogenic origin) or by anthropogenic activities. In most cases, heavy metals occur naturally within the subsurface aquifers and their presence in hydrothermal environments is due to both recent or ancient volcanic activity (gas fumes) and sedimentary rocks [7]. Metal levels in the atmosphere are due to wind erosion processes, volcanic emissions, forest fires and volatilization processes, the latter due mainly to biological transformations such as methylation by some microorganisms. The effect of the anthropogenic activities is been progressively observed due to the increment in population, industrialization and urbanization [1, 3] and is the major source of metal enrichment in soils. Heavy metals such as lead, arsenic, cadmium and mercury are discharged from industrial operations and/or agricultural activities, like application of agrochemical agents for plague control, or long-term usage of sewage sludge in agricultural fields [8].

Heavy metals exist both in bioavailable and non-bioavailable forms; this availability depends on physic-chemical properties of the environment: pH, redox potential, salinity, temperature and extent of aeration [9]. Under aerobic conditions, metals are usually found as soluble cationic forms, while under anaerobic conditions, sulfide or carbonate metal derivatives do precipitate as insoluble forms. At low pH, metal bioavailability increases due to its free ionic molecular species, while at high pH it decreases due principally to the formation of insoluble mineral of phosphate and carbonate derivatives [10]. Therefore, assessment of heavy metal bioavailability helps to evaluate the impact of metals on microorganisms from soils and to predict the application of technologies that could be used to clean up metals from the polluted soils [11].

Heavy Metal Remediation Using Microorganisms: Bioremediation

The removal of toxic heavy metals from industrial wastewaters has been performed for several decades. Physical and chemical processes such as oxidation and reduction, chemical precipitation, filtration, electrochemical treatment, evaporation, ion-exchange and reverse osmosis processes are widely available, but their cost-effectiveness is limited and they are difficult to apply in undeveloped countries. Moreover, they are not completely effective, leaving behind trace quantities of the metal [12] and producing frequently toxic sludge (or other waste products) that require safe and expensive post-treatment. Therefore, over the years the knowledge in science and technology, in parallel to industrial revolution, has facilitated the development of several alternative bioremediation strategies in order to overcome the pollution of damaged environments.

At present, bioremediation is the most effective controlling tool to manage polluted environments and their further recovery [10]. This is an alternative that offers the possibility to neutralize or render harmless various contaminants, such as heavy metals, using the biological activity of the living beings. Compared to other methods, bioremediation is considered a more promising, efficient and generally less expensive way for cleaning up contaminated environments, mainly soil and water [13, 14]. Bioremediation uses biological agents, mainly eukaryotic microorganisms, such as yeast or fungi to clean up contaminated environments [15]; under the name of phytoremediation authors have applied aquatic or terrestrial plants to achieve metal concentration from contaminated water or soil [16].

However, bacterial remediation has not been profusely applied, even when specific advantages for the use of bacteria as metals sorbent can be addressed: small size, broad metabolic capacity, large number of available genome sequences that can be exploited, simplicity and low cost of the processes [17]. Some bioremediation strategies have been developed so far in bacteria, and they could be combined to obtain a higher accumulation yield: i) expression of chelating agents (proteins) that retain metals inside the cell; ii) extracellular sequestration by the expression of chelating proteins that retain metals associated to the cell wall; iii) deletion of genes involved in metal efflux; and iv) expression of membrane channels or porins that increase the metal uptake [18]. Metals bind to cellular molecules by means of the functional groups, which are well characterized on the cell walls and include imidazole, thiol, amino, phosphate, carboxyl, sulfate, thioether, phenol, carbonyl and hydroxyl among others [19].

Nevertheless, there are relatively few studies describing the use of bacteria for heavy metals remediation based on the points described above, although biosorption-chelation or bioretention-accumulation processes are the most common. The latter can be understood as an active, metabolically driven process where cells are viable, whereas the former would be a passive metabolically independent system [20]. For microbial chelation, bacterial strains expressing exopolysaccharides (EPS; naturally secreted or by means of genetic manipulation) have been used. Members from *Bacillus* or from the cyanobacteria group (among others) are frequently exopolysaccharide overproducers; the negative charges of the EPS associated to the outer membrane are frequently used for the passive biosorption of different metals [21].

Microorganisms have different resistance mechanisms to tolerate the toxic effects of heavy metals. By cellular induction of specific metal-regulatory proteins, the resistance mechanism can facilitate the bioretention/accumulation of toxic metals inside the cell withstanding its toxic effects. Some well-studied metal binding proteins are metallothioneins (MTs). MTs are low molecular weight proteins (3.5 to14 kDa), non-ribosomal synthesized cysteine-rich peptides that contain the reactive thiol group (-SH). Heavy metals such as cadmium, lead, zinc, and copper can induce the expression of MTs in certain bacteria. Members from cyanobacteria (*Anabaena*, *Oscillatoriabrevis, Synechococcus,* etc.) and other bacterial strains (*Pseudomonas*, *Bacillus*, etc.) have been reported to encode MTs to maintain cytosolic metal homeostasis [22].

In some bacterial strains, combinations of different strategies for metal remediation were successful, such as the case of arsenic retention by *Escherichia coli* [17], or the use of *Corynebacterium glutamicum* as it would be described later [18].

MECHANISMS OF RESISTANCE TO METALLIC STRESS IN *CORYNEBACTERIUM GLUTAMICUM*

Coryneform bacteria are a diverse group of Gram-positive actinomycetes that are broadly distributed in the environment; they can be found in mucous and skin of humans and animals, and in samples of soil and water. They can be divided as pathogenic and non-pathogenic species, with *Corynebacterium diphtheriae* [23] and *Corynebacterium glutamicum* as representative members from the different groups, respectively. More than 50 *Corynebacterium* species from the 84 validly published (May 2010) [24], have been described as medically relevant species that cause occasional infections in humans. *C. atypicum, C. urealyticum, C. jeikeium, C. nigricans,* and so on, are some examples of species responsible of significant and occasionally fatal diseases, particularly in immune-compromised patients [25]. On the other hand, these species are resistant to different antibiotics due to the indiscriminate use and abuse of antibiotics in human and animal medicine [26].

The non-pathogenic *Corynebacterium* species are characterized by their ubiquity and the production of primary (nucleotides, amino acids) and secondary metabolites (antibiotics and antitumorals). Microbiological characterization of contaminated areas with metals such as mercury, lead, cadmium, arsenic, gold, zinc or cobalt has revealed the presence of a great number of bacteria [27, 28, 29], most of them from actinobacteria, and mainly from the genus *Corynebacterium* [29]. These evidences suggested that *Corynebacterium* has been adapted to "contaminated" environments, developing specific defense mechanisms to cope with antibiotics or heavy metal exposure [28, 30], in contrast to other very studied prokaryotic and eukaryotic model organisms, such as *E. coli* and *Saccharomyces cerevisiae*, which are naturally less exposed to heavy metals concentrations. *C. glutamicum* is also a safe bacterium to handle [31], its genome has been completely sequenced [32] and it was used as model of biological research for several decades [33]; therefore *C. glutamicum* strains could serve as a bioremediation tool by accumulation of large amounts of heavy metals and to be used as an efficient indicator of the presence of heavy metals.

Resistance Genes for Mercury, Cadmium, Lead, Copper and Zinc in *C. glutamicum*

Mercury (Hg) is a very toxic heavy metal for life; it is ranked at sixth position among the top ten hazardous elements [34]. It exists in nature mainly as cinnabar ores and enters into aquatic environment through leaching and washing of soils sediments and rocks by rain [35]. The mercury pollution in marine sediments and its effects through bioaccumulation in the food chain are one of the major environmental and human health problems. Mercury accumulates in kidney, neurological tissue and liver and exhibits noxious effects, including neurotoxicity, nephrotoxicity and gastrointestinal toxicity [36]. Mercury species [inorganic-mercurials (Hg^{2+}) and organomercurials (methylmercury, MeHg)] when they pass through biological membranes form covalent bonds with low molecular weight thiols (LMW-thiols) and strongly binds to thiol groups of the cysteine residues of proteins. The releasing of these mercury ions modifies the LMW-thiols metabolism and provokes cellular damaging [37], increases lipid peroxidation and affects the antioxidant activities of the cells [36].

C. glutamicum cells could survive and grow at mercury concentrations ranging from 2-15 μM, whereas 20 μM led to growth arrest [30]. This growth/resistance is due to the presence of chromosomal mercury resistance genes in *C. glutamicum*. The genetic resistance system, named "*mer* operon," is in fact the best known bacterial mercury resistance system that transforms mercury toxic targets into volatile non-toxic forms [35]. *mer* operons are widely distributed among a large community of microbes which means a horizontal gene transfer, reflecting the importance and compatibility of this mechanism in nature [35]. Bioinformatic analyses reflect that *C. glutamicum* present a unique *merA* gene (cg3339) in its genome that encodes a mercuric reductase enzyme. The mercury resistance mechanism in *C. glutamicum* is therefore limited to enzymatic detoxification of a unique type of mercury compounds: the inorganomercurials.

When Hg^{2+} passes through the biological membranes of *C. glutamicum*, a proteomic response is achieved and results in dramatic changes of the protein synthesis [28]. Thirteen proteins were upregulated in the presence of 10 μM of mercury, and were classified into three different functional classes: (i) enzymes with antioxidant properties; (ii) proteins involved in transport systems and (iii) miscellaneous proteins, including those involved in inositol metabolism or Zn-dependent alcohol dehydrogenases. However, the production of reactive oxygenic species (ROS) is also known to be one of the major mechanisms of the toxicities exerted by heavy metals [38], and also by mercury [35]. ROS, which are produced by heavy metals as a result of oxidative stress, cause severe damage to biological macromolecules because they bind to lipids, proteins and DNA and may lead to cytotoxicity, mutagenicity and carcinogencity. Fanous et al. (2008) indicated that *C. glutamicum* cells responded to the mercury oxidative stress by upregulation of the corresponding antioxidant proteins, such as thioredoxin, thioredoxin reductase, superoxide dismutase, and even LMW-thiols. In the actinomycetes group, mycothiol (MSH) is the major LMW-thiol identified so far. MSH is even more resistant than glutathione (GSH) to heavy metals ion-catalyzed oxidation, being the latter the major LMW-thiol in prokaryotic and eukaryotic organisms, whereas MSH is the LMW-thiol antioxidant used by *C. glutamicum* [28].

Cadmium (Cd) and lead (Pb) have no known biological function in bacteria and when they enter into cells they disturb the normal functioning of the organism. Their toxicity comes from their ability to mimic other biologically important metals, most notably calcium, iron, and zinc [39], which are functioning as cofactors in many enzymatic reactions and are considered as essential elements for all living organisms.

Lead is a ubiquitous toxic metal detectable practically in all phases of the inert environment and biological systems [39]. It is frequently used as an additive in the fabrication of batteries, metal alloys, gasoline, paints and ceramic products; however its use for caulking and pipe solder has been dramatically reduced in recent years because of health concerns. Ingestion of contaminated food and drinking water is the most common source of lead exposure in humans. Relatively little is known about lead resistance in bacteria. Some bacterial lead-resistance mechanisms involve the intra- and extracellular binding of inorganic

Pb^{2+}; the molecular mechanism responsible for the formation of lead precipitates in bacteria are not known, however better characterized lead-resistance mechanisms include active efflux pumps, such as P-type ATPases, which transport ions against the concentration gradient using energy provided by ATP hydrolysis [40]. *Cupriavidus metallidurans* CH34 contained the most studied lead resistance mechanism, based on the *pbr* operon (*pbrTRABCD*) [40], which also conferred zinc and cadmium resistance. Up to now, we did not found any lead resistant gene in the *C. glutamicum* genome.

Cadmium is an element of great concern as exposure to this element can cause both chronic and acute health effects in living organisms. Cadmium occurs naturally in the earth's crust and in ocean water [39]. The average terrestrial abundance of cadmium is 0.09-0.18 μM, whereas in ocean waters it ranges from 0.04 to 0.98 μM. Cadmium and its derivative compounds are carcinogenic to humans and are classified as Group 1 of risk by the International Agency for Research on Cancer (IARC) [39]. *C. glutamicum* is resistant to 10 μM cadmium, whereas this cadmium concentration is lethal for *E. coli* [28]. The cadmium resistance mechanism in *C. glutamicum* is still unclear, but the exposure of *C. glutamicum* cells to low cadmium concentrations resulted in dramatic changes of the protein patterns [28]. Thus when *C. glutamicum* was treated with 5 μM of cadmium, 35 proteins were induces and 8 proteins were repressed, which were classified into eight different functional classes: i) enzymes with antioxidant properties; ii) proteins involved in repair and biogenesis of Fe-S cluster proteins; iii) molecular chaperones; iv) proteases; v) proteins involved in terpenoid biosynthesis; vi) proteins involved in lipoate synthase; vii) proteins involved in energy metabolism; and viii) other proteins.

Copper (Cu) is a trace element essential to living organisms. It plays a catalytic role in enzymes, especially in oxidation/reduction reactions, due to its ability to change between the oxidized (Cu^{2+}) and reduced form (Cu^{+}) [41]. However, at the same time, copper has toxic effects on cellular functions when present in large amounts, because it can trigger the ROS response or lead to sulfhydryl depletion [42]. Hence, the amount of copper ions inside the cell must be accurately regulated to prevent deprivation as well as high toxic copper concentrations. In *C. glutamicum* the growth rates were reduced with the addition of more than 20 μM of copper; the addition of 500 μM copper completely inhibited the growth of *C. glutamicum* [42]. The copper resistance system of *C. glutamicum* has been recently studied and a dual model copper resistance mechanism has been proposed [42]. *C. glutamicum* presents two genes (*copR* and *copS*; cg3285 and cg3284, respectively) as well as the up- and downstream genes (cg3286-cg3289 and cg3283-cg3281) that are particularly important for the adaption to copper excess conditions. Under aerobic conditions copper is present in the "periplasmic" space in the oxidation states Cu^{+}/Cu^{2+}. The two-component system CopRS recognizes a high copper concentration and activated two operons containing the putative copper resistance genes for a copper oxidase, CopO (cg3287), and the copper export ATPase, CopB (cg3281); the function of the rest of the genes is still unknown [42]. As a result, CopB export the excess of the intracellular Cu^{+} to outside and CopO detoxify Cu^{+} by its oxidization to the less toxic Cu^{2+}. On the other hand, the copper specific regulator CsoR can sense elevated intracellular copper concentrations, promoting the transcription of the copper export ATPase CtrV (cg0464), which is also responsible for the cellular copper release [42].

Zinc (Zn) is considered an essential nutrient for all living organisms, because of its crucial role in several enzymatic reactions and in proteins structures; however, zinc is toxic to living cells when is present in excess, and therefore zinc uptake, efflux, storage and metabolism must be tightly regulated in bacteria [43]. The regulation of the zinc homeostasis in *C. glutamicum* requires various types of enzymes encoded in different operons, conforming the Zur regulon [43], however only two zinc inducible transporter genes *zrf* (cg1447) and *zra* (cg1448) have been studied in depth [44]. Both genes seem to be responsible for the export of zinc excess out of cells: Zrf encodes a member of the cation diffusion facilitator (CDF) family, whereas Zra encodes a putative metal-translocating P-type ATPase [44]. It is noteworthy that the *zrf* gene in *C. glutamicum* is involved in both, cobalt and zinc resistance [45].

It is most than likely that *C. glutamicum,* a soil bacterium very abundant in contaminated soils had developed highly specific defense mechanisms to cope with heavy metal exposure along the evolution; however only the arsenic resistance mechanism has been genetically and biochemically analyzed and applied in bioremediation assays.

THE ARSENIC RESISTANCE MODEL IN *C. GLUTAMICUM:* THE *ARS* OPERON

Arsenic (As) is an extremely toxic metalloid that, when is present in high concentrations, severely threatens the biota and human health, and whose toxicity accounts for the first position of the Substance Priority List of the Superfund Amendments and Reauthorization Act (http://www.atsdr.cdc.gov).

Arsenic contamination of soil, water, and air is a global growing environmental problem due to leaching from geological formations, the burning of fossil fuels, wastes generates by the gold mining industry present in uncontrolled landfills, and improper agriculture or medical uses [18].

The ubiquity of arsenic in the environment has led to the appearance of several arsenic defense mechanisms in microbes. The most frequent, important and studied arsenic resistant mechanism is the presence of arsenic resistant genes organized in operones called ars operones [46, 47]; they are more frequent even than the genes of the biosynthesis of amino acids, such as tryptophan [48].

C. glutamicum is one of the most arsenic-resistant microorganisms described to date: up to 12 mM arsenite (As^{III}) and 60 mM arsenate (As^{V}) [49]. Genes involved in arsenic resistance are located on the C. glutamicum bacterial chromosome, forming two separated operones named ars1 and ars2 and two orphan genes distant from each other in the chromosome (acr3-3, an hypothetical arsenite permease, and arsC4, an arsenate reductase) (Figure 1).

Both ars operons showed a similar genetic structure of three compounds (arsR-acr3-arsC), but operon ars1 contained an additional gene (arsC1'). The existence of several arsenic resistant operons in the genome of a single bacterial specie is not frequent, but it has been also identified in bacteria such as Pseudomonas putida [50], Ochrobactrum tritici [51] and Herminiimonas arsenicoxydans [52].

However, the identification of additional genes in the ars operon is more common [47]. The genes of the C. glutamicum ars operons encode the transcriptional repressors (ArsR1 and ArsR2), which control the expression of rest of the genes of the ars operons; the arsenite permeases Acr3-1 and Acr3-2 are arsenite efflux pumps that remove As^{III} out of the cell; the arsenate reductases ArsC1 and ArsC2 reduce As^{V} to As^{III} using mycothiol/mycorredoxin-1 redox system (Figure 1) [49].

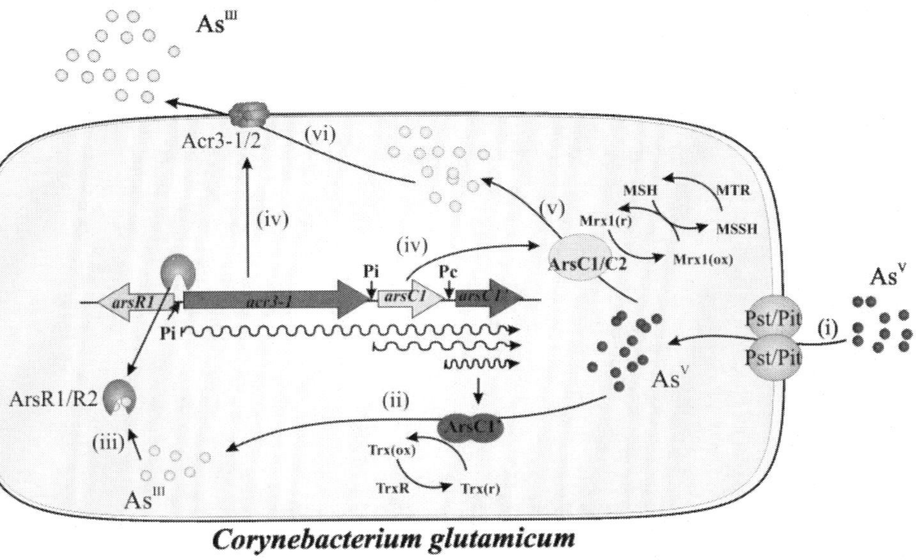

Figure 1. The arsenic detoxification model in *C. glutamicum* and genes involved in arsenic resistance in *C. glutamicum.* (i) As^{V} enters the cell by phosphate transports systems (Pst/Pit); (ii) the constitutive expression of the Trx-dependent ArsC1' allows immediate but controlled reduction of As^{V} to As^{III}; (iii) the As^{III} binds to ArsR repressors (ArsR1/R2) and structural changes makes them releasing from the *ars1* and *ars2* operons: (iv) transcription of genes from the *ars* operons; (v) the inducible arsenate reductases ArsC1/C2 are reducing As^{V} to As^{III} using the MSH/Mrx1-dependent reduction system; and (vi) the inducible Acr3 arsenite permeases (Acr3-1/2) on the membrane are pumping As^{III} out of the cell [Modified from 64].

Arsenic Uptake Systems in *C. glutamicum*

All cells possess mechanisms that control the level of essential metals that exist therein, such as zinc, or toxic metalloids as arsenic [53]. These mechanisms are necessary as some essential metals, when they are in excess, become toxic to the cell. Arsenic can act paradoxically as a toxic carcinogen and a chemotherapeutic agent for various diseases. To understand and prevent the toxicity and carcinogenesis exerted by arsenic, is important to know how arsenic enter into the cells [54], as well as knowing the chemical nature of the metalloid in the extracellular environment.

Two inorganic arsenic species are generally found in soils and in surface waters: As^{III} and As^{V}. Under oxigenic conditions, As^{V} is the most prevalent form. Its toxicity can be explained by the resemblance of its tetrahedral oxyanion structure (AsO_4^{3-}) to phosphate (PO_4^{3-}), and therefore it uncouples conversions in intermediary metabolism, e.g. glycolysis or oxidation phosphorylation [55]. In *C. glutamicum*, as in other microorganisms such as *E. coli* and *Bacillus subtilis* or eukaryotic organisms like *S. cerevisiae* or *Arabidopsis thaliana*, As^{V} enters through the phosphate transport systems [54]. In *C. glutamicum* two phosphate transport systems have been found: Pit (Phosphate Inspecific Transport) and Pst (Phosphate

Specific Transport), and both systems seem to be involved in As^V uptake (Figure 1), but the degree of involvement of the Pit and Pst systems in As transport is unknown [56].

On the other hand, As^{III} is more toxic than As^V, disturbing redox homeostasis because of its ability to react with free thiols [15]. In aqueous solution and at neutral pH, As^{III} usually enters bacterial cells through aquaglyceroporins (GlpF). This reflects the fact that As^{III} is protonated to form arsenic hydroxide [$As(OH)_3$], which structurally resembles glycerol [$C_3H_5(OH)_3$] [57]. No *glpF* homologues were found in the *C. glutamicum* genome [18], which is in agreement with the recent observation that *C. glutamicum* is unable to grow using glycerol as the unique carbon source [58]. As^{III} uptake has also been correlated to unspecific sugar transport, such as the hexose permeases described in yeast [59]. The genome of *C. glutamicum* contains more than ten genes coding for putative hexose permeases; interruption and overexpression analyses of two of them in *C. glutamicum* did not show differences in the As^{III} incorporation related to control strains [data unpublished]. Nowadays, the As^{III} uptake system in *C. glutamicum* is unknown, but it can be claimed that it does not take place through GlpF.

Mechanisms for Arsenic Release in *C. glutamicum*

Efflux of As^{III} from cells is often the primary arsenic resistance mechanism. It is catalyzed by members of three families of unrelated As^{III} transporters: i) the ATP-binding cassette transporters of the multidrug resistance-associated protein superfamily; ii) ArsB permeases, which, historically, was the first arsenite permease to be identified [57]; iii) the arsenite permease family Acr3, which is a member of the bile/arsenite/riboflavin transporter (BART) superfamily that includes members in bacteria, archaea, and fungi and is more widely distributed than ArsB [60].

C. glutamicum has two functional arsenite permeases belonging the Acr3 family, Acr3-1 and Acr3-2 (Figure 1). The presence of several gene copies for As^{III} detoxification is not infrequent [47], but it allows organisms, such as *C. glutamicum*, to respond to high concentrations of As^{III}. The Acr3-1 protein conferred almost 80% of arsenite permease activity, and Acr3-2 stands for the remaining 20% [61].

The energy source used by *C. glutamicum* Acr3 proteins for As^{III} efflux is coupled to the electrochemical proton gradient established basically by NADH respiration [61]. Acr3 proteins could be considered an metalloid-proton exchangers with $As(OH)_3/H^+$ antiporter activity, because the dissipation of only one of the individual component of the electrochemical proton gradient (electrical or chemical component), does not inhibit the As^{III} transport [61].

Both Acr3 proteins, Acr3-1 and Acr3-2, have a limited substrate specificity, they are involved in As^{III} efflux but not in other heavy metal efflux. Acr3-1 and Acr3-2 are 72% identical to each other with a putative topology of ten transmembrane (TM) domains and several conserved residues in different TM segments. Using a putative Acr3-1 topological map, the participation of six conserved residues in the activity of Acr3-1 was examined, but only the conserved cysteine 129 (C^{129}) and the glutamate 305 (G^{305}) are involved in the active transport of As^{III} [61].

Therefore, Acr3 proteins from *C. glutamicum* are antiporters that catalyze arsenite-proton exchange with Cys and Glu residues involved in efflux [61].

Arsenate Reduction Mechanisms in *C. glutamicum*

In bacteria, the main arsenic detoxification mechanism is based on the reduction of intracellular As^{V} to As^{III} by cytoplasmic arsenate reductase enzymes (ArsCs). ArsCs are small cytoplasmic redox enzymes intensely studied during the last two decades [62]. For As^{V} reduction, ArsC proteins make use of electrons coming from the cellular redox system; electrons are transferred from NADPH by the sequential involvement of different thiolate nucleophiles. Based on these reduction mechanisms, four different ArsC classes can be defined [62]: i) the thioredoxin (Trx)/thioredoxin reductase (TrxR)-dependent class; ii) the gluthation (GSH)/glutaredoxine (Grx)-coupled class; iii) the trypanothione/trypanothione reductase class; and iv) the mycothiol (MSH)/mycoredoxin (Mrx)-dependent class.

As indicated above, there are three *arsC* genes that code for three functional arsenate reductases (ArsC1, ArsC2 and ArsC1'; Figure 1) in the genome of *C. glutamicum*, being ArsC1 and ArsC2 the most relevant arsenate reductases in *C. glutamicum* and playing a key role in the defense against As^{V} toxicity [63]. They are single-cysteine monomeric enzymes with a completely new enzyme mechanism for the reduction of As^{V} in which the electron transfer is coupled to mycothiol (MSH) and mycoredoxin1 (Mrx1) and its expression is only observed in the presence of As^{III} (Figure 1). A clear link between the production of MSH and the level of As^{V} resistance has been observed in *C. glutamicum* [63]. In contrast, ArsC1' is a three-cysteine protein containing a homodimer that is constitutively expressed at very low levels from its own constitutive promoter (Figure 1) [64]. ArsC1' is very efficient in reducing As^{V} to As^{III} using electrons coming from the thioredoxin pathway (Figure 1).

All together, *C. glutamicum* has evolved to survive As^{V} stress using two different mechanisms, which have to work in concert to pump the produced As^{III} out of the cell by the Acr3 proteins (Figure 1). Linkage to distinct arsenate reductase mechanisms is not usually observed in bacteria, so it looks that *C. glutamicum* has evolved towards a most optimal way to survive arsenic stress in an ever-changing environment [64].

BIOREMEDIATION OF ARSENIC SPECIES USING *C. GLUTAMICUM*

C. glutamicum is one of the most arsenic-resistant microorganisms described up to date and many of the molecular biology techniques developed during the last fifteen years have being used to obtain engineered *C. glutamicum* mutants strains able to speciate and/or remove arsenic from contaminated environments [18, 56].

Any strain used for arsenic bioremediation should show increased uptake of As^{V} or As^{III} and reduced the efflux of As^{III}. In addition, arsenic (As^{V} or/and As^{III}) present in the cell must be complexed to enhance the cellular tolerance.

Arsenate Bioremediation

In order to obtain *C. glutamicum* mutant strains able to accumulate As^{V}, the As^{V} uptake in *C. glutamicum* has to be increased, later on reduced to As^{III}, and removed from the cell; alternatively, As^{V} must be complexed.

As mentioned previously, As^{V} enters the cell through the specific (Pst) or inspecific (Pit) phosphate transporters (Figure 1), Therefore, the As^{V} entrance/accumulation would increase if *C. glutamicum* is grown in media with a reduced level of phosphate and without any arsenate reduction activity.

Efficient As^{V} biocontainers in *C. glutamicum* were obtained by the removal of both arsenate reductase genes (*arsC1* and *arsC2*; ArsC1-C2 mutant) or the genes *mshA* and *mshC* involved in MSH synthesis (MshA and MshC mutants).

None of these mutant strains were able to reduce As^{V} to As^{III} and they specificaly accumulated As^{V} in amounts that were from 27.50 to 30.00 μmol/g (27- to 30-fold higher than the As^{V} accumulated by wild type strain; Table 1) [65]; furthermore, these mutants are the most efficient bacterial As^{V} biocontainers described to date, accumulating 9-10 times more As^{V} than *E. coli* [65].

However, a specific method to increase the cellular concentration of As^{V} in *C. glutamicum* by cloning the genes involved in poly-P biosynthesis, *ppk2A* and *ppk2B*, in a multicopy plasmid, were not be useful to accumulate As^{V} [65].

Arsenite Bioremediation

In order to obtain *C. glutamicum* strains able to accumulate As^{III}, the following properties have to be considered: i) the As^{III} efflux has to be reduced; ii) the As^{III} uptake in *C. glutamicum* has to be increased; and iii) As^{III} must to be complexed inside the cell. As^{III} extrusion systems were removed to obtain effective As^{III} biocontainers that accumulated 7.2- and 4-fold higher than *C. glutamicum* wild strain in presence of As^{III} [65], and 15- fold higher when were assayed in the presence of As^{V} (Table 1) [56].

In bacteria, As^{III} uptake takes place through aquaglyceroporins (GlpF), but the corresponding genes are absent in the *C. glutamicum* genome [65], so we attempted to increase the cellular As^{III} uptake/accumulation by the heterologous expression in *C. glutamicum* of *glpF* genes from two actinobacterias *Corynebacterium diphtheriae* NCTC 13129 (CdGlpF) and *Streptomyces coelicolor* A3(2) (ScGlpF). As^{III} accumulation assays in the *C. glutamicum* engineered strains expressing *CdglpF* gene and without arsenite permease pumps (Acr3-1 and Acr3-2) showed an increased As^{III} retention of 11 folds more than *C. glutamicum* wild type strain (Table 1) [65].

This result provides evidence for the involvement of GlpF in As^{III} uptake and a more effective expression of *CdglpF* than of *ScglpF* in terms of arsenic accumulation in *C. glutamicum.*

To analyze the third property, overexpression of specific chelators to increase the As^{III} concentration in *C. glutamicum,* we attempted to clone the genes encoding ArsR1 and ArsR2 regulators from *C. glutamicum* and the ArsD (metallo-chaperone) from *E. coli* in a multicopy plasmid.

The resulting strains (*C. glutamicum* 2Δars + ArsR1/ArsR2/ArsD) were unable to accumulate more As^{III} than the original strain *C. glutamicum* 2Δars (Table 1) [65].

In addition, the overexpression of arsR1 and arsR2 genes in C. glutamicum was apparently toxic, as the growth and the As^{III} retention of this strain was delayed in the presence of As^{III}.

Table 1. Arsenic accumulation amounts in different *C. glutamicum* mutant strains

	Arsenic added[a]	Strains	Accumulation amount (μmol/ g DCW) ± SD[b]	Times more of arsenic retained related to wild type strain[c]
C. glutamcium As^V biocontainers	As^V	*C. glutamicum* wild type	0.98±0.10	--
		C. glutamicum ArsC1-C2	28.50 ±0.34	28
		C. glutamicum MshA	27.50±0.75	27
		C. glutamicum MshC	30.00±1.00	30
C. glutamcium As^{III} biocontainers	As^{III}	*C. glutamicum* wild type	0.50±0.11	--
		C. glutamicum 2Acr3	3.60±0.17	7.2
		C. glutamicum 2Δars	1.94±0.10	4
		C. glutamicum 2Δcr3 + GlpFCd	5.56±0.30	11
		C. glutamicum 2Δcr3 + GlpFSc	5.01±0.29	10
		C. glutamicum 2Δars + GlpFCd	3.77±0.40	7.5
		C. glutamicum 2Δars + GlpFSc	2.87±0.20	5.7
		C. glutamicum 2Δars + ArsR1	1.11±0.20	2.2
		C. glutamicum 2Δars + ArsR2	1.27±0.30	2.5
		C. glutamicum 2Δars + ArsD	2.00±0.25	4
	As^V	*C. glutamicum* 2Acr3	7.44±0.32	15

DCW, dry cell weight; SD, standard deviation.

[a] The arsenic specie added in the assay.

[b] Using "resting cells" exposed for 2 h to As^V (0.1 mM) or As^{III} (1 mM). SD was determined from three independent experiments.

[c] Folds of arsenic that the strain retains more compared with wild type *C. glutamicum* ATCC 13032 in the same conditions.

Conclusion and Future Prospects

Remediation techniques for the removal of toxic compounds from the environment (frequently associated to the human activities) are urgently needed taking into account the development of transformation industries during the two last centuries. The global requirement for a solution to the environmental pollution problem includes remediation techniques such as the use of chemical methods (chemical chelation), biological tools to convert the toxic substances in environmentally healthy (biotransformation) and its contention when conversion in a less toxic form is rather impossible (biocontention). The latter is required for heavy metals released to the environment (mostly by anthropogenic mechanisms), and when bioremediation using plants and/or microorganisms are considered the best choice, due to its multiple advantages, compared to other remediation protocols.

Environmental heavy metal pollution is one of the most frequent events in nature; in some cases the contaminated areas are far away from the waste discharge places affecting to many organisms, to their interactions and the evolution of those organisms. It is clear that heavy metal pollution has provided an opportunity for some species to evolve toward tolerant populations either in prokaryotes or in eukaryotes.

In addition to the described arsenic bioremediation strategies, several future research orientations are suggested. Research on toxic effects of heavy metals in certain organisms has received considerable attention, mainly when the metal toxicity is high (mercury, arsenic, cadmium, and so on), but this needs further exploration. Certain authors have described the important effect of sub-lethal metal presence in fresh water environments, which could uncover previously unrecognized heavy metal pollution impacts such as immunomodulation. This was studied in rainbow trout, where doses of cadmium, mercury, or zinc clearly affected the immune system function [66].

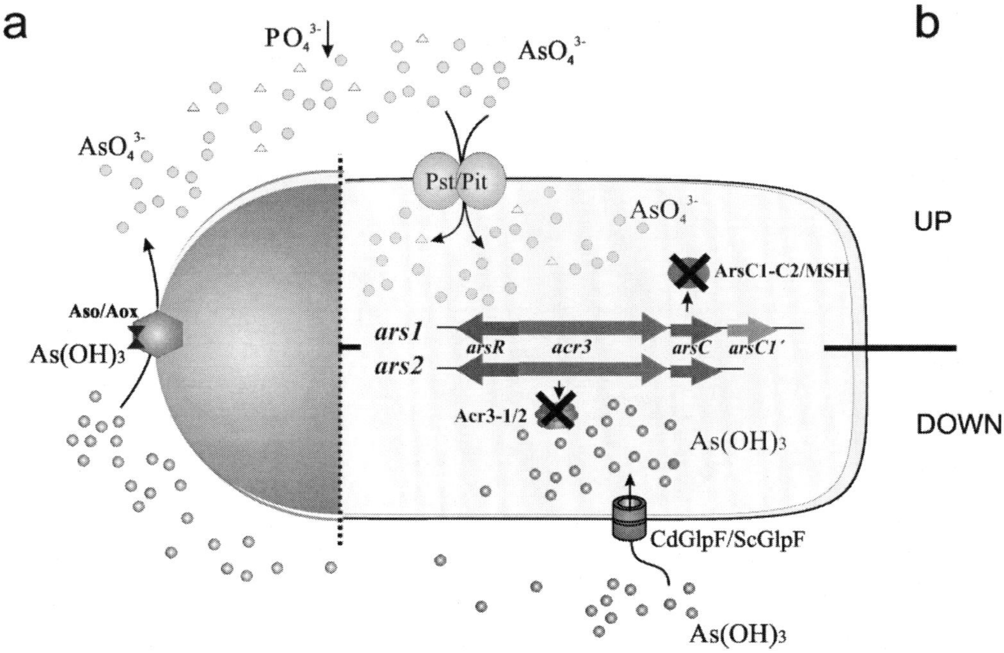

Figura 2. Proposed microbial consortium system for the bioremediation of inorganic arsenic species. a) In aquatic environments and at neutral pH, some species of bacteria (e.g., *Ochrobactrum tritici*) use periplasmic arsenite oxidases (Aso/Aox proteins) to oxidize arsenite [$As(OH)_3$] to arsenate (AsO_4^{3-}), which provides an energy source. b) When phosphate levels (PO_4^{3-}) in the environment are low, engineered *C. glutamicum* strains can use the Pst/Pit system to incorporate As^V. Strains lacking functional arsenate reductases (ArsC1/C2) or unable to synthesize MSH (UP) accumulate As^V; mutant strains lacking arsenite permeases activities (Acr3-1/2) and expressing heterologous GlpF proteins (CdGlpF/ScGlpF) (DOWN) incorporate both arsenic species, but only As^{III} is finally accumulated [Modified from 65].

Furthermore, the mechanisms of interaction between bacteria and metals are very complex and a large number of factors are involved in this process. Indeed, the abovementioned researches point out that metal bioremediation depends on: i) chemical and morphological features of the microbial cells; ii) chemical characteristics of the exopolimers surrounding the cells; iii) the chemical and physical properties of metals; iv) metal interactions with certain cellular compounds (MTs, LMW-thiols, etc.), if present; finally, v) the device and conditions utilized in the recovery treatment. The remarkable interest in using bioremediation for the recovery of noxious metals and the increasing public perception of the importance of green biotechnologies in the treatment of solid or liquid wastes from industries, power plants and incineration plants, containing substantial amounts of toxics might give the opportunity to develop new industrial processes.

Combinations of new remediation approaches can result very useful to accomplish the ultimate goal, the achievement of an optimal biocontainer. In the case of arsenic as pollutant two bacterial species have been used as bioaccumulation prototypes, *E. coli* and *C. glutamicum*. In both cases the amount of the arsenic accumulated depended on the arsenic specie (As^{V} or As^{III}) and how the original strain was genetically manipulated. In certain situations we can imagine the involvement of different bacterial species (a bacterial consortium) for the remediation of a metal contaminated environment. In the case of arsenic, the bacterial consortium could be based on bacteria able to transform one toxic metal specie (As^{III}) to the less toxic (As^{V}), such as *Ochrobactrum tritici* [67], and bacteria able to accumulate the just converted As^{V}, like engineered *C. glutamicum* mutant strains (mutants ArsC1-C2, Msh or 2Acr3) (Figure 2). This remediation approach would offer a novel and efficient bio-tool for arsenic bioremediation using a two-step system for arsenic detoxification.

ACKNOWLEDGMENTS

AFV was recipient of postdoctoral fellowship from the Junta de Castilla y León (Spain). This work was supported by grants the Junta de Castilla y León (LE326U14).

REFERENCES

[1] Valls M, de Lorenzo V. Exploiting the genetic and biochemical capacities of bacteria for remediation of heavy metal pollution. *FEMS Microbiol. Rev*. 2002;26:327-338.
[2] Filali BK, Taoufik J, Zeroual Y, Dzairi FZ, Talbi M, Blaghen M. Waste water bacterial isolates resistant to heavy metals and antibiotics. *Current Microbiol.* 2000;41:151-156.
[3] Issazadch K, Jahanpour N, Pourghorbanali F, Racisi G, Faekhondeh J. Heavy metals resistance by bacterial strains. *Annals Biol. Research* 2013;4:60-63.
[4] Haferburg G, Kothe E. Microbes and metals: interactions in the environment. *J. Basic Microbiol.* 2007;47:453-467.
[5] Rosen BP. Biochemistry of arsenic detoxification. *FEBS Lett.* 2002;529:86-92.
[6] Emsley J. Nature's Building Blocks: An A-Z Guide to the Elements. Oxford: Oxford University Press; 2001.
[7] Lièvremont D, Bertin PN, Lett MC. Arsenic in contaminated waters: Biogeochemical cycle, microbial metabolism and biotreatment processes. *Biochimie* 2009;91:1229-1237.
[8] Rosen BP. Transport and detoxification systems for transition metals, heavy metals and metalloids in eukaryotic and prokaryotic microbes. *Comp. Biochem. Physiol. Mol. Integr.* Physiol. 2002;133:689-693.
[9] Brown G, Foster AL, Ostergren JD. Mineral surfaces and bioavailability of heavy metals: a molecular-scale perspective. *Proc. Nat. Acad. Sci.* 1999;96:3388–3395.

[10] Ahemad M, Khan MS. Pesticide interactions with soil microflora: importance in bioremediation. In: Microbes and microbial technology: agricultural and environmental applications. Springer. New York: Ahmad I, Ahmad F, Pichtel J; 2011.
[11] Khan MS, Zaidi A, Wani PA, Oves M. Role of plant growth promoting rhizobacteria in the remediation of metal contaminated soils. *Environ. Chem. Lett.* 2009;7:1–19.
[12] Chwirka JD, Thompson BM, Stomp JM. Removing arsenic from groundwater. *J. Am. Water Works Assoc.* 2000;92:79–88.
[13] Kamaludeen SPBK, Arunkumar KR, Avudainayagam S, Ramasamy K. Bioremediation of chromium contaminated environments. *Indian J. Exp. Biol.* 2003;41:972–985.
[14] Ahemad M. Implication of bacterial resistance against heavy metals in bioremediation: a review. *IIOABJ* 2012;3:39-46.
[15] Tsai SL, Singh S, Chen W. Arsenic metabolism by microbes in nature and the impact on arsenic remediation. *Curr. Opin. Biotechnol.* 2009;20:659–667.
[16] Wojas S, Clemens S, Skłodowska A. Arsenic response of AtPCS1- and CePCS-expressing plants-effects of external As(V) concentration on As-accumulation pattern and NPT metabolism. *J. Plant Physiol.* 2010;167:169–175.
[17] Tsai SL, Singh S, Da Silva NA, Chen W. Co-expression of *Arabidopsis thaliana* phytochelatin synthase and *Treponema denticola* cysteine desulfhydrase for enhanced arsenic accumulation. *Biotecnol. Bioeng.* 2012;109:605–608.
[18] Mateos LM, Ordóñez E, Letek M, Gil JA. *Corynebacterium glutamicum* as a model bacterium for the bioremediation of arsenic. *Int. Microbiol.* 2006;9:207-215.
[19] Javanbakht V, Alavi SA, Zilouei H. Mechanisms of heavy metal removal using microorganisms as biosorbent. *Water Sci. Technol.* 2014;69:1775-1787.
[20] Volesky B, Holan ZR. Biosorption of heavy metals. *Biotechnol. Prog.* 1995;11:235–250.
[21] De Philippis R, Colica G, Micheletti E. Exopolysaccharide-producing cyanobacteria in heavy metal removal from water: molecular basis and practical applicability of the biosorption process. *Appl. Microbiol. Biotechnol.* 2011;92:697-707.
[22] Blindauer CA, Harrison MD, Robinson AK, Parkinson JA, Bowness PW, Sadler PJ, Robinson NJ, Multiple bacteria encode metallothioneins and SmtA-like fingers. *Mol. Microbiol.* 2002;45:1421–1432.
[23] Costa J, Michel J, Rappuoli R, Murphy J. Restriction map of corynebacteriophages βc and βvir and physical localization of the diphtheria *tox* operon. *J. Bacteriol.* 1981;148:124-130.
[24] Bergey's Manual of Systematic Bacteriology. Volume 5.
[25] Funke G, Bernard KA. Coryneform gram-positive rods. In P. R. Murray, E. J. Baron, J. H. Jorgensen, M. L. Landry, and M. A. Pfaller, editor. Manual of clinical microbiology, ASM Press, Washington; 2007.
[26] Ortiz-Pérez A, Martín-De-Hijas NZ, Esteban J, Fernández-Natal M, García-Cía JI, Fernández-Roblas R. High frequency of macrolide resistance mechanisms in clinical isolates of *Corynebacterium* species. *Microb. Drug Resist* 2010;16:273-277.
[27] Margesin R, Płaza GA, Kasenbacher S. Characterization of bacterial communities at heavy-metal-contaminated sites. *Chemosphere* 2011;82:1583-1588.
[28] Fanous A, Hecker M, Görg A, Parlar H, Jacob F. *Corynebacterium glutamicum* as an indicator for environmental cobalt and silver stress-a proteome analysis. *J. Environ. Sci. Health B* 2010;45:666-675.

[29] Lee D, Ho K, Chen Y. Degradation of cresols by phenol-acclimated aerobic granules. *Appl. Microbiol. Biotechnol.* 2011;89:209-215.

[30] Fanous A, Weiss W, Görg A, Jacob F, Parlar H. A proteome analysis of the cadmium and mercury response in *Corynebacterium glutamicum. Proteomics* 2008;8:4976-4986.

[31] Meiswinkel TM, Rittman D, Lindner SN, Wendisch VF. Crude glycerol-based production of amino acids and putrescine by *Corynebacterium gluta*micum. *Bioresour. Technol.* 2013;145:254–258.

[32] Kalinowski J, Bathe B, Bartels D, Bischoff N, Bott M, Burkovski A, Dusch N, Eggeling L, Eikmanns BJ, Gaigalat L, Goesmann A, Hartmann M, Huthmacher K,Krämer R, Linke B, McHardy AC, Meyer F, Möckel B, Pfefferle W, Pühler A, Rey DA, Rückert C, Rupp O, Sahm H, Wendisch VF, Wiegräbe I, Tauch A. The complete *Corynebacterium glutami*cum ATCC 13032 genome sequence and its impact on the production of L-aspartate-derived amino acids and vitamins. *J. Biotechnol.* 2003;4:5-25.

[33] Mateos LM, Pisabarro A, Pátek M, Malumbres M, Guerrero C, Eikmanns BJ, Sahm H, Martín JF. Transcriptional analysis and regulatory signals of the *hom-thrB* cluster of *Brevibacterium lactofermentum. J. Bacteriol.* 1994;176:7362-7371.

[34] Nascimento AM, Chartone-Souza E. Operon *mer*: bacterial resistance to mercury and potential for bioremediation of contaminated environments. *Genet. Mol. Res.* 2003;31:92-101.

[35] Mathema VB, Thakuri BC, Sillanpää M. Bacterial *mer* operon-mediated detoxification of mercurial compounds: a short review. *Arch. Microbiol.* 2011;193:837-844.

[36] Ercal N, Gurer-Orham H, Aykin-Burns N. Toxic metal and oxidative stress part I: Mechanisms involved in metal induced oxidative damage. *Curr. Topn. Med. Chem,* 2001 1, 529-539.

[37] Hulberg B, Anderson A, Isaksson A. Interaction of metals and thiols in cell damage and glutathione distribution: Potentiation of mercury toxicity by dithiothreitol. *Toxicology* 2001 156, 93-100.

[38] Stohs SJ, Bagchi D. Oxidative mechanisms in the toxicity of metal-ions. *Free Radic. Biol. Med.* 1995;18:321–336.

[39] Sharma B, Singh S, Siddiqi NJ. Biomedical Implications of Heavy Metals: Induced Imbalances in Redox Systems. *BioMed. Res. Internat* 2014;http://dx.doi.org/10.1155/2014/640754.

[40] Hynninen A, Touzé T, Pitkänen L, Mengin-Lecreulx D, Virta M. An efflux transporter PbrA and a phosphatase PbrB cooperate in a lead-resistance mechanism in bacteria. *Mol. Microbiol.* 2009;74:384-394.

[41] Teramoto H, Inui M, Yukawa H. *Corynebacterium glutamicum* CsoR acts as a transcriptional repressor of two copper/zinc-inducible P_{1B}-type ATPase operons. *Biosci. Biotechnol. Biochem.* 2012;76:1952-1958.

[42] Schelder S, Zaade D, Litsanov B, Bott M, Brocker M. The two-component signal transduction system CopRS of *Corynebacterium glutamicum* is required for adaptation to copper-excess stress. *PLoS One* 2011;6:e22143. doi: 10.1371/journal.pone.0022143.

[43] Schröder J1, Jochmann N, Rodionov DA, Tauch A. The Zur regulon of *Corynebacterium glutamicum* ATCC 13032. *BMC Genomics* 2010;doi: 10.1186/1471-2164-11-12.

[44] Teramoto H, Inui M, Yukawa H. *Corynebacterium glutamicum* Zur acts as a zinc-sensing transcriptional repressor of both zinc-inducible and zinc-repressible genes involved in zinc homeostasis. *FEBS J.* 2012;279:4385-4397.

[45] Takeno S, Nakamura M, Fukai R, Ohnishi J, Ikeda M. The Cgl1281-encoding putative transporter of the cation diffusion facilitator family is responsible for alkali-tolerance in *Corynebacterium glutamicum. Arch. Microbiol.* 2008;190:531-538.

[46] Oremland RS, Stolz JF. The ecology of arsenic. *Science* 2003;300:939-944.

[47] Páez-Espino D, Tamames J, de Lorenzo V, Cánovas D. Microbial responses to environmental arsenic. *Biometals* 2009;22:117-130.

[48] Silver S, Phung LT. A bacterial view of the periodic table: genes and proteins for toxic inorganic ions. *J. Ind. Microbiol. Biotechnol.* 2005;32:587-605.

[49] Ordóñez E, Letek M, Valbuena N, Gil JA, Mateos LM. Analysis of genes involved in arsenic resistance in *Corynebacterium glutamicum* ATCC 13032. *App. Environ. Microbiol.* 2005;71:6206-6215.

[50] Cánovas D, Cases I, de Lorenzo V. Heavy metal tolerance and metal homeostasis in *Pseudomonas putida* as revealed by complete genome analysis. *Environ. Microbiol.* 2003;5:1242-1256.

[51] Branco R, Francisco R, Chung AP, Vasconcelos-Morais P. Identification of an Aox system that requires cytochrome c in the highly arsenic-resistant bacterium *Ochrobactrum tritici* SCII24. *Appl. Environ. Microbiol.* 2009;75:5141–5147.

[52] Muller D, Medigue C, Koechler S, Barbe V, Barakat M, Talla E, Bonnefoy V, Krin E, Arsene-Ploetze F, Carapito C. A tale of two oxidation states: Bacterial colonization of arsenic-rich environments. *PLoS Genetics* 2007;3:e53.

[53] Finney LA, O'Halloran TV. Transition metal speciation in the cell: insights from the chemistry of metal ion receptors. *Science* 2003;300:931-936.

[54] Fu H, Jiang X, Rosen BP. Metalloid transport systems. In: Biological chemistry of arsenic, antimony and bismuth. Eds. Sun H, Winley J. (2011).

[55] Rosen BP, Ajees AA, McDermott TR. Life and death with arsenic. Arsenic life: an analysis of the recent report. "A bacterium that can grow by using arsenic instead of phosphorus." *Bioessays* 2011;33:350-357.

[56] Feo JC, Ordóñez E, Letek M, Castro MA, Muñoz MI, Gil JA, Mateos LM, Aller AJ Retention of inorganic arsenic by coryneform mutant strains. *Water Res.* 2007;41:531–542.

[57] Meng YL, Liu Z, Rosen BP. As(III) and Sb(III) uptake by GlpF and efflux by ArsB in *Escherichia coli. J. Biol. Chem.* 2004;279:18334–18341.

[58] Meiswinkel TM, Rittmann D, Lindner SN, Wendisch VF. Crude glycerol-based production of amino acids and putrescine by *Corynebacterium glutamicum. Bioresour. Techno* 2013;145:254-258.

[59] Liu Z, Boles E, Rosen BP. Arsenic trioxide uptake by hexose permeases in *Saccharomyces cerevisiae. J. Biol. Chem.* 2004;279:17312-17318.

[60] Achour A. R., Bauda P., Billard P. Diversity of arsenite transporter genes from arsenic-resistant soil bacteria. *Res. Microbiol.* 2007;158:128-137.

[61] Villadangos AF, Fu HL, Gil JA,Messens J, Rosen BP, Mateos LM Efflux permease CgAcr3-1 of *Corynebacterium glutamicum* is an arsenite-specific antiporter. *J. Biol. Chem.* 2012;287:723–735.

[62] Messens J., Silver S. Arsenate reduction: Thiol cascade chemistry with convergent evolution. *J. Mol. Biol.* 2006;362:1-17.

[63] Ordóñez E, Van Belle K, Roos G, De Galan S, LetekM, Gil JA,Wyns L, Mateos LM, Messens J Arsenate reductase, mycothiol, and mycoredoxin concert thiol/disulfide exchange. *J. Biol. Chem.* 2009;284:15107–15116.

[64] Villadangos AF, Van Belle K, Wahni K, Dufe VT, Freitas S, Nur H, De Galan S, Gil JA, Collet JF, Mateos LM, Messens J. *Corynebacterium glutamicum* survives arsenic stress with arsenate reductases coupled to two distinct redox mechanisms. *Mol. Microbiol.* 2011;82:998–1014.

[65] Villadangos AF, Ordóñez E, Pedre B, Messens J, Gil JA, Mateos LM. Engineered coryneform bacteria as a bio-tool for arsenic remediation. Appl. Microbiol. Biotechnol. 2014;DOI 10.1007/s00253-014-6055-2.

[66] Sanchéz-Dardon J, Voccia I, Hontela A, Chilmonczyk S, Dunier M, Boermans H, Blakley B, Fournier M. Immunomodulation by heavy metals tested individually or in mixtures in rainbow trout (*Oncorhynchus mykiss*) exposed *in vivo*. *Environ. Toxicol. Chem.* 1999;18:1492–1497.

[67] Branco R, Francisco R, Chung AP, Vasconcelos-Morais P. Identification of an Aox system that requires cytochrome c in the highly arsenic-resistant bacterium *Ochrobactrum tritici* SCII24. *Appl. Environ. Microbiol.* 2009;75:5141–5147.

INDEX

B

C

D

E

F

G

H

I

J

K

L

M

N

O

P

T

U

V

W

Z